MODELING METHODS AND PRACTICES IN SOIL AND WATER ENGINEERING

Innovations in Agricultural and Biological Engineering

MODELING METHODS AND PRACTICES IN SOIL AND WATER ENGINEERING

Edited by
Balram Panigrahi, PhD
Megh R. Goyal, PhD, PE

APPLE ACADEMIC PRESS

Apple Academic Press Inc. | Apple Academic Press Inc.
3333 Mistwell Crescent | 9 Spinnaker Way
Oakville, ON L6L 0A2 | Waretown, NJ 08758
Canada | USA

©2016 by Apple Academic Press, Inc.

First issued in paperback 2021

Exclusive worldwide distribution by CRC Press, a member of Taylor & Francis Group
No claim to original U.S. Government works

ISBN 13: 978-1-77463-602-2 (pbk)
ISBN 13: 978-1-77188-326-9 (hbk)

Library and Archives Canada Cataloguing in Publication

Modeling methods and practices in soil and water engineering/edited by Balram Panigrahi, PhD, and Megh R. Goyal, PhD, PE.

(Innovations in agricultural and biological engineering)
Includes bibliographical references and index.
Issued in print and electronic formats.
ISBN 978-1-77188-326-9 (hardcover).--ISBN 978-1-77188-327-6 (pdf)
1. Soil mechanics--Mathematical models. 2. Water resources development--Mathematical models. 3. Crops and water--Mathematical models. I. Panigrahi, Balram, author, editor II. Goyal, Megh Raj, editor III. Series: Innovations in agricultural and biological engineering

TA710.M63 2016   627   C2016-901547-5   C2016-901548-3

Library of Congress Cataloging-in-Publication Data

Names: Panigrahi, Balram, editor. | Goyal, Megh Raj, editor.
Title: Modeling methods and practices in soil and water engineering / editors: Balram Panigrahi, PhD, Megh R. Goyal, PhD, PE.
Description: Oakville, ON ; Waretown, NJ : Apple Academic Press, [2016] | Includes index.
Identifiers: LCCN 2016009310 (print) | LCCN 2016010785 (ebook) | ISBN 9781771883269 (hardcover : alk. paper) | ISBN 9781771883276 ()
Subjects: LCSH: Soil mechanics--Mathematical models. | Water resources development--Mathematical models. | Crops and water--Mathematical models.
Classification: LCC TA710 .M585 2016 (print) | LCC TA710 (ebook) | DDC 627--dc23
LC record available at http://lccn.loc.gov/2016009310

Apple Academic Press also publishes its books in a variety of electronic formats. Some content that appears in print may not be available in electronic format. For information about Apple Academic Press products, visit our website at **www.appleacademicpress.com** and the CRC Press website at **www.crcpress.com**

CONTENTS

List of Contributors .... *vii*

List of Abbreviations .... *xi*

List of Symbols .... *xv*

Preface 1 by Balram Panigrahi .... *xvii*

Preface 2 by Megh R. Goyal .... *xxiii*

Foreword by M. Kar .... *xxv*

Warning/Disclaimer .... *xxvii*

About the Editors .... *xxix*

Book Messages .... *xxxi*

*List of Other Books on Soil and Water Engineering
by Apple Academic Press Inc.* .... *xxxiii*

Editorial .... *xxxv*

**PART 1: MODELING METHODS IN SOIL AND
WATER ENGINEERING** 1

1. **Groundwater Recharge Estimation Using
   Physical-Based Modeling** 3

   Zijian Wang, Kibreab A. Assefa, Allan D. Woodbury, and
   Hartmut M. Holländer

2. **Constraints in Rainfall-Runoff Modeling Using
   Artificial Neural Network** 31

   M. U. Kale, M. M. Deshmukh, S. B. Wadatkar, and A. S. Talokar

3. **Soil Water Balance Simulation of Rice Using
   Hydrus-2D and Mass Balance Model** 41

   Laxmi Narayan Sethi and Sudhindra Nath Panda

4. **A Multi-Model Ensemble Approach for
   Stream Flow Simulation** 71

   Dwarika Mohan Das, R. Singh, A. Kumar, D. R. Mailapalli,
   A. Mishra, and C. Chatterjee

5.   **Multi-Criteria Analysis for Groundwater**
     Structures in Hard Rock 103
     Ranu Rani Sethi, Ashwani Kumar, Madhumita Das, and S. K. Jena

6.   **Hydrologic Modeling of Watersheds Using**
     Remote Sensing, GIS and AGNPS 131
     Susanta Kumar Jena and Kamlesh Narayan Tiwari

PART 2: RESEARCH INNOVATIONS IN SOIL
AND WATER ENGINEERING 167

7.   **Social-Based Exploratory Assessment of**
     Sustainable Urban Drainage Systems (SUDS) 169
     Hohuu Loc, K. N. Irvine, and Nirakar Pradhan

8.   **Climate Change Impacts on Planning and**
     Management of Water Resources 187
     G. B. Sahoo and S. G. Schladow

9.   **Remote Sensing and GIS Applications for**
     Water Resources Planning in Micro-Watershed 215
     P. K. Rout, J. C. Paul, and B. Panigrahi

PART 3: IRRIGATION MANAGEMENT OF CROPS 279

10. **Water Productivity of Rice Under Deficit Irrigation** 281
     Kajal Panigrahi

11. **Performance of Plantation Crops Under**
     Conservation Trenches 301
     C. R. Subudhi

12. **Water Requirement of Crops** 311
     A. Dalei, C. R. Subudhi, and B. Panigrahi

     Appendices *359*

     Index ....

LIST OF CONTRIBUTORS

Kibreab A. Assefa, PhD
Professor, Department of Civil Engineering, University of Manitoba, Winnipeg, Manitoba R3T 5V6, Canada

C. Chatterjee, PhD
Professor, Department of Agricultural and Food Engineering, Indian Institute of Technology, Kharagpur, 721302, West Bengal, India

A. Dalei
Ex-Post Graduate Student, Department of Soil and Water Conservation Engineering, College of Agricultural Engineering and Technology, Orissa University of Agriculture and Technology, Bhubaneswar, Odisha, India

Dwarika Mohan Das
PhD Scholar, Department of Agricultural and Food Engineering, Indian Institute of Technology, Kharagpur, 721302, West Bengal, India, E-mail: dwarika.dmd@gmail.com, Phone: 7548929957

Madhumita Das, PhD
Principal Scientist, Indian Institute of Water Management (ICAR), Opposite Rail Vihar, Chandrasekarpur, Bhubaneswar-751023, Odisha, India

M. M. Deshmukh, PhD
Professor, Department of Irrigation and Drainage Engineering, Dr. Panjabrao Deshmukh Krishi Vidyapeeth, Krishinagar PO, Akola (MS) 444104, India

Megh R. Goyal, PhD, PE
Retired Professor in Agricultural and Biomedical Engineering, University of Puerto Rico – Mayaguez Campus; and Senior Technical Editor-in-Chief in Agriculture Sciences and Biomedical Engineering, Apple Academic Press Inc., PO Box 86, Rincon, PR 00677, USA. E-mail: goyalmegh@gmail.com

Hartmut M. Holländer, PhD
Professor, Department of Civil Engineering, University of Manitoba, Winnipeg, Manitoba R3T 5V6, Canada. E-mail: hartmut.hollaender@umanitoba.ca

K. N. Irvine, PhD
Professor, National Institute for Education, Nanyang Technological University, Nanyang Walk, Singapore

Susanta K. Jena, PhD
Principal Scientist, Indian Institute of Water Management (ICAR), Opposite Rail Vihar, Chandrasekarpur, Bhubaneswar-751023, Odisha, India, Phone: +91 9437221616, E-mail: skjena_icar@yahoo.co.in

M. U. Kale, PhD
Professor, Department of Irrigation and Drainage Engineering, Dr. Panjabrao Deshmukh Krishi Vidyapeeth, Akola 444104 (MS), India. E-mail: kale921@gmail.com

A. Kumar, PhD
Professor, Department of Agricultural and Food Engineering, Indian Institute of Technology, Kharagpur, 721302, West Bengal, India

Ashwani Kumar, PhD
Ex-Director, ICAR-Indian Institute of Water Management, Opposite Rail Vihar, Chandrasekarpur, Bhubaneswar-751023, Odisha, India

HoHuu Loc, PhD
Professor, Water Engineering and Management Program, Asian Institute of Technology, Pathumthani, 12120, Thailand; Research Center for Environmental Quality Management, Kyoto University. 1–2 Yumihama, Otsu, 520–0811, Japan, E-mail: ho.huu.45z@st.kyoto-u.ac.jp

D. R. Mailapalli, PhD
Professor, Department of Agricultural and Food Engineering, Indian Institute of Technology, Kharagpur, 721302, West Bengal, India

A. Mishra, PhD
Professor, Department of Agricultural and Food Engineering, Indian Institute of Technology, Kharagpur, 721302, West Bengal, India

Sudhindra Nath Panda, PhD
Professor, Department of Agricultural and Food Engineering, Indian Institute of Technology, Kharagpur-721302, India, Phone: +91 9434009156; E-mail: snp@iitkgp.ernet.in

Balram Panigrahi, PhD
Professor and Head, Department of Soil and Water Conservation Engineering, College of Agricultural Engineering and Technology, Orissa University of Agriculture and Technology, Bhubaneswar, Odisha, India. E-mail: kajal_bp@yahoo.co.in

Kajal Panigrahi
MTech Student, Department of Civil Engineering, National Institute of Technology, Rourkela, Odisha, India, Phone +91 8895997206, E-mail: kajalpanigrahi@yahoo.in

J. C. Paul, PhD
Associate Professor, Department of Soil and Water Conservation Engineering, College of Agricultural Engineering and Technology, Orissa University of Agriculture and Technology, Bhubaneswar, Odisha, India. Phone: +91 9437762584, E-mail: jcpaul66@gmail.com

Nirakar Pradhan, PhD
PhD Scholar, Environmental Engineering and Management Program, Asian Institute of Technology, Pathumthani, 12120, Thailand

P. K. Rout
Ex-M. Tech. Student, Department of Soil and Water Conservation Engineering, College of Agricultural Engineering and Technology, Orissa University of Agriculture and Technology, Bhubaneswar, Odisha, India

G. B. Sahoo, PhD, PE, D.WRE
Professor, Department of Civil and Environmental Engineering and Tahoe Environmental Research Center, University of California, One Shields Avenue, Davis, CA 95616. E-mail: gbsahoo@ucdavis.edu, Telephone: (530) 752–1755, Fax: (530) 752–7872

S. G. Schladow, PhD
Professor, Department of Civil and Environmental Engineering, University of California, One Shields Avenue, Davis, CA 95616, USA

Laxmi Narayan Sethi, PhD
Associate Professor, Department of Agricultural Engineering, Assam University (A Central University), Silchar-788011, Assam, India, Phone: +91 9864372058, E-mail: lnsethi06@gmail.com

Ranu Rani Sethi, PhD
Senior Scientist, ICAR-Indian Institute of Water Management, Opposite Rail Vihar, Chandrasekarpur, Bhubaneswar-751023, Odisha, India. Phone: +91 8763667446. E-mail: ranurani@yahoo.com

R. Singh, PhD
Professor, Department of Agricultural and Food Engineering, Indian Institute of Technology, Kharagpur, 721302, West Bengal, India

C. R. Subudhi, PhD
Associate Professor, Department of Soil and Water Conservation Engineering, College of Agricultural Engineering and Technology, Orissa University of Agriculture and Technology, Bhubaneswar-751003, Odisha, India, Phone: +91 9437645234, E-mail: rsubudhi5906@gmail.com

A. S. Talokar, PhD
Professor, Department of Irrigation and Drainage Engineering, Dr. Panjabrao Deshmukh Krishi Vidyapeeth, Akola 444104 (MS), India

Kamlesh Narayan Tiwari, PhD
Professor, Department of Agricultural and Food Engineering, Indian Institute of Technology, Kharagpur, West Bengal-721302, India, Email: kamlesh@agfe.iitkgp.ernet.in

S. B. Wadatkar, PhD
Professor, Department of Irrigation and Drainage Engineering, Dr. Panjabrao Deshmukh Krishi Vidyapeeth, Akola 444104 (MS), India

Zijian Wang, PhD
Professor, Department of Civil Engineering, University of Manitoba, University of Manitoba Winnipeg, Manitoba, R3T 5V6, Canada

Allan D. Woodbury, PhD
Professor, Department of Civil Engineering, University of Manitoba, Winnipeg, Manitoba, R3T 5V6, Canada

LIST OF ABBREVIATIONS

AAP	Apple Academic Press Inc.
AAT	Arc Attribute Table for Line
ABE	Agricultural and Biological Engineering
ABEs	Agricultural and Biological Engineers
AE	Agricultural Engineering
AET	Actual Evapotranspiration
AGNPS	Agricultural Nonpoint Source Pollution
AIS	All India Soil
AMC	Antecedent Moisture Conditions
API	Antecedent Precipitation Index
AWBM	Australian Water Balance Model
BASINS	Better Assessment Science Integrating Point and Nonpoint Sources
BCs	Boundary Conditions
BMP	Best Management Practices
BOD	Biological Oxygen Demand
BS	Brier Score
CAET	College of Agricultural Engineering and Technology
CDF	Cumulative Distribution Function
CGS	Critical Growth Stage
CN	Curve Number
COD	Chemical Oxygen Demand
CRM	Coefficient of Residual Mass
CRPS	Continuous Rank Probability Score
CS	Continuously Submerged
CWN	Canadian Water Network
DAS	Days After Sowing
DEM	Digital Elevation Model
DLM-WQ	Dynamic Lake Model with Water Quality
DMIP	Distributed Model Inter-Comparison Project
DON	Dissolved Organic Nitrogen

DOP	Date of Pass
DYRESM	DYnamic REservoir Simulation Model
EI	Energy Intensity
EPA	Environmental Protection Agency
EPS	Ensemble Streamflow Prediction
ER	Effective Rainfall
ESA	Endangered Species Act
FCC	Flood Control Centre
FCCs	False Color Composites
FDAR	Five Day Antecedent Rainfall
FEM	Finite Element Method
GFDL	Geophysical Fluid Dynamics Laboratory
GIS	Geographical Information System
GUI	Graphical User Interface
HEC-HMS	Hydrologic Modeling System
HEPEX	Hydrologic Ensemble Prediction Experiment
HG	Hargreaves
HGM map	Hydrogeomorphological Map
HSPF	Hydrologic Simulation Program – FORTRAN
IIT	Indian Institute of Technology
IWDP	Integrated Watershed Development Project
L-M	Levenberg-Marquardt algorithm
LID	Low Impact Development
LP	Linear Programming
LS	Lateral Seepage
LTIMP	Lake Tahoe Interagency Monitoring Program
LU/LC	Land Use/Land Cover
LUCHEM	Land Use Change on Hydrology by Ensemble Modeling
LUS	Land Use Survey
MAD	Manageable Allowable Deficit
MAE	Mean Absolute Error
ME	Mean Error
ME	Modeling Efficiency
MLP-ANN	Multi Layer Perceptron-Artificial Neural Network Technique
MME	Multi-Model Ensembles

MSE	Mean Square Error
NCDC	National Climatic Data Center
NDEP	Nevada Division of Environmental Protection
NGVD	National Geodetic Vertical Datum
NH	National Highways
NRSA	National Remote Sensing Agency
NSE	Nash Sutcliffe Efficiency
NWDPRA	National Watershed Development Project for Rainfed Areas
OFR	On-Farm Reservoir
OFT	Overland Flow Time
OUAT	Orissa University of Agriculture and Technology
P-T	Priestley-Taylor
PE	Prediction Efficiency
PET	Potential Evapotranspiration
PON	Particulate Organic Nitrogen
PTF	Pedotransfer Functions
PWP	Permanent Wilting Point
R-R	Rainfall-Runoff
RBP	Resilient Back Propagation
RLL	Rainfall Runoff Library
RMSE	Root Mean Square Error
RPS	Rank Probability Score
RPSS	Rank Probability Skill Score
RS	Remote Sensing
RS	Reproductive Stage
SAC	Space Application Center
SAT	Saturation
SCC	Surface Condition Constant
SDSS	Spatial Decision Support System
SI	Supplemental Irrigation
SIMHYD	Simplified Daily Conceptual Rainfall-Runoff
SMAR	Soil Moisture and Accounting Model
SNOTEL	SNOwpack TELemetry
SUDS	Sustainable Urban Drainage Systems
SWAT	Soil and Water Assessment Tool

SWCE	Soil and Water Conservation Engineering
SYI	Sediment Yield Index
TCID	Truckee-Carson Irrigation District
TERC	Tahoe Environmental Research Center
TNAU	Tamil Nadu Agricultural University
USLE	Universal Soil Loss Equation
VGM	van Genuchten-Mualem model
VP	Vertical Percolation
WR	Water Requirement
WSI	Water Saving Irrigation
WUE	Water Use Efficiency

PREFACE 1 BY BALRAM PANIGRAHI

Mathematical models in the field of soil and water engineering have become essential tools for the planning, development, and management of land and water resources. They are increasingly used to analyze quantity and quality of stream flow, groundwater and soil water, and different water resources management activities. Their application in the fields of soil and water engineering has expanded the horizon of innovative research. *Modeling Methods and Practices in Soil and Water Engineering*, a volume in the new book series *Innovations in Agricultural and Biological Engineering*, discusses the development of useful models and their applications in soil and water engineering. The book contains 12 chapters spreading over three parts: Modeling Methods in Soil and Water Engineering, Research Innovations in Soil and Water Engineering, and Irrigation Management of Crops.

The first section covers various modeling methods, including groundwater recharge estimation, rainfall-runoff modeling using artificial neural networks, development and application of a water balance model and a HYDRUS-2D model for cropped fields, a multi-model approach for stream flow simulation, multi-criteria analysis for construction of groundwater structures in hard rock terrains, hydrologic modeling of watersheds using remote sensing, and GIS and AGNPS.

The effects of climate change on water resources planning and management and sustainable urban drainage systems are discussed in the book's second section. A specialized chapter deals with applications of remote sensing and GIS for watershed planning and management.

Irrigation water for agriculture is diminishing day by day, which affects rice production in many rice-growing countries. The third section addresses productivity of rice under deficit irrigation and suggests different water-saving irrigation techniques to help in obtaining optimum yield. Finally methods to estimate water requirement of different crops and performance of plantation crops in degraded watersheds under different conservation trenches are discussed.

The book will serve as a reference manual for graduate and undergraduate students of agriculture and agricultural, civil, and biological engineering and will also be valuable for those who teach, practice, and research soil and water conservation methods for agriculture.

The contributions by the authors of different chapters of this book are very valuable and without their support this book would have not been published successfully. The authors are well experts in their fields. Their names are mentioned in each chapter and also separately in the list of contributors. The readers are requested to offer constructive suggestions that may help to improve the next edition.

I take the opportunity to offer my heartfelt obligations to Distinguished Professor Megh R. Goyal "Father of Irrigation Engineering of 20th Century in Puerto Rico" and editor of this book who has benevolently given me an opportunity to serve as a lead editor of this book. We both thank all the editorial staff of Apple Academic Press Inc. for making every effort to publish this book. I express my deep obligations to my family, friends and colleagues for their help and moral support during preparation of the book.

Balram Panigrahi, PhD
Editor
December 31 of 2015

PREFACE 2 BY MEGH R. GOYAL

Due to increased agricultural production, irrigated land has increased in the arid and sub-humid zones around the world. Agriculture has started to compete for the water use with industries, municipalities and other sectors. This increasing demand along with increments in water and energy costs have made it necessary to develop new innovative technologies for the adequate management of natural resources. The intelligent use of soil and water for crops requires understanding of evapotranspiration processes and use of efficient irrigation methods under limited resources.

Our planet will not have enough potable water for a population of >10 billion persons in 2115. The situation will be further complicated by multiple factors that will be adversely affected by the global warming. The http://www.un.org/waterforlifedecade/scarcity.shtml indicates, "Water scarcity already affects all continents. Around 1.2 billion people, or almost one-fifth of the world's population, live in areas of physical scarcity, and 500 million people are approaching this situation. Another 1.6 billion people, or almost one quarter of the world's population, face economic water shortage (where countries lack the necessary infrastructure to take water from rivers and aquifers). Water scarcity is among the main problems to be faced by many societies and the World in the 21st century. Water use has been growing at more than twice the rate of population increase in the last century, and, although there is no global water scarcity as such, an increasing number of regions are chronically short of water. Water scarcity is both a natural and a human-made phenomenon. There is enough freshwater on the planet for seven billion people but it is distributed unevenly and too much of it is wasted, polluted and unsustainably managed." The crisis is rampant.

I have been involved in Soil and Water Conservation Engineering (SWCE) since 1971. I know what the cooperating authors have emphasized in this book volume. I am a staunch supporter of preserving our natural resources. The updated seventh edition of "*Soil and Water Conservation Engineering by http://www.asabe.org*" emphasizes engineering design of soil and water conservation practices and their impact on the environment, primarily air and water quality. Other books on SWCE advocate the same. Importance of wise use of our natural resources has been taken up seriously by Universities, Institutes/Centers, Government Agencies and Non-Government Agencies. With an example in the next paragraph, I conclude that the agencies and departments in SWCE have contributed to the ocean of knowledge.

At Orissa University of Agriculture and Technology (http://www.caet.org.in/caet/), the Department of SWCE is well equipped with soil and water conservation, irrigation and drainage and hydraulics laboratories, which have facilities of modern systems and equipment's like Neutron Probe moisture meter, Pressure plate apparatus, Digital stage level recorder, Micro Irrigation systems, different types of hydraulic pumps, models and prototypes of chute and drop spillway, models to study ground water flow dynamics, Geo-resistivity meter, etc. The department has also working co-ordination with various Government organizations and industries to help students to gain practical knowledge. Besides, the facility of testing and evaluation of various pumps, sprinklers and drippers are available. The SWCE has rendered its yeomen service to the farming community through programs related to irrigation and drainage water management, ground water and wells, SWE, Hi-tech irrigation systems and watershed management. Besides, the department has undertaken research component of World Bank funded project, "National Watershed Development Project for Rainfed Areas (NWDPRA)," and Integrated Watershed Development Project (IWDP) throughout the Orissa state at different watersheds, Improvement of hydraulic structures in Mahanadi Delta irrigation command, Rehabilitation of degraded watersheds, etc. Similar and specialized description of SWCE programs is available throughout the world.

Our book also contributes to the ocean of knowledge on SWCE. Agricultural and Biological Engineers (ABEs) with expertise in SWCE work to better understand the complex mechanics of natural resources, so that they can be used efficiently and without degradation. ABEs determine crop water requirements and design irrigation systems. They are experts in agricultural hydrology principles, such as controlling drainage, and they implement ways to control soil erosion and study the environmental effects of sediment on stream quality. Natural resources engineers design, build, operate and maintain water control structures for reservoirs, flood-ways and channels. They also work on water treatment systems, wetlands protection, and other water issues. While making call for chapters for a book volume on SWCE, we mentioned to the prospective authors following focus areas:

- Academia to industry to end user loop in soil and water engineering
- Aquaculture engineering
- Biological engineering in SWE
- Biotechnology applications in SWE
- Climate change and its impact on SWE
- Design in irrigation and drainage systems
- Drainage principles, management, practices
- Education in SWE: curricula/scope/opportunities
- Energy potential in SWE
- Environment engineering
- Extension methods in SWE
- Flood damage in crop production
- Flow through porous media
- Global warming due ill effects of SWCE
- Ground water and tube-wells: principles, management, and practices
- Groundwater simulation for sustainable agriculture,
- Human factors engineering in SWE
- Hydrologic applications in SWE
- Irrigation principles, management, practices
- Management of water resources
- Nanotechnology applications in SWE
- Natural resources engineering and management
- Principles of hydraulics in SWE
- Robot engineering in SWE
- Simulation, optimization and computer modeling
- Society and natural resources
- Soil and water engineering
- Waste management engineering

Therefore, I conclude that scope of SWCE is wide enough and focus areas may overlap one another. The mission of this book volume is to serve as a reference manual for graduate and under graduate students of agricultural, biological and civil engineering; horticulture, soil science, crop science and agronomy. I hope that it will be a valuable reference for professionals that work with soil and water management; for professional training institutes, technical agricultural centers, irrigation centers, Agricultural Extension Service, and other agencies. I cannot guarantee the information in this book series will be enough for all situations.

After my first textbook, *Drip/Trickle or Micro Irrigation Management* by Apple Academic Press Inc., and response from international readers, Apple Academic Press Inc. published for the world community the ten-volume series on *Research Advances in Sustainable Micro Irrigation*, edited by Megh R. Goyal.

At 49th annual meeting of Indian Society of Agricultural Engineers at Punjab Agricultural University during February 22–25 of 2015, a group of ABEs convinced me that there is a dire need to publish book volumes on focus areas of agricultural and biological engineering (ABE). This is how the idea was born for a new book series titled "Innovations in Agricultural and Biological Engineering." Here we present the volume titled *Modeling Methods and Practices in Soil and Water Engineering*.

My longtime colleague, Dr. Balram Panigrahi, joins me as a Lead Editor of this volume. Dr. Panigrahi holds exceptional professional qualities in addition to Professor and Head for Department of Soil and Water Conservation Engineering in College of Agricultural Engineering & Technology (CAET) at Orissa University of Agriculture & Technology, Bhubaneswar, India. His contribution to the contents and quality of this book has been invaluable.

We will like to thank editorial staff, Sandy Jones Sickels, Vice President, and Ashish Kumar, Publisher and President, at Apple Academic Press, Inc., for making every effort to publish the book when the diminishing water resources are a major issue worldwide. Special thanks are due to the AAP Production Staff. We request that the reader offer us your constructive suggestions that may help to improve the next edition.

I express my deep admiration to my family for understanding and collaboration during the preparation of this book. As an educator, there is a piece of advice to one and all in the world: "*Permit that our almighty God, our Creator and excellent Teacher, irrigate the life with His Grace of rain trickle by trickle, because our life must continue trickling on… and Get married to your profession"*

—Megh R. Goyal, PhD, PE
Senior Editor-in-Chief
December 31 of 2015

FOREWORD

Soil and water are the two vital natural resources of the world. These two natural resources are fast depleting. With increasing population pressure coupled with urbanization and industrialization, these resources are begin further exploited. These resources must be conserved and maintained carefully for sustainable crop production. Innovation researches are very much essential for conservation and development of these natural resources. Now-a-days, a number of models are used in the field of soul and water engineering. These models help in optimum utilization of resources and at the same time help in proper planning and management of decision support systems.

In this context, This volume is a commendable work by the authors. I congratulate Dr. Balam Panigrahi, Professor Dr. Megh R. Goyal, "Father of Irrigation Engineering of 20th Century in Puerto Rico". The book contains useful chapters dealing with different aspects of soil and water engineering, including groundwater, surface water, water requirement of crops, remote sensing and GIS application in watershed planning and management, irrigation and drainage. The book will serve as an invaluable resource for graduate and undergraduate students of agriculture, agricultural, biological and civil engineering and all deal with other natural resources engineering. I believe that the book will be helpful for all teaching and farming community, practicing engineers, research scientists, soil conservationists, agronomists, planners, managers and all policy makers dealing with soil and water conservation.

I wish the publication a great success.

—M. Kar

Vice Chancellor, Orissa University of Agriculture & Technology,
Bhubaneswar, Odhisa, India

WARNING/DISCLAIMER

PLEASE READ CAREFULLY

The goal of this compendium is to guide the world engineering community on how to efficiently design for economical crop production. The reader must be aware that dedication, commitment, honesty, and sincerity are the most important factors in a dynamic manner for success.

The editor, the contributing authors, the publisher, and the printer have made every effort to make this book as complete and as accurate as possible. However, there still may be grammatical errors or mistakes in the content or typography. Therefore, the contents in this book should be considered as a general guide and not a complete solution to address any specific situation in irrigation. For example, one size of irrigation pump does not fit all sizes of agricultural land and to all crops.

The editor, the contributing authors, the publisher, and the printer shall have neither liability nor responsibility to any person, any organization or entity with respect to any loss or damage caused, or alleged to have caused, directly or indirectly, by information or advice contained in this book. Therefore, the purchaser/reader must assume full responsibility for the use of the book or the information therein.

The mention of commercial brands and trade names is only for technical purposes. It does not mean that a particular product is endorsed over another product or equipment not mentioned. The editors, cooperating authors, educational institutions, and the publisher Apple Academic Press Inc. do not have any preference for a particular product.

All weblinks that are mentioned in this book were active on December 31 of 2015. The editors, the contributing authors, the publisher, and the printing company shall have neither liability nor responsibility, if any of the weblinks is inactive at the time of reading of this book.

OTHER BOOKS ON
SOIL AND WATER CONSERVATION
ENGINEERING
BY APPLE ACADEMIC PRESS, INC.

Management of Drip/Trickle or Micro Irrigation
Megh R. Goyal, PhD, PE, Senior Editor-in-Chief

Evapotranspiration: Principles and Applications for Water Management
Megh R. Goyal, PhD, PE, and Eric W. Harmsen, Editors

Book Series: Research Advances in Sustainable Micro Irrigation
Senior Editor-in-Chief: Megh R. Goyal, PhD, PE
    Volume 1: Sustainable Micro Irrigation: Principles and Practices
    Volume 2: Sustainable Practices in Surface and Subsurface Micro Irrigation
    Volume 3: Sustainable Micro Irrigation Management for Trees and Vines
    Volume 4: Management, Performance, and Applications of Micro Irrigation Systems
    Volume 5: Applications of Furrow and Micro Irrigation in Arid and Semi-Arid Regions
    Volume 6: Best Management Practices for Drip Irrigated Crops
    Volume 7: Closed Circuit Micro Irrigation Design: Theory and Applications
    Volume 8: Wastewater Management for Irrigation: Principles and Practices
    Volume 9: Water and Fertigation Management in Micro Irrigation
    Volume 10: Innovation in Micro Irrigation Technology

Book Series: Innovations and Challenges in Micro Irrigation
Senior Editor-in-Chief: Megh R. Goyal, PhD, PE
    Volume 1: Principles and Management of Clogging in Micro Irrigation

Volume 2: Sustainable Micro Irrigation Design Systems for
               Agricultural Crops: Methods and Practices
Volume 3: Performance Evaluation of Micro Irrigation Management:
               Principles and Practices
Volume 4: Potential Use of Solar Energy and Emerging Technologies
               in Micro Irrigation
Volume 5: Micro Irrigation Management: Technological Advances
               and Their Applications

Book Series: Innovations in Agricultural and Biological Engineering
Senior Editor-in-Chief: Megh R. Goyal, PhD, PE

Dairy Engineering: Advanced Technologies and Their Applications
Editors: Murlidhar Meghwal, PhD, Megh R. Goyal, PhD, PE, and
Rupesh S. Chavan, PhD

**Developing Technologies in Food Science: Status, Applications, and
Challenges**
Editors: Murlidhar Meghwal, PhD, and Megh R. Goyal, PhD, PE

Emerging Technologies in Agricultural Engineering
Editor: Megh R Goyal, PhD, PE

**Engineering Practices for Agricultural Production and Water
Conservation: An Interdisciplinary Approach**
Editors: Megh R. Goyal, PhD, PE, and R. K. Sivanappan, PhD

Flood Assessment: Modeling and Parameterization
Editors: Eric W. Harmsen, PhD, and Megh R. Goyal, PhD

Food Engineering: Emerging Issues, Modeling, and Applications
Editors: Murlidhar Meghwal, PhD, and Megh R. Goyal, PhD

**Food Process Engineering: Emerging Trends in Research and Their
Applications**
Editors: Murlidhar Meghwal, PhD, and Megh R. Goyal, PhD, PE

Modeling Methods and Practices in Soil and Water Engineering
Editors: BalramPanigrahi, PhD, and Megh R. Goyal, PhD, PE

**Soil and Water Engineering: Principles and Applications of
Modeling**
Editors: BalramPanigrahi, PhD, and Megh R. Goyal, PhD, PE

**Soil Salinity Management in Agriculture: Technological Advances
and Applications**
Editors: S. K. Gupta, PhD, CE, and Megh R. Goyal, PhD, PE

ABOUT EDITOR

 Dr. Balram Panigrahi is an agricultural engineer with specialization in soil and water engineering. Dr. Panigrahi is presently Professor and Head of the Department of Soil and Water Conservation Engineering (SWCE) at the College of Agricultural Engineering and Technology (CAET), Orissa University of Agriculture and Technology (OUAT), in Bhubaneswar, India. He also served as Chief Scientist of the Water Management Project and Associate Director of Research in Regional Research Station of OUAT.

Dr. Panigrahi has published about 180 technical papers in different international and national journals and conference proceedings. He has written several book chapters, practical manuals, and monographs. He has also written two textbooks in the field of irrigation engineering. He has been awarded with 14 gold medals and awards, including the Jawaharlal Nehru Award for best post-graduate research in the field of natural resources management by Indian Council of Agricultural Research, New Delhi; the Samanta Chandra Sekhar Award for best scientist in the state of Odisha, India; and the Gobinda Gupta Award as outstanding engineer of the state of Odisha, given by the Institution of Engineers (India), Odisha state center. He has also received a Japanese Master Fellowship for pursuing a master of engineering study at the Asian Institute of Technology, Thailand. In addition to being the editor and an editorial board member of several journals, Dr. Panigrahi is a reviewer for many journals. He is the member of a number of professional societies at national and international levels. He has chaired several international and national conferences both in India and abroad. With 26 years of of teaching and research experience, he has guided several PhD and many MTech students. Dr. Panigrahi's research interests include irrigation and drainage engineering, water management in rainfed and irrigated commands, and modeling of irrigation systems.

He obtained his BTech in agricultural engineering from Orissa University of Agriculture and Technology (OUAT), Bhubaneswar, Odisha, India, and his Master of Engineering in water resources engineering from the Asian Institute of Technology, Thailand. He was awarded his PhD in agricultural engineering from the Indian Institute of Technology, Kharagpur, India.

ABOUT SENIOR EDITOR-IN-CHIEF

Megh R. Goyal, PhD, PE, is a Retired Professor in Agricultural and Biomedical Engineering from the General Engineering Department in the College of Engineering at University of Puerto Rico–Mayaguez Campus; and Senior Acquisitions Editor and Senior Technical Editor-in-Chief in Agriculture and Biomedical Engineering for Apple Academic Press Inc.

He has worked as a Soil Conservation Inspector and as a Research Assistant at Haryana Agricultural University and Ohio State University. He was first agricultural engineer to receive the professional license in Agricultural Engineering in 1986 from College of Engineers and Surveyors of Puerto Rico. On September 16, 2005, he was proclaimed as "Father of Irrigation Engineering in Puerto Rico for the twentieth century" by the ASABE, Puerto Rico Section, for his pioneer work on micro irrigation, evapotranspiration, agroclimatology, and soil and water engineering. During his professional career of 45 years, he has received many prestigious awards. A prolific author and editor, he has written more than 200 journal articles and textbooks and has edited over 35 books. He received his BSc degree in engineering from Punjab Agricultural University, Ludhiana, India; his MSc and PhD degrees from Ohio State University, Columbus; and his Master of Divinity degree from Puerto Rico Evangelical Seminary, Hato Rey, Puerto Rico, USA.

BOOK MESSAGES

"Conservation of soil and water resources is urgently needed to save our planet from degradation. Agricultural engineers can help to alleviate these crises. The editors of this book volume have contributed a drop in the ocean. It is our ethical duty to educate our fraternity on this topic."

—Miguel A Muñoz, PhD, Ex-President of University of Puerto Rico, and Professor and Soil Scientist

"I believe that this innovative book on soil and water engineering will aid educators throughout the world."

—A. M. Michael, PhD, Former Professor/Director, Water Technology Centre, IARI; Ex-Vice-Chancellor, Kerala Agricultural University, Trichur, Kerala

"In providing this resource in soil and water engineering, Balram Panigrahi and Megh R. Goyal, as well as the Apple Academic Press, are rendering an important service to the conservationists."

—Gajendra Singh, PhD, Ex-President of ISAE; Former Deputy Director General (Engineering) of Indian Council of Agricultural Research, and Former Vice-President/Dean/Professor and Chairman, Asian Institute of Technology, Thailand

"Water is increasingly scarce and extremely valuable resource, without which sustainable development is impossible. Agriculture is the largest water-using sector worldwide. The gross irrigated area in the world has increased from 94 M-ha in 1950's to about 280 M-ha at present. Most of the areas are in the developing countries, especially in India, the gross irrigated area is more than 100 M-ha. Research in water resources, quantity, quality of water, water management in agriculture including drip irrigation is taken up seriously by the scientists especially for Rice, fruits, vegetables, cotton, banana, sugarcane plantations crops, etc. According to the FAO (1990), 60% of the water supplied for irrigation goes unused and leads to water logging and salinization.

Hence, water requirements for various crops are worked out in surface irrigation, sprinkler and drip irrigation methods to use water efficiently. There are two strategies, which are used to meet challenges: (i) Supply management; and (ii) Demand management. To solve the problems related to water management, water should be considered as an economic asset. The increase in the value of water, demand management will become more important than supply management. This book volume addresses emerging technologies in SWCE."

—R. K. Sivanappan, PhD, Former Professor and Dean, College of Agricultural Engineering & Technology, Tamil Nadu Agricultural University (TNAU), Coimbatore, India

"The emerging technologies have potential to conserve water that can facilitate timely sowing of crops under the delayed monsoon situation that has occurred in 2014 and provide solutions to monsoon worries. Agricultural Engineers need to provide leadership opportunities for in water resources and water management sector, water resources, irrigation, soil conservation, watersheds, environment and energy for stability of agriculture and in turn stable growth of Indian economy. This book volume is an asset in this path."

—V. M. Mayande, PhD, Former Vice Chancellor, Dr. Panjabrao Deshmukh Krishi Vidyapeeth & MAFSU, Nagpur Former President, Indian Society of Agricultural Engineers

EDITORIAL

Apple Academic Press Inc., (AAP) is publishing this volume, *Modeling Methods and Practices in Soil and Water Engineering*, as part of our AAP book series titled Innovations in Agricultural and Biological Engineering. Over a span of 8–10 years, Apple Academic Press Inc., will publish subsequent volumes in the specialty areas defined by the American Society of Agricultural and Biological Engineers (www.asabe.org). The mission of this series is to provide knowledge and techniques for agricultural and biological engineers (ABEs). The series aims to offer high-quality reference and academic content in Agricultural and Biological Engineering (ABE) that is accessible to academicians, researchers, scientists, university faculty, and university-level students and professionals around the world. The following material has been edited/modified and reproduced below ["*Goyal, Megh R., 2006. Agricultural and Biomedical Engineering: Scope and Opportunities. Paper Edu_47 at the Fourth LACCEI International Latin American and Caribbean Conference for Engineering and Technology (LACCEI' 2006): Breaking Frontiers and Barriers in Engineering: Education and Research by LACCEI University of Puerto Rico – Mayaguez Campus, Mayaguez, Puerto Rico, June 21–23*"]:

WHAT IS AGRICULTURAL AND BIOLOGICAL ENGINEERING (ABE)?

"*Agricultural Engineering (AE) involves application of engineering to production, processing, preservation and handling of food, fiber, and shelter. It also includes transfer of technology for the development and welfare of rural communities,*" according to www.isae.in. "*ABE is the discipline of engineering that applies engineering principles and the fundamental concepts of biology to agricultural and biological systems and tools, for the safe, efficient and environmentally sensitive production, processing, and management of agricultural, biological, food, and natural resources systems,"* according to www.asabe.org. *"AE is the branch of engineering involved with the design of farm machinery, with soil management, land development, and mechanization and automation of livestock farming, and with the efficient planting, harvesting, storage, and processing of farm commodities,"* definition by: http://dictionary.reference.com/browse/agricultural+engineering.

"AE incorporates many science disciplines and technology practices to the efficient production and processing of food, feed, fiber and fuels. It involves disciplines like mechanical engineering (agricultural machinery and automated machine systems), soil science (crop nutrient and fertilization, etc.), environmental sciences (drainage and irrigation), plant biology (seeding and plant growth management), animal science (farm animals and housing) etc.," by: http://www.ABE.ncsu.edu/academic/agricultural-engineering.php.

"According to https://en.wikipedia.org/wiki/Biological_engineering: *"BE (Biological engineering) is a science-based discipline that applies concepts and methods of biology to solve real-world problems related to the life sciences or the application thereof. In this context, while traditional engineering applies physical and mathematical sciences to analyze, design and manufacture inanimate tools, structures and processes, biological engineering uses biology to study and advance applications of living systems."*

SPECIALTY AREAS OF ABE

Agricultural and Biological Engineers (ABEs) ensure that the world has the necessities of life including safe and plentiful food, clean air and water, renewable fuel and energy, safe working conditions, and a healthy environment by employing knowledge and expertise of sciences, both pure and applied, and engineering principles. Biological engineering applies engineering practices to problems and opportunities presented by living things and the natural environment in agriculture. BA engineers understand the interrelationships between technology and living systems, have available a wide variety of employment options. *"ABE embraces a variety of following specialty areas,"* www.asabe.org. As new technology and

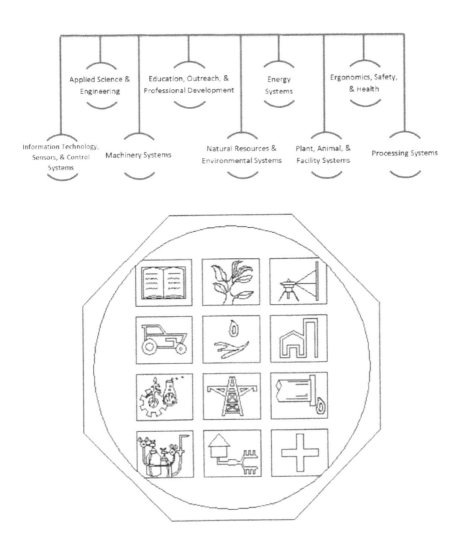

information emerge, specialty areas are created, and many overlap with one or more other areas.

1. **Aquacultural Engineering**: ABEs help design farm systems for raising fish and shellfish, as well as ornamental and bait fish. They specialize in water quality, biotechnology, machinery, natural resources, feeding and ventilation systems, and sanitation. They seek ways to reduce pollution from aquacultural discharges, to

reduce excess water use, and to improve farm systems. They also work with aquatic animal harvesting, sorting, and processing.

2. **Biological Engineering** applies engineering practices to problems and opportunities presented by living things and the natural environment.

3. **Energy:** ABEs identify and develop viable energy sources—biomass, methane, and vegetable oil, to name a few—and to make these and other systems cleaner and more efficient. These specialists also develop energy conservation strategies to reduce costs and protect the environment, and they design traditional and alternative energy systems to meet the needs of agricultural operations.

4. **Farm Machinery and Power Engineering**: ABEs in this specialty focus on designing advanced equipment, making it more efficient and less demanding of our natural resources. They develop equipment for food processing, highly precise crop spraying, agricultural commodity and waste transport, and turf and landscape maintenance, as well as equipment for such specialized tasks as removing seaweed from beaches. This is in addition to the tractors, tillage equipment, irrigation equipment, and harvest equipment that have done so much to reduce the drudgery of farming.

5. **Food and Process Engineering:** Food and process engineers combine design expertise with manufacturing methods to develop economical and responsible processing solutions for industry. Also food and process engineers look for ways to reduce waste by devising alternatives for treatment, disposal and utilization.

6. **Forest Engineering**: ABEs apply engineering to solve natural resource and environment problems in forest production systems and related manufacturing industries. Engineering skills and expertise are needed to address problems related to equipment design and manufacturing, forest access systems design and construction; machine-soil interaction and erosion control; forest operations analysis and improvement; decision modeling; and wood product design and manufacturing.

7. **Information and Electrical Technologies Engineering** is one of the most versatile areas of the ABE specialty areas, because it is applied to virtually all the others, from machinery design to soil

testing to food quality and safety control. Geographic information systems, global positioning systems, machine instrumentation and controls, electromagnetics, bioinformatics, biorobotics, machine vision, sensors, spectroscopy: These are some of the exciting information and electrical technologies being used today and being developed for the future.

8. **Natural Resources:** ABEs with environmental expertise work to better understand the complex mechanics of these resources, so that they can be used efficiently and without degradation. ABEs determine crop water requirements and design irrigation systems. They are experts in agricultural hydrology principles, such as controlling drainage, and they implement ways to control soil erosion and study the environmental effects of sediment on stream quality. Natural resources engineers design, build, operate and maintain water control structures for reservoirs, floodways and channels. They also work on water treatment systems, wetlands protection, and other water issues.

9. **Nursery and Greenhouse Engineering**: In many ways, nursery and greenhouse operations are microcosms of large-scale production agriculture, with many similar needs – irrigation, mechanization, disease and pest control, and nutrient application. However, other engineering needs also present themselves in nursery and greenhouse operations: equipment for transplantation; control systems for temperature, humidity, and ventilation; and plant biology issues, such as hydroponics, tissue culture, and seedling propagation methods. And sometimes the challenges are extraterrestrial: ABEs at NASA are designing greenhouse systems to support a manned expedition to Mars!

10. **Safety and Health:** ABEs analyze health and injury data, the use and possible misuse of machines, and equipment compliance with standards and regulation. They constantly look for ways in which the safety of equipment, materials and agricultural practices can be improved and for ways in which safety and health issues can be communicated to the public.

11. **Structures and Environment:** ABEs with expertise in structures and environment design animal housing, storage structures, and greenhouses, with ventilation systems, temperature and humidity

controls, and structural strength appropriate for their climate and purpose. They also devise better practices and systems for storing, recovering, reusing, and transporting waste products.

CAREER IN AGRICULTURAL AND BIOLOGICAL ENGINEERING

One will find that university ABE programs have many names, such as biological systems engineering, bioresource engineering, environmental engineering, forest engineering, or food and process engineering. Whatever the title, the typical curriculum begins with courses in writing, social sciences, and economics, along with mathematics (calculus and statistics), chemistry, physics, and biology. Student gains a fundamental knowledge of the life sciences and how biological systems interact with their environment. One also takes engineering courses, such as thermo-dynamics, mechanics, instrumentation and controls, electronics and electrical circuits, and engineering design. Then student adds courses related to particular interests, perhaps including mechanization, soil and water resource management, food and process engineering, industrial microbiology, biological engineering or pest management. As seniors, engineering students team up to design, build, and test new processes or products.

For more information on this series, readers may contact:

Ashish Kumar, Publisher and President	Megh R. Goyal, PhD, PE
	Book Series Senior
Sandy Sickles, Vice President	Editor-in-Chief
Apple Academic Press, Inc.,	*Innovations in Agricultural and*
Fax: 866-222-9549; E-mail:	*Biological Engineering*
ashish@appleacademicpress.com	E-mail: goyalmegh@gmail.com
http://www.appleacademicpress. com/publishwithus.php	

PART I

MODELING METHODS IN SOIL AND WATER ENGINEERING

CHAPTER 1

GROUNDWATER RECHARGE ESTIMATION USING PHYSICAL-BASED MODELING

ZIJIAN WANG, KIBREAB A. ASSEFA, ALLAN D. WOODBURY, and HARTMUT M. HOLLÄNDER

Department of Civil Engineering, University of Manitoba, Winnipeg, Manitoba R3T 5V6, Canada. E-mail: hartmut.hollaender@umanitoba.ca

CONTENTS

1.1   Introduction 4
1.2   Methods and Materials 7
    1.2.1   Study Area 7
    1.2.2   Data 8
    1.2.3   Laboratory Measurements 10
1.3   Vadose Zone Modeling 11
    1.3.1   Governing Equations 11
    1.3.2   Boundary Conditions 14
    1.3.3   Calibration 15
    1.3.4   Sensitivity Analysis 16
1.4   Results 17
    1.4.1   Calibration 17
    1.4.2   Sensitivity Analysis 17

    1.4.3   Recharge.. 20
1.5   Discussions 22
1.6   Conclusions.. 25
1.7   Summary.. 25
Keywords 26
Acknowledgements... 27
References.. 27

1.1   INTRODUCTION

Groundwater protection strategies are conventionally based on the assess-ment of aquifer and well vulnerability. The hydrogeological characteristics of a specific groundwater supply system are quantified through the consider-ation of the intrinsic physical characteristics of the subsurface materials and on average annual, steady-state hydrologic conditions, including ground-water recharge. Seasonal climatic variability produces extreme changes in both the spatial and temporal distribution of recharge. Under consideration of the decreasing surface water resources available worldwide, and the real-ization of the importance of groundwater, there is no doubt that recharge estimation becomes more and more important for sustainable groundwater management.

Estimation of groundwater recharge, which is fundamentally transient and spatially variable [6] can be a difficult task [41] and their estimates are often connected with a high degree of uncertainty. There are six general approaches noted in the literature for recharge estimation:

   (i)   direct measurements,
  (ii)   chemical methods,
 (iii)   geophysical methods,
  (iv)   derivation from the water budget,
   (v)   analytical analysis of the groundwater table fluctuation, and
  (vi)   numerical and analytical methods related to infiltration processes.

Direct measurements as well as chemical methods tend to depend on point measurements using lysimeters, isotopes, heat and other tracers.

Direct measurements can also be obtained on the catchment scale by stream gauging [30, 41]. However, these two methods (i and ii) are susceptible to measurement errors and spatial variability, and are often limited by their cost [27]. Geophysical methods (iii above) attempt to differentiate between the changes in soil moisture content using an associated physical property. For example, Ganz et al. [19] used electrical resistivity tomography to monitor a recharge experiment over 50 minutes near Hanover, Germany. Such a method needs sophisticated equipment and creates substantial data, which have to be inferred from soil moisture. This results in relatively short observation periods and long processing times. Recharge may be derived from the water budget (iv above) if other components in the hydrologic cycle are known with sufficient accuracy. Budget-based methods are often used for studies where meso to large-scale area recharge is needed on a gridded basis (e.g., the river Elbe basin, [29]). However, these estimations are highly uncertain at the local scale [51]. Observations of groundwater table fluctuations (v above) are another widely applied method in basins where wells are available [32, 33]. This method suffers from uncertainty due to imprecision in porosity estimation especially when working in (semi-) conducting aquifers [23]. Indirect analytical and numerical (vi above) methods intend to predict the soil water movement. The initial implementation of this method occurred before the 1990s, e.g., Gardner [20] showed that, for water contents above field capacity, the discharge out of a soil column behaves, in good approximation, proportional to the square of water content. Another commonly used method to estimate recharge is physical-based vadose zone modeling [18, 26, 46]. These studies pointed out that vadose zone modeling is effective in the estimation of the "true" recharge.

This study uses an indirect analytical and numerical method (iv above) to predict recharge. In order to determine reliable recharge estimations using physical-based vadose zone modeling, accurate weather, vegetation, and soil data, such as soil temperature and soil water content are needed [9, 21, 22]. Specifically, the HYDRUS-1D finite element code was developed to solve the Richards equation in vertical direction. This code is widely used to simulate one-dimensional water movement in variable saturated media [44]. Assefa and Woodbury [6] and

Chen, Willgoose [12] pointed out that the HYDRUS-1D can provide acceptable results of the infiltration rate and accumulative infiltration at different scales. Most soil moisture measurements are carried out in the uppermost part of the vadose zone, especially at the root zone, due to the feasible depths of the sensors [39]. The limited installation depth indirectly addresses uncertainties in distinguishing between deep percolation and recharge due to the lack of information below the root zone [31, 37, 39]. Installing sensors to measure soil moisture content as a necessary step for recharge estimation is critical.

Authors believe that the best results can be achieved by using remote sensor techniques to measure all required field data [25, 47] since the technique is: (i) a relatively inexpensive and rapid method of acquiring up-to-date information (ii) capable of obtaining data from remotely accessible areas (iii) able to reduce travel between observation sites and working stations, and (iv) compatible with computer devices and software, e.g. combined with GIS [4]. An automated data recorder can be installed with a weather station, which records detailed climate data (e.g., air temperature, precipitation, wind speed, solar radiation) and thus, adds valuable data to any project. The additional transfer of data using a cellular network allows frequent retrieval of the observed data and early warnings on malfunctions of the sensors. This total data package allows for numerical modeling using the observed soil moisture for calibration, while simulating the recharge with climate data as input. Of major importance for the acceptance of such a method are cost and robustness.

This study shows how data from a low cost remote sensor weather station with additional soil moisture and soil temperature sensors from short-term observations can be used for robust recharge predictions. In order to calculate the relevant long-term recharge estimates, data from a short-term measurement campaign are temporally upscaled and added to additional weather data from Environment Canada.

Finally, authors have applied the aforementioned monitoring-modeling approach to the Abbotsford aquifer. An intensive agricultural use of nutrients and a high annual precipitation can trigger leaching, which is a key factor for endangering the groundwater quality within this important trans-boundary aquifer.

1.2 METHODS AND MATERIALS

1.2.1 STUDY AREA

This study was conducted in Southern Abbotsford, British Columbia, Canada (Figure 1.1). Most of the Abbotsford area is situated over the Abbotsford-Sumas Aquifer. The Abbotsford-Sumas Aquifer is a trans-boundary aquifer, which spans the border between British Columbia and Western Washington and provides drinking water for more than 100,000 people in both areas [49]. The whole aquifer covers 260 km^2 and groundwater in British Columbia flow from northeast to southwestern.

The annual average temperature recorded from 1971 to 2000 in Abbotsford was 10°C, and the average daily maximum and minimum

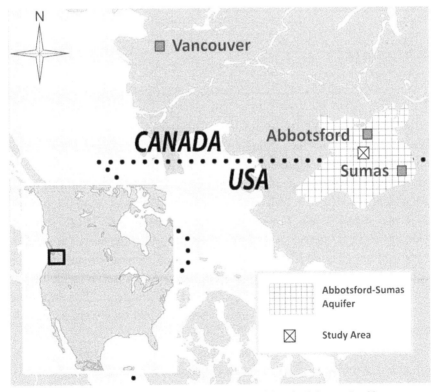

FIGURE 1.1   Location of the study area and of the Abbotsford-Sumas Aquifer.

temperatures were 14.7°C and 5.3°C, respectively [15]. The climate in the study area is oceanic, being influenced by the Pacific Ocean. The region has mild and moist winters, and majority of precipitation falls in the hydrologic winter period; approximately 70% of the total precipitation occurs during the October-March period. The mean annual precipitation is 1507 mm (1984–2013). Most of it (96.5%) is contributed by rainfall and the rest by snowfall [15].

The soil type is a sandy soil with a medium permeability. The groundwater table at the outer boundaries of the aquifer is only 0–5 m below the surface. In the central portion, where the study area is located, the groundwater table is at least 30 m below the surface [2]. This region has the largest agriculture production and heaviest concentration of agriculture-related goods in British Columbia [11]. The dominant agricultural crop is red raspberry (*Rubus idaeus L.*) with significant areas in forage grass (*Dactylis glomerata L.*) and pasture [53].

1.2.2 DATA

A state-of-art HOBOTM U30 weather station at (49.010441 *N*, –122.33256 E) was installed in April 2012. This weather station contains sensors, which are capable of recording soil-related data and weather data. The weather station provides "plug-and-play" smart sensors for measuring soil moisture, soil temperature and climate data, including precipitation, air temperature, wind speed, atmospheric pressure, relative humidity and solar radiation (Table 1.1). The recording time step was set to 30 minutes to receive detailed data. The soil temperature sensors and FDR (Frequency Domain Reflectometry), which observed soil moisture, were installed at two locations (S1 and S2) in the near vicinity of the weather station. Three sensors were installed vertically at each location to observe soil temperature and soil moisture at 10 cm, 37 cm and 100 cm depth. The data were stored both internally and on a cell network so that they were accessible via the internet.

The cost of the weather station, including the additional soil sensors was about CAD 4,890. In addition, API [5] reported that the average relative accuracy and cost of groundwater numerical models are moderate to high (low cost: less than US$10,000; moderate cost: US$ 10,000 ~ US$ 50,000;

TABLE 1.1   Specifications of Sensors

Parameter	Instrument	Installation	Range	Accuracy
Barometric Pressure	S-BPB-CM50	1 m height	600–1070 mbar	±3.0 mbar
Rain fall	S-RGB	1 m height	0–127 mm/h	±1% <20 mm/h
Solar Radiation (Spectral Range):	S-LIB-M003 300 to 1100 nm	2 m height	0–1280 W/m^2	±10 W/m^2
Air temperature	S-THB-M008	2 m height	–40–75°C	±0.13°C
Relative humidity	S-THB-M008	2 m height	0–100%	±2.5%
Soil temperature	S-TMB-M006	10 cm depth	–40–100°C	±0.2°C
Water content	S-SMC-M005	10 cm depth	0–0.55 m^3/m^3	±3.1%
Wind speed	S-WSA-M003	2 m height	0–45 m/s	±1.1 m/s

high cost: greater than US\$ 50,000 (US\$ 1 ≈ CAD 1.36 in year 1996), respectively. The annual cost for cellular telemetry is CAD 300. With the consideration of economic development and price inflation, the total cost of CAD 5,190 at the present time has to be considered as low cost. Additionally, one day each was required for installing and dismantling of the weather station and one day for maintenance and installing of an additional soil moisture sensor at 100 cm depth (15th August 2012). Each field trip needed two extra days for traveling. This results in a total time of about nine working days for the weather station.

The average temperature recorded by the HOBO™ weather station was 10°C; maximum and minimum temperature were measured as 25°C and –3°C, respectively in the observation period from 18th April 2012 to 19th March 2013. Nine days had average daily air temperatures below 0°C during the observation period. The available precipitation sensors do not allow for the differentiation between snow and rain (Table 1.1). Therefore, any precipitation during the observation period could be snow but on average 96.5% of the precipitation in the region is rain [15]. Due to the oceanic climate nearly 60% of the total precipitation (1375.8 mm) within the observation period was contributed in the hydrologic winter period, from 1st November 2012 to 19th March 2013 (date of dismantling).

Additional weather data of the Abbotsford A climate station was provided by Environment Canada [15] to determine a long term estimate of recharge. The Abbotsford A climate station is located about 2 km northwest from the installed HOBO™ weather station. Daily maximum temperature, minimum temperature, total precipitation, wind speed and solar radiation from 1984 to 2010 (27 years) recorded at the Abbotsford A (Abbotsford Airport) climate station were available for this location. The mean temperature was 10°C, and the annual mean precipitation was recorded as 1529 mm/year with a standard deviation σ of 223 mm/year in the 27-year period.

1.2.3 LABORATORY MEASUREMENTS

One soil sample was collected from each S1 and S2. These two soil samples (S1 and S2) represented the entire depth covered by the three sensors (depth of 36 inches, almost 1 m). A standard soil core sampling kit was used to extract the soil samples during the disassembly of the weather station. Two laboratory measurements were conducted to determine the properties of soil samples: i) sieving analyzes and ii) constant head permeability test. These two tests were applied to each soil sample.

The sieve analysis was carried out according to ASTM C 136–06 [7] using an Oscillatap ML-4330. TS Sieve Shaker with sieves at 6 different sizes: 4.75 mm, 2.00 mm, 0.85 mm, 0.42 mm, 0.18 mm and 0.075 mm. The resulting data were applied to the USDA textural classification to determine the sand [48], silt and clay composition of all samples (Table 1.2).

TABLE 1.2   Soil Texture and Classification

Soil sample depth	Sand, %	Silt, %	Clay, %	Soil texture
S1 0"–6"	87.1	4.9	8.0	Sand
S1 10"–16"	55.2	41.8	3.0	Sandy loam
S1 18"–24"	58.0	39.0	3.0	Sandy loam
S2 0"–6"	63.2	31.8	5.0	Sandy loam
S2 10"–16"	66.2	32.4	1.3	Sandy loam
S2 18"–24"	55.9	22.1	22.0	Sandy clay loam
S2 30"–38"	84.1	5.9	10.0	Loamy sand

The permeability testing was conducted in accordance with ASTM D 2434–68 [8] and can be applied to granular, non-cohesive soils with higher conductivities. All soil samples were tested three times at five different hydraulic gradients. The average value of hydraulic conductivities with a standard deviation was used for each sample soil during the numerical modeling (Table 1.3).

1.3 VADOSE ZONE MODELING

1.3.1 GOVERNING EQUATIONS

Physical-based vadose zone modeling was applied to estimate recharge, using HYDRUS-1D, version 4.16 [44] as a modeling tool. There are five governing equations used in simulating heat, moisture transport, and in the estimation of recharge in the study area. First of all, using the Richards' equation for variably saturated water and convection-dispersion type equations, HYDRUS-1D numerically solves heat and moisture transport for a given soil [44]. There are three basic formulations of Richards' equation: head-based formulation, saturation formulation and mixed formulation [10]. Head-based formulations tend to large mass balance errors, and the saturation-based formulations are limited with discontinues nature of moisture content. To minimize mass balance errors without reducing modeling capability near saturation, the mixed form of Richards' equation [Eq. (1)] which combines head-based and saturation-based formulations is applied by HYDRUS-1D by using Galerkin-type linear finite element schemes [10, 44].

TABLE 1.3   Hydraulic Conductivities of Soil Samples

Soil sample depth	K_s cm/d	σ cm/d
S1 0"–6"	27.8	0.7
S1 10"–16"	7.0	0.2
S1 18"–24"	29.5	1.1
S2 0"–6"	96.4	8.8
S2 10"–16"	105.4	7.3
S2 18"–24"	36.5	1.5
S2 30"–38"	108.8	6.2

$$\frac{\partial \theta}{\partial t} = \frac{\partial}{\partial z}\left[K(\psi)\left(\frac{\partial \psi}{\partial z}\right) - S \right]$$

(1)

where, θ is the volumetric water content [L^3L^{-3}], ψ is the pressure head [L], t is the time [T], z is the elevation [L], S is the sink term [$L^3L^{-3}T^{-1}$] and $K(\psi)$ [LT^{-1}] is an unsaturated hydraulic conductivity function of ψ and of the saturated hydraulic conductivity K_s [LT^{-1}].

Secondly, the van Genuchten-Mualem (VGM) model [35, 50] was chosen as an indirect method for studying water retention behavior on the given soil samples. Hydraulic retentivity and conductivity functions are one category of indirect methods, which can be formulated from empirical nonlinear regression equations, and/or methods with more physical foundations [35, 50]. The van Genuchten-Mualem model is defined in Eqs. (2) and (3), and consists of eight parameters: pressure head h [L], saturated water content θ_s [L^3L^{-3}], residual water content θ_r [L^3L^{-3}], and shape empirical parameters α [L^{-1}], n [–] and m [–], saturated hydraulic conductivity K_s [LT^{-1}] and S_e [–] is the effective saturation.

$$\theta(h) = \begin{cases} \theta_r + \dfrac{\theta_s - \theta_r}{\left[1+|\alpha h^n| \right]^m} & h < 0 \\ \theta_s & h \geq 0 \end{cases}$$

(2)

$$K(h) = K_s S_e^L \left[1 - \left(1 - S_e^{\frac{1}{m}} \right)^m \right]^2$$

(3)

Besides K_s, which was measured during the laboratory measurement period, actual measurements of soil hydraulic properties are time-consuming, complex, and rather costly. Therefore, the other four parameter within the VGM model were estimated using ROSETTA [42]. ROSETTA determines pedotransfer functions (PTF) to predict the water retention parameters according to van Genuchten and the saturated hydraulic conductivity by using the textural distribution as gathered in the laboratory from undisturbed soil samples as input (Table 1.3).

TABLE 1.4 VGM Parameters Using Pedotransfer Functions (PTF)

Soil sample depth	θ_r m³/m³	θ_s m³/m³	α 1/cm	n	K_s cm/d
S1 0"–6"	0.05	0.37	0.03	2.07	28.0
S1 10"–16"	0.03	0.41	0.02	1.45	7.0
S1 18"–24"	0.03	0.41	0.02	1.43	30.0
S2 0"–6"	0.03	0.39	0.03	1.41	96.4
S2 10"–16"	0.03	0.41	0.04	1.43	105.0
S2 18"–24"	0.06	0.39	0.02	1.36	36.5
S2 30"–38"	0.05	0.37	0.03	1.80	108.8

The Chung and Horton equation [13] is also dependent on the soil texture, and is defined in Eq. (3). The Chung and Horton equation is included in the HYDRUS-1D code and is used to estimate thermal conductivity. Table 1.2 shows that the main component of the soil is sand, therefore in this study, parameters were taken from the default values based on the soil class of sand.

$$\lambda_0(\theta) = b_1 + b_2\theta + b_3\theta^{0.5} \tag{4}$$

where, θ is the volumetric water content [L³L⁻³], and b_1, b_2 and b_3 are empirical parameters [MLT³K⁻¹].

The effect of evapotranspiration and root water uptake to the water distribution in the vadose zone is represented by Feddes-type uptake functions [17]. The Feddes functions estimate the flux due to root uptake as a function of potential transpiration and the pressure head. Since pasture is the vegetation type at the weather station, the default parameters from the database, which is included in the HYDRUS-1D code were used. The potential evapotranspiration was calculated by Penman-Monteith equation [3]. The methods requires climate data as input, such as daily mean temperature, wind speed, relative humidity and solar radiation, all of these data were recorded by HOBO™ weather station.

$$\lambda ET = \frac{\Delta(R_n - G) + \rho_a C_p \dfrac{e_s - e_a}{r_a}}{\Delta + \gamma(1 + \dfrac{r_s}{r_a})} \tag{5}$$

where, R_n [MT^{-3}] is the net radiation, G [MT^{-3}] is the soil heat flux, $(e_s - e_a)$ [$ML^{-1}T^{-2}$] represents the vapor pressure deficit of the air, ρ_a [ML^{-3}] is the mean air density at constant pressure, C_p [$L^2T^{-2}K^{-1}$] is the specific heat of the air, Δ[$ML^{-1}T^{-2}K^{-1}$] represents the slope of the saturation vapor pressure temperature relationship, γ [$ML^{-1}T^{-2}K^{-1}$] is the psychrometric constant, and r_s and r_a [TL^{-1}] are the (bulk) surface and aerodynamic resistances.

1.3.2 BOUNDARY CONDITIONS

Boundary conditions (BCs) can be either system-dependent or system-independent. In this study, BCs were related to the external situation, such as variable precipitation, the distinct temperature difference between summer and winter, and the existing soil moisture conditions. The upper BC in the study area was defined as "atmospheric BC with surface runoff" in order to address the high precipitation volume and available daily meteorological data. With the potential of surface ponding, the BC was obtained by limiting the absolute value of the flux by the following two conditions, prescribed flux or prescribed pressure head [16]:

$$\left| -k\frac{\partial \varphi}{\partial x} - K \right| \leq E \text{ at } x = L, \text{ and} \qquad (6)$$

$$\varphi_A \leq \varphi \leq \varphi_S \text{ at } x = L \qquad (7)$$

where, E is defined as the maximum potential rate of infiltration or evapotranspiration under the given atmospheric conditions [LT^{-1}], φ_A and φ_S and are minimum and maximum pressure head at the soil surface [L]. However, when one of the end points of Eq. (7) [44] is reached, a prescribed head boundary condition will be used to calculate the actual surface flux. Similarly, when any point results in exceeding the maximum potential rate of infiltration or evapotranspiration described in Equation 6, the potential rate will be used as a prescribed flux boundary which will be considered as saturation excess overland flow [6].

Using "free drainage" as the lower BC is a situation often occurs in field studies of water flow and drainage in the vadose zone. This lower boundary condition is most appropriate for situations where the water table lies far below the domain of interest [44]. In the central part of the study area, the depth to the groundwater table is larger 30 m (refer to Study Area). Therefore, the lower BC is considered free drainage at a depth of 2 m since this is already below the root zone and an effective capillary rise.

1.3.3   CALIBRATION

Under section Governing Equations, the prediction of the VGM parameters using ROSETTA is described. However, ROSETTA is developed based on a set of soil samples obtained from the USA and European countries. Thus, it could cause inaccurate predictions when applied on the soil samples from anywhere else, such as Canada [44]. For these reasons, the HYDRUS-1D parameter estimation module [44] was used at both sites to improve estimates for the purpose of providing an authentic description of soil properties. Since large amount of parameters involved for each layer of soil, only some of them are really verifiable, such as hydraulic conductivity, which we obtained from laboratory measurement, and heat transport parameters determined by soil texture from HYDRUS-1D database. The VGM parameter θ_r, α, and n were estimated in this study. As a consequence, these hydraulic parameters were used as calibration parameter versus the observed soil moisture and soil temperature data at different depths.

Various performance criteria were used in the calibration of HYDRUS-1D vadose zone models to determine and minimize the misfit between observations and model simulations. In this study, three performance criteria were applied to the calibration results. Firstly, ME (Mean Error) is the most commonly used method to test if the simulations are over- or under- predicted (Equation 8) [24].

$$ME = \frac{1}{n}\sum_{i=1}^{n}(S_i - O_i)$$ (8)

where, n is the observation days [T], S_i is the simulation value at the ith day, and O_i is the observation value at the ith day. The RMSE (Root Mean Square Error) is a measure of the calibration accuracy (Equation 9) [24].

$$\text{RMSE} = \sqrt{\frac{\sum_{i=1}^{n}(S_i - O_i)^2}{n}} \tag{9}$$

Zero is a perfect value for the ME and RMSE analysis, which means simulations can perfectly match with observations. The third method is the Nash–Sutcliffe Efficiency NSE [36], which is used to assess the predictive power of hydrological models.

$$\text{NSE} = 1 - \frac{\sum_{i=1}^{n}(O_i - S_i)^2}{\sum_{i=1}^{n}(O_i - \overline{O_i})^2} \tag{10}$$

where, $\overline{O_i}$ is the mean observed value.

NSE can range from $-\infty$ to 1. An efficiency of 1 corresponds to a perfect match of modeled value to the observed data while values of 0 shows that the modeling results are as good as the mean of the predictor.

1.3.4 SENSITIVITY ANALYSIS

A sensitivity analysis is typically required to evaluate model performance in terms of the importance, strength and relevance of the input parameters in determining the variation in the output [40] and to identify sensitive parameters as a way of screening parameters for calibration. One key objective of this study is to develop a robust model to estimate recharge. Therefore, in this study, the sensitivity analysis was used to test the robustness of the simulated results. This was accomplished by determining the influence of the VGM model inputs on the recharge estimate. The sensitivity of the recharge and the soil moisture content were evaluated, which were the actual calibration measures.

The sensitivity analysis is limited to soil moisture as a reference, since calibration showed that the model outputs are much more sensitive to changes in soil moisture than soil temperature. The various parameters in the VGM model were varied to determine their influence on the calculated water retention and recharge amount. The key parameters involved in the VGM model were chosen to estimate the robustness of the model: empirical shape parameters α and n and saturated hydraulic conductivity.

One of the simplest and most common approaches is that of changing one-factor-at-a-time to determine the effect produced on the output [14]. Therefore, all other parameters were fixed at the baseline by changing one parameter at a time. To increase the comparability of the results, the calibration result was kept at the baseline value and the VGM parameter α and n were individually changed by ±5%, ±10% and ±25%. K_s were changed by ± σ.

1.4 RESULTS

1.4.1 CALIBRATION

The soil moisture content was calibrated at 10 cm, 37 cm and 100 cm depth and the soil temperature at 10 cm and 37 cm depth for each site (Figures 1.2(a–c) and 1.3(a-b)). Comparing the simulated soil moisture and soil temperature using the estimated VGM parameters with those estimated from the ROSETTA derived parameters showed that the model was successfully adjusted as demonstrated by low ME, low RMSE and close-to-1 NSE for both soil moisture and soil temperature (Table 1.5). All of these measures represent values, which are much beyond the calibration standard for hydrological models [34]. The difference in soil moisture and soil temperature from both sites by means of RMSE was between 2 and 4% and between 1.06°C and 1.91°C, respectively, and by means of NSE ranged from 0.67 to 0.88 and from 0.92 to 0.96, respectively (Table 1.5). MEs showed that there was not a tendency for over- or under-prediction since the values were near zero.

1.4.2 SENSITIVITY ANALYSIS

The VGM parameters α, n, and K_s were evaluated further during the sensitivity analysis. Overall, the same importance at each depth was identified. Taking analysis results at 100 cm depth as an example, the empirical shape parameter n (Figure 1.4) showed the highest sensitivity on soil moisture estimation, followed by α then K_s.

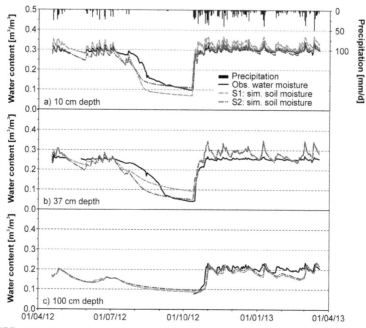

FIGURE 1.2 Calibration of the soil water contents: Top (figure a) at 10 cm, Center (figure b) at 37 cm, and Bottom (figure c) at 100 cm depth.

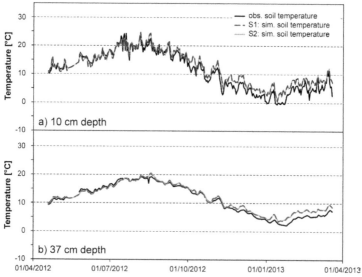

FIGURE 1.3 Calibration of the soil temperature: Top (figure a) at 10 cm, and Bottom (figure b) at 37 cm depth.

TABLE 1.5 Calibration Performance Testing by RMSE, AE and NSE (Calibration Reference: Soil Moisture)

	Parameter	S1			S2		
Soil moisture	Depth [cm]	10	37	100	10	37	100
	RMSE [m³/m³]	0.038	0.034	0.030	0.024	0.040	0.020
	AE [m³/m³]	0.002	0.015	−0.021	−0.009	0.000	−0.010
	NSE [–]	0.70	0.76	0.77	0.88	0.67	0.83
Soil temperature	RMSE [°C]	1.90	1.07		1.91	1.06	
	AE [°C]	1.28	0.59		1.24	0.59	
	NSE [–]	0.90	0.96		0.92	0.96	

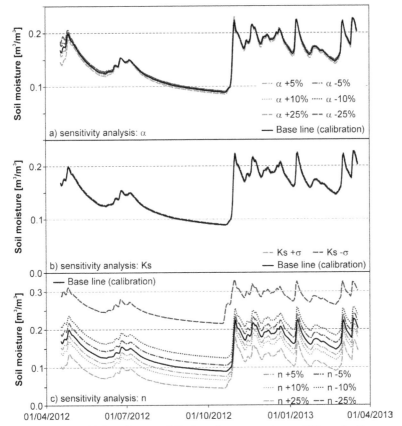

FIGURE 1.4   Sensitivity of n, α, and K_s for soil water content at 100 cm depth.

Taking soil sample S1 as an example, the sensitivity analysis showed that for the parameter n the recharge values changed by –0.9%/+1.2%, –1.5%/+2.9%, and –2.3%/+12% if the n value was changed by 5%, 10% and 25%, respectively. Similarly, changes were received for α and K_s, the order of sensitivity of parameters for groundwater recharge was: n, α, and K_s. Comparing the sensitivity of α with the sensitivity of n, α was slightly less sensitive. In addition, all changes were relatively small, and the changes as measured percent change were nearly constant which approximately resulted in a linear behavior of the changes. Final cumulated recharge was stable at ± 5% of the baseline (Figure 1.5).

1.4.3   RECHARGE

The total precipitation recorded by the HOBO™ weather station during the observation period (from 18th April 2012 to 19th March 2013) was 1375.8 mm. This was similar to the Environment Canada observation (1360.4 mm) at the Abbotsford Airport weather station over the same period. The groundwater recharge was simulated as 863 mm and 816 mm for S1 and S2 respectively (Figure 1.6).

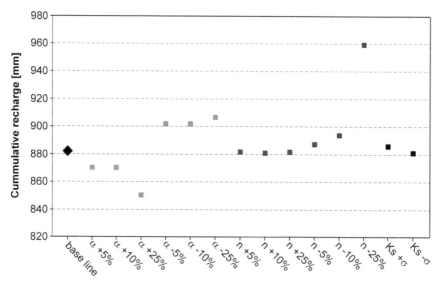

FIGURE 1.5   Impact of parameter changes on the cumulative recharge.

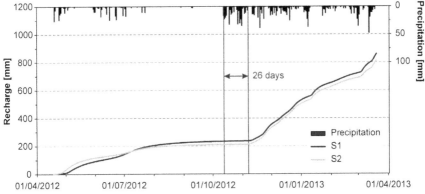

FIGURE 1.6   Simulated recharge for the observed period.

The ratio of recharge-precipitation was calculated as 63% and 59% for S1 and S2, respectively. Figure 1.6 also presents a 26-day delay between the beginning of the rains (October 12th) and the effective start of the recharge (November 7th) based on the lower BC at 2 m depth. The lag time increases with the larger depth of the lower BC.

To evaluate the behavior of recharge at lower depth, the lower BC was extended to a depth of 30 m. The recharge estimate for the long-term period was almost the same using a 30 m and 2 m lower BC depth. The recharge estimate for the shorter period resulted in larger difference due to the initial condition of the soil moisture, which was set to field capacity. However, all the modeling results below the 2 m depth for the lag time are vague, due to insufficient field data. For the long-term study period, 1984–2010 (27 years). The annual average precipitation was calculated as 1529 mm year with σ of 223 mm. The resulting recharge was estimated in the 27-years period to 848 mm/year with σ of 206 mm/year and 859 mm/year having σ of 208 mm/year for the S1 and S2, respectively (Figure 1.7). Model results suggest that the ratio of annual recharge-precipitation varied from 43% to 69%.

Monthly recharge statistics showed that the maximum recharge occurred in December or January with a mean value of 177 mm/month with σ of 82 mm/month and mean value of 185 mm/month with σ of 72 mm/month, respectively (Figure 1.8). According to the monthly mean recharges, there was a nearly linear decrease in recharge rate of 21 mm/month from January to October and an increase in recharge rate of 57 mm/month from October to December.

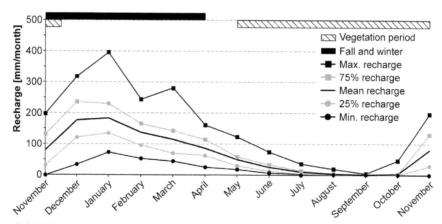

FIGURE 1.7   Monthly estimated recharge distribution in long-term period.

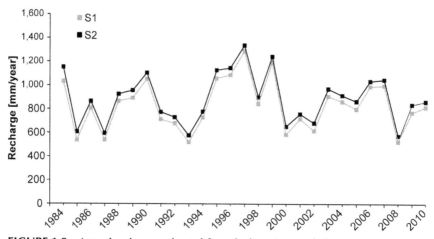

FIGURE 1.8   Annual recharge estimated from the long-term period.

1.5   DISCUSSIONS

Soil moisture and soil temperature measurements contain standard measurement errors. The overall error in soil moisture estimation expressed by RMSE (1–2%, Table 1.5) had the same magnitude as the measurement error of the soil moisture sensors (3.1%, Table 1.1). However, the additional calibration parameter, soil temperature, showed a RMSE of 1.1 to 1.9°C

(Table 1.5) which was larger than the measurement error of the temperature sensors (0.2°C, Table 1.1). Since the soil moisture drives the water flow within the vadose zone and, therefore, the recharge estimation, the adopted method is able to predict the recharge robustly without increasing any errors due to the measurement. Therefore, the uncertainty in the prediction of recharge is not increased by the modeling process but depends on the quality of the climate observations. Also the NSE and the ME verify the good agreement between estimated values and measured data.

In contrast to the prediction quality over the whole period, there was a significant difference in the predicted and observed soil moisture at 37 cm depth starting in October 2012. The simulated soil moisture at 37 cm depth (Figure 1.2b) regularly changed with precipitation events and showed approximate linear changes with precipitation amount. The observed soil moisture did not show this sensitive behavior to the precipitation; the soil moisture was kept at 0.25 m³/m³ ± 0.01 m³/m³. This may be the result of a mismatch between the scale of the soil moisture measurement and of the numerical result: while the field measurement integrates the soil moisture content of the neighboring soil, the numerical model explicitly determines the soil moisture of a cell with a thickness of 1 cm. Additionally, the heterogeneity of the soil has to be taken into consideration. The obtained soil data at the two sites may not reflect the general heterogeneity in the area. However, this general issue of heterogeneity is addressed by the use of two soil columns. Although any impact of heterogeneity cannot be negated, in this case, there is strong evidence that the behavior of the observed soil moisture was not due to the heterogeneity because simulated soil moisture at both sites is very similar.

Soil water contents (Figure 1.2a–c) showed the largest variation near the soil surface. With increasing depth, the variation became smaller. This has two reasons: (i) the root density was coupled with the root water uptake. So the transpiration effects were decreasing with increasing depth, and (ii) the unsaturated hydraulic conductivity decreased non-linearly as soil moisture content decreased. This resulted also in lower variations with increasing soil depth.

From the sensitivity analysis, the empirical shape parameter n was found to be most sensitive. This agrees with findings by Abbasi, Jacques [1]. However, Figure 1.5 indicates that no matter how much each

VGM parameter (α, n, and K_s) was changed; the cumulative recharge was not significantly changed.

The annual recharge estimated on the two sites over the 27-year period was very similar: the annual average recharge calculation revealed for S1: 848 mm/year with σ of 206 mm/year and S2: 859 mm/year with σ of 208 mm/year. The estimate on the long-term annual recharge (Figure 1.7) underlined the robustness of the chosen method: The average recharge values agreed with 851–900 mm/year estimated by Scibek and Allen [43] in the Abbotsford region. Furthermore, the variation in the estimated recharge between the two sites was within the accuracy of the measurement.

The calculated ratio of annual recharge to precipitation ranged from 43% to 69%, which agrees with the study by Kohut [28]. Kohut reported that annual recharge range between 37% and 81% of the mean annual precipitation in the Abbotsford area. This interval indicates that the total amount of recharge through the vadose zone is controlled by the total amount of local precipitation. Figure 1.7 illustrates the effect of wet and dry years on recharge estimation. For instance, the amount of recharge in 1997 (S1: 1328 mm/year and S2: 1355 mm/year) almost doubled that in 1993 (587 mm/year and 595 mm/year respectively) with regard to the total precipitation (1997: 1999 mm/year, 1993: 1170 mm/year).

According to the monthly recharge statistic chart (Figure 1.7), the majority of recharge was contributed in the hydrological winter period from November to April (Figure 1.8), especially in the months of December (178 mm/month with σ of 82 mm/month) and January (185 mm/month with σ of 72 mm/month), which agrees with Piteau Associates Engineering Ltd. [38]'s report on the onsite climate conditions. On the contrary, only small amounts of recharge were simulated during the vegetation period (May to November). Therefore, based on the degree of land use in the study area and nitrogen loading from agricultural production [52], Abbotsford recharge system shows a high potential for groundwater contamination in hydrological winter period. Late summer fertilization, which contributes 33% of the yearly fertilization amount [45] is not recommended due to a rapid increase of recharge at the end of summer, with an average increase of 60 mm/month occurs from October to December. This recharge could result in the leaching of nutrients, such as nitrate, which are left in the root zone and have not been absorbed by the crops.

1.6 CONCLUSIONS

The study shows how data from a low cost remote sensor weather station with additional soil moisture and soil temperature sensor measurements from a short-term observation period can be used for robust recharge predictions. The method has a high accuracy of recharge estimation since the estimated recharge from the observation period (April 2012–March 2013) as well from the long-term period (1984–2010) agrees with recent studies in the same area. At this point, this method implements the practice of saving time and cost from using short-term data instead of long-term measurements. The main advantage of using remote sensor technique is that data can be obtained from difficult or even inaccessible areas. This method can overcome the accessibility problems and upload the data to user's device as scheduled.

The implemented vadose zone model, HYDRUS-1D, which only uses soil information and climate data as input, allows for cost-effective, efficient and robust recharge estimates. Results strongly suggest using temporally and spatially variant recharge in groundwater management, since recharge varies between S1 and S2. In other words, increasing the comprehensiveness of soil data measurements (multiple points) will better reflect the recharge behavior of an aquifer model. In addition, due to its physical-based parameters, this method can be applied to other locations. The HYDRUS-1D parameter estimation function, ROSETTA was used to make preliminary estimates of the VGM parameters, which resulted in a significantly shorter calibration time. The VGM parameters are often difficult to estimate. The parameter estimation allowed for limited experimental requirements and reduced time and costs. Other parameters can be easily obtained experimentally, such as soil texture and hydraulic conductivity; the authentic results prove that high-quality recharge estimation can be achieved by strictly applying ASTM standards. As a consequence, the final calibration performance was much beyond the normal standard for vadose zone modeling.

1.7 SUMMARY

Groundwater recharge estimation using physical-based modeling and data from a low-cost weather station with remote sensor technique is a critical development for sustainable groundwater management. A study was conducted to explore the feasibility and robustness a physical-based numerical method to estimate groundwater recharge on sandy soils. The study area is located in Southern Abbotsford, British Columbia, Canada. Recharge was simulated using the Richard's based vadose zone hydrological model, HYDRUS-1D. The required meteorological data were collected using a HOBO™ weather station for a short observation period (about 1 year), and an existing weather station (Abbotsford Airport) for long-term study purpose (27 years). Soil moisture and temperature was collected at 2 sites (S1 and S2) near the HOBO™ weather station.

Model performance was evaluated by using observed soil moisture and soil temperature data obtained from subsurface remote sensors. Recharge estimates from the short observation period, based on half-hourly meteorological data were 863 mm and 816 mm for S1 and S2 respectively. The annual average recharge estimates based on a time series of 27 years using daily data were 848 mm/year and 859 mm/year for S1 and S2 respectively. The relative ratio of annual recharge-precipitation varied from 43% to 69%. From a monthly recharge perspective, the majority (80%) of recharge due to precipitation occurred during the hydrologic winter period.

Peak recharge, up to 400 mm/month occurred in December and January. On the contrary, only 20% of annual average recharge was generated during the vegetation period. Overall, by comparing and contrasting with recent studies, the results indicate that the method is a robust, cost-efficient and effective method to reliably estimate recharge.

KEYWORDS

- **Abbotsford**
- **ASTM standard**
- **calibration**
- **field work**
- **heat transfer**
- **hydraulic conductivity**

- **HYDRUS-1D**
- **inverse modeling**
- **laboratory measurements**
- **long-term estimate**
- **low cost**
- **numerical modeling**
- **pedotransfer function**
- **physical-based modeling**
- **recharge estimation**
- **robust prediction**
- **ROSETTA**
- **sensitivity analysis**
- **short-term observation**
- **simulation**
- **soil moisture**
- **vadose zone**
- **van Genuchten-Mualem model**
- **weather station**

ACKNOWLEDGEMENTS

This research was supported by the Canadian Water Network (CWN). Mr. Lynch and Mr. Wang gave tremendous help while supporting this work in the laboratory and at the field site.

REFERENCES

1. Abbasi, F., Jacques, D., Simunek, J., Feyen, J., and van Genuchten, M. T., 2003. Inverse estimation of soil hydraulic and solute transport parameters from transient field experiments: Heterogeneous soil. *Transactions of the ASAE*, 46:1097–1111.
2. Abbotsford-Sumas Aquifer International Task Force, 2014. *Management of Specific Aquifers*. http://www.env.gov.bc.ca/wsd/plan_protect_sustain/groundwater/aquifers/absumas.html.

3. Allen, R. G., Pereira, L. S., Raes, D., and Smith, M., 1998. *Crop evapotranspiration. Guidelines for computing crop water requirements.* Irrigation and Drainage Paper 56. Rome: FAO. pages 300.

4. Anbazhagan, S., Ramasamy, S. M., and Gupta, S. D., 2005. Remote sensing and GIS for artificial recharge study, runoff estimation and planning in Ayyar basin, Tamil Nadu, India. *Environmental Geology*, 48:158–170.

5. API, 1996. *Estimation of Infiltration and Recharge for Environmental Site Assessment.* American Petroleum Institute: Albuquerque, New Mexico. p. 204.

6. Assefa, K. A. and Woodbury, A. D., 2013. Transient, spatially varied groundwater recharge modeling. *Water Resources Research*, 49:4593–4606.

7. ASTM, 2006. *ASTM C136–06. Standard test method for sieve analysis of fine and coarse aggregates.* American Society for Testing and Materials: West Conshohocken, Pennsylvania, PA 19428, USA.

8. ASTM, 2006. *ASTM D2434–68. Standard Test Method for Permeability of Granular Soils (Constant Head).* American Society for Testing and Materials: West Conshohocken, PA.

9. Bormann, H., Holländer, H. M., Blume, T., Buytaert, W., Chirico, G. B., Exbrayat, J. F., Gustafsson, D., Hölzel, H., Kraft, P., Krauße, T., Nazemi, A., Stamm, C., Stoll, S., Blöschl, G., and Flühler, H., 2011. Comparative discharge prediction from a small artificial catchment without model calibration: Representation of initial hydrological catchment development. *Die Bodenkultur: Journal for Land Management, Food and Environment*, 62:23–29.

10. Celia, M. A., Bouloutas, E. T., and Zarba, R. L., 1990. A General Mass-Conservative Numerical-Solution for the Unsaturated Flow Equation. *Water Resources Research*, 26:1483–1496.

11. Chamber of Commerce, 2008. *The Economic Impact of Agriculture in Abbotsford* in *Strengthening Farming Report.* Chamber of Commerce, Ministry of Agriculture and Lands.

12. Chen, M., Willgoose, G. R., and Saco, P. M., 2014. Spatial prediction of temporal soil moisture dynamics using HYDRUS-1D. *Hydrological Processes*, 28:171–185.

13. Chung, S. O. and Horton, R., 1987. Soil heat and water flow with a partial surface mulch. *Water Resources Research*, 23:2175–2186.

14. Czitrom, V., 1999. One-Factor-at-a-Time Versus Designed Experiments. *The American Statistician*, 53:126–131.

15. Environment Canada, 2013. *Abbotsford daily data report.* http://climate.weather.gc.ca.

16. Feddes, R. SA., Bresler, E., and Neuman, S. SP., 1974. Field-Test of a Modified Numerical-Model for Water Uptake by Root Systems. *Water Resources Research*, 10:1199–1206.

17. Feddes, R. A., Kowalik, P. J., and Zaradny, H., 1978. *Simulation of field water use and crop yield.* Simulations MonograPhone Wageningen: Pudoc. p. 189.

18. Franke, O. L., Reilly, T. E., Pollock, D. W., and LaBaugh, J. W., 1998. *Estimating Areas Contributing Recharge to Wells*, USGS. U.S. Government Printing Office: Denver, CO, US. p. 14.

19. Ganz, C., Bachmann, J., Noell, U., Duijnisveld, W. H. M., and Lamparter, A., 2014. Hydraulic Modeling and in situ Electrical Resistivity Tomography to Analyze Ponded Infiltration into a Water Repellent Sand. *Vadose Zone Journal*, 13.

20. Gardner, W.R., 1962. Approximate Solution of a Non-Steady-State Drainage Problem. *Soil Science Society of America Journal*, 26:129–132.
21. Holländer, H. M., Blume, T., Bormann, H., Buytaert, W., Chirico, G. B., Exbrayat, J. F., Gustafsson, D., Hölzel, H., Kraft, P., Stamm, C., Stoll, S., Blöschl, G., and Flühler, H., 2009. Comparative predictions of discharge from an artificial catchment (Chicken Creek) using sparse data. *Hydrology and Earth System Sciences*, 13:2069–2094.
22. Holländer, H.M., Bormann, H., Blume, T., Buytaert, W., Chirico, G. B., Exbrayat, J. F., Gustafsson, D., Hölzel, H., Krauße, T., Kraft, P., Stoll, S., Blöschl, G., and Flühler, H., 2014. Impact of modelers' decisions on hydrological a priori predictions. *Hydrology and Earth System Sciences*, 18:2065–2085.
23. Holländer, H.M., Mull, R., and Panda, S. N., 2009. A concept for managed aquifer recharge using ASR-wells for sustainable use of groundwater resources in an alluvial coastal aquifer in Eastern India. *Physics and Chemistry of the Earth, parts A/B/C*, 34:270–278.
24. Hyndman, R. J. and Koehler, A. B., 2006. Another look at measures of forecast accuracy. *International Journal of Forecasting*, 22:679–688.
25. Jackson, T. J., 2002. Remote sensing of soil moisture: implications for groundwater recharge. *Hydrogeology Journal*, 10:40–51.
26. Jimenez-Martinez, J., Skaggs, T. H., van Genuchten, M. T., and Candela, L., 2009. A root zone modeling approach to estimating groundwater recharge from irrigated areas. *Journal of Hydrology*, 367:138–149.
27. Jyrkama, M. I., Sykes, J. F., and Normani, S. D., 2002. Recharge estimation for transient ground water modeling. *Ground Water*, 40:638–648.
28. Kohut, A.P., 1987. *Ground Water Supply Capability Abbotsford Upland*. W. M. B. B.C. Ministry of Environment. p. 16.
29. Kunkel, R. and Wendland, F., 2002. The GROWA98 model for water balance analysis in large river basins—the river Elbe case study. *Journal of Hydrology*, 259:152–162.
30. Lerner, D. N., Issar, A. S., and Simmers, I., 1990. *Groundwater Recharge: A Guide to Understanding and Estimating Natural Recharge*. Hannover: Verlag Heinz Heise. p. 345.
31. Leterme, B., Mallants, D., and Jacques, D., 2012. Sensitivity of groundwater recharge using climatic analogs and HYDRUS-1D. 16:2485–2497.
32. Meinzer, O. E., 1923. *The occurrence of groundwater in the United States with a discussion of principles*. US Geol. Surv. Water-Supply. p. 321.
33. Meinzer, O. E. and Stearns, N. D., 1929. *A study of groundwater in the Pomperaug Basin, Conn. with special reference to intake and discharge*. US Geol. Surv. Water-Supply. p. 73–146.
34. Moriasi, D. N., Arnold, J. G., Van Liew, M. W., Bingner, R. L., Harmel, R. D., and Veith, T. L., 2007. Model evaluation guidelines for systematic quantification of accuracy in watershed simulations. *Transactions of the ASABE*, 50:885–900.
35. Mualem, Y., 1976. A new model for predicting the hydraulic conductivity of unsaturated porous media. *Water Resources Research*, 12:513–522.
36. Nash, J. E. and Sutcliffe, J. V., 1970. River flow forecasting through conceptual models part I — A discussion of principles. *Journal of Hydrology*, 10:282–290.
37. Ochoa, C. G., Fernald, A. G., Guldan, S. J., and Shukla, M. K., 2007. Deep percolation and its effects on shallow groundwater level rise following flood irrigation. Transactions of the ASABE, 50:73–81.

38. Piteau Associates Engineering Ltd., 2006. *Hydrogeological investigation for ground-water supply development, North Abbotsford, B.C.* City of Abbotsford.

39. Rimon, Y., Dahan, O., Nativ, R., and Geyer, S., 2007. Water percolation through the deep vadose zone and groundwater recharge: Preliminary results based on a new vadose zone monitoring system. *Water Resources Research*, 43.

40. Saltelli, A., Ratto, M., Andres, T., Campolongo, F., Cariboni, J., Gatelli, D., Saisana, M., and Tarantola, S., 2008. *Global Sensitivity Analysis. The Primer.* Chichester, England: John Wiley & Sons. p. 292.

41. Scanlon, B. R., Healy, R., and Cook, P. G., 2002. Choosing appropriate techniques for quantifying groundwater recharge. *Hydrogeology Journal*, 10:18–39.

42. Schaap, M. G., Leij, F. J., and van Genuchten, M. T., 2001. ROSETTA: A computer program for estimating soil hydraulic parameters with hierarchical pedotransfer functions. *Journal of Hydrology*, 251:163–176.

43. Scibek, J. and Allen, D. M., 2005. *Numerical groundwater flow model of the Abbotsford-Sumas Aquifer, Central Fraser Lowland of BC, Canada and Washington State, US.* Ottawa: Environment Canada. p. 203.

44. Simunek, J., van Genuchten, M. T., and Sejna, M., 2008. *The HYDRUS-1D Software Package for Simulating the Movement of Water, Heat, and Multiple Solutes in Variably Saturated Media, Version 4.0.* Department of Environmental Sciences, University of California Riverside, Riverside, California, USA. p. 296.

45. Spectrum Analytic Inc., *A Guide to Fertilizing Raspberries and Other Brambles.* Washington C. H., Ohio. p. 8.

46. Timlin, D., Starr, J., Cady, R., and Nicholson, T., 2003. *Comparing Ground-Water Recharge Estimates Using Advanced Monitoring Techniques and Models*, USDA Agriculture, Editor.: Beltsville, MD. p. 57.

47. Tweed, S. O., Leblanc, M., Webb, J. A., and Lubczynski, M. W., 2007. Remote sensing and GIS for mapping groundwater recharge and discharge areas in salinity prone catchments, southeastern Australia. *Hydrogeology Journal*, 15:75–96.

48. USDA, 1987. *Soil textural soil classification study guide.* United States Department of Agriculture, Washington, D.C. U.S. p. 48.

49. USGS, 1999. Hydrogeology, Ground-Water Quality, and Sources of Nitrate in Lowland Glacial Aquifers of Whatcom County. *Washington and British Columbia, Canada.* 98–4195.

50. van Genuchten, M. T., 1980. A closed-form equation for predicting the hydraulic conductivity of unsaturated soils. *Soil Science Society of America Journal*, 44:892–898.

51. Varni, M. R. and Usunoff, E. J., 1999. Simulation of regional-scale groundwater flow in the Azul River basin, Buenos Aires Province, Argentina. *Hydrogeology Journal*, 7:180–187.

52. Zebarth, B. J., Hii, B., Liebscher, H., Chipperfield, K., Paul, J. W., Grove, G., and Szeto, S. Y., 1998. Agricultural land use practices and nitrate contamination in the Abbotsford Aquifer, British Columbia, Canada. *Agriculture Ecosystems and Environment*, 69:99–112.

53. Zebarth, B. J., Paul, J. W., and Van Kleeck, R., 1999. The effect of nitrogen management in agricultural production on water and air quality: evaluation on a regional scale. *Agriculture Ecosystems and Environment*, 72:35–52.

CHAPTER 2

CONSTRAINTS IN RAINFALL-RUNOFF MODELING USING ARTIFICIAL NEURAL NETWORK

M. U. KALE, M. M. DESHMUKH, S. B. WADATKAR, and A. S. TALOKAR

Department of Irrigation and Drainage Engineering, Dr. Panjabrao Deshmukh Krishi Vidyapeeth, Akola 444104 (MS), India; E-mail: kale921@gmail.com

CONTENTS

2.1 Introduction 32
2.2 Methodology 33
    2.2.1 Structure of ANN model (MLP Network) 33
    2.2.2 Working of Multi-layer Perceptron (MLP) Network 34
2.3 Discussion 34
    2.3.1 Determination of Number of Hidden Layers 34
    2.3.2 Determination of Number of Hidden Nodes in Hidden Layers 35
    2.3.3 Selection of Training Algorithm 35
    2.3.4 Selection of Activation Function 36
    2.3.5 Selection of Stopping Criteria 36
    2.3.6 Input Parameters 37
2.4 Conclusions 37
2.5 Summary 38

Keywords 38
References 39

2.1 INTRODUCTION

Usually rainfall-runoff (R-R) information is required to provide basic information for reservoir management in a multipurpose water system optimization framework. The relationship between rainfall and subsequent inflow to reservoir (R-R) is an extremely complex and difficult problem involving many variables, which are interconnected in a very complicated way. For prediction of this relationship, various models (such as knowledge driven, i.e., conceptual and deterministic models; and data driven, i.e., artificial neural network and fuzzy logic, etc.) have been developed over the years. Conceptual and deterministic models are designed to simulate the physical mechanism that determines the hydrological cycle. This involves the physical laws of water transfer and the parameters associated with the characteristics of the catchment area [12]. Such models require sophisticated mathematical tools, a significant amount of calibration data, and some degree of expertise and experience with the model. Data driven models do not provide any information on the physical laws of the hydrologic processes.

Among data driven models, *Artificial Neural Network* (ANN) has recently gained popularity as an emerging and challenging computational technology. It offers advantages over conventional modeling, including the ability to handle large amounts of high variability data from dynamic and non-linear systems, especially when the underlying physical relationships are not fully understood. Other associated benefits include improved opportunities to provide estimates of prediction confidence through comprehensive bootstrapping techniques. ANN models have proved to be very useful for river flow forecasting, where main concern is accurate predictions of runoff (i.e., inflow to the reservoir) [3, 10, 17].

ANN is probably the most successful machine learning technique with flexible structure, which is capable of identifying complex non-linear relationship between input and output data without attempting to reach understanding as to the nature of the phenomena. But selection of variables/parameters in

ANN structure is more complex. The multi-layer perceptron (MLP) ANN is most widely used type of ANN in hydrological modeling [19].

This chapter discusses constraints in prediction of Inflow to reservoir using *Multi Layer Perceptron-Artificial Neural Network (MLP-ANN) Technique*.

2.2 METHODOLOGY

2.2.1 STRUCTURE OF ANN MODEL (MLP NETWORK)

ANNs are flexible mathematical structures that are capable of identifying complex non-linear relationships between input and output data sets. The neural network consists of three main layers: input layer, connecting input information to the network (not carrying out any computation); one or more hidden layers, acting as intermediate computational layers; and an output layer, producing the final output.

A neural network consists of a large number of simple processing elements that are variously called neurons, units, cells, or nodes. Each neuron is connected to other neurons by means of direct communication links, each with an associated weight that represents information being used by the net to solve a problem. The network usually has one or more layers of processing units (hidden layers) where each processing unit in each layer is connected to all processing units in the adjacent layers. General structure of MLP feed-forward network is as follows (Figure 2.1).

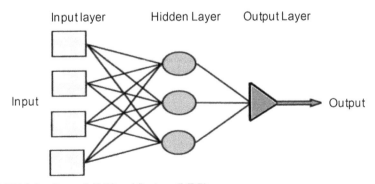

FIGURE 2.1   General ANN architecture (MLP).

2.2.2 WORKING OF MULTI-LAYER PERCEPTRON (MLP) NETWORK

Input parameters consist of any perceptible signals encoded as vectors. The vector is multiplied by the weight matrices of the neural network. The results are summed and an additional value, a so-called bias (*b*), is commonly added to this value. The resulting net input (*net*) is transformed by a transfer function into an activation value of the neuron. This activation value is then propagated to subsequent neurons. If the product exceeds the node's thresholds, the node is "activated," and it conveys a signal. Nodes are interconnected in layers and networks to form neural networks. Specific patterns of input vectors elicit specific outputs from the network. Ideal values of neural network's weights and threshold values are found via relatively simple optimization techniques. After the network's weights and thresholds are optimized, training is stopped. When the trained network is presented with new input, it mimics the patterns that are present between parameters in the real world.

There are no specific rules for determination of number of hidden layers and hidden neurons in each layer, selection of learning rule, activation function, training criteria and input parameters, as well. The literature was reviewed thoroughly to find out a way to overcome these constraints and are discussed in detail in this chapter.

2.3 DISCUSSION

2.3.1 DETERMINATION OF NUMBER OF HIDDEN LAYERS

The major concern of an ANN structure is to determine the appropriate number of hidden layers and number of neurons in each layer. There is no systematic way to establish a suitable architecture, and the selection of the appropriate number of neurons is basically problem specific. Hornik et al. [11] proved that a single hidden layer network containing a sufficiently large number of neurons can be used to approximate any measurable functional relationship between the input and the variables to any desired accuracy. De Villars and Barnard [2] showed that an ANN comprised of two hidden layers tends to be less robust and converges with less accuracy than its single hidden layer counterpart. Furthermore, some studies indicate that the benefits of using a second hidden layer are marginal to the rainfall-runoff modeling problem [1, 15]. Dowson and Wilby [8] stated that an ANN with only one hidden layer is enough to represent the nonlinear relationship between rainfall and the corresponding runoff or inflow to the reservoir. Taking recognizes of above studies, a single hidden layer should be used to develop R-R model.

2.3.2 DETERMINATION OF NUMBER OF HIDDEN NODES IN HIDDEN LAYERS

The performance of the ANN depends on the number of nodes in the hidden layer. There are some algorithms, including pruning and constructive algorithms, to determine an 'optimum' number of neurons in hidden layer(s) during training. However, a trial and error procedure using different number of neurons is still be preferred choice of most users [1, 16, 20]. As no specific guidelines exist for choosing the optimum number of hidden nodes for a given problem, this network parameter should be optimized using a combination of empirical rules and trial and error procedure.

2.3.3 SELECTION OF TRAINING ALGORITHM

The ANN training is fundamentally a problem of non-linear optimization, which minimizes the error between the network output and the target output by repeatedly changing the values of ANN's connection weights according to a predetermined algorithm. Paradigms of supervised learning include error-correction learning. An important issue concerning supervised learning is the problem of error convergence, i.e., the minimization of error between the desired and computed unit values. The aim is to determine a set of weights, which minimizes the error. That's why a training algorithm is needed to solve a neural network problem. Since there are so many types of algorithm available for training a network, selection of an algorithm that converges fast and find global minimum without trapped in local minima.

Among various algorithms reported in literature Levenberg-Marquardt (L-M) algorithm is much more robust and outperformed other algorithms (e.g., variable learning rate and momentum (BPvm), Resilient Back Propagation (RBP), Polak-Ribiere, etc.) in terms of accuracy and convergence speed [18]. Levenberg-Marquardt algorithm is a quasi-Newton method that proved to be quickest and was less easily trapped in local minima [5]. Hence, Levenberg-Marquardt (L-M) algorithm should be used for R-R modeling application.

2.3.4   SELECTION OF ACTIVATION FUNCTION

Activation functions are needed for introducing non linearity into the network and it is non linearity that makes multilayer networks so powerful. The exact form of the activation function is not critical, as long as it is bounded and it increases monotonically [13]. Various transfer functions reported in literature are sigmoidal types (logistic and hyperbolic tangent function), hard limit transfer functions (bounded to 0 or 1), linear, polynomial, rational function (ratios of polynomial) and Fourier series (sum of cosines). In literature, most commonly used transfer functions for rainfall-runoff (R-R) modeling are sigmoidal type transfer functions in the hidden layers and linear transfer functions in the output layer due to its advantage in extrapolation beyond the range of the training data [6, 12, 14, 20].

2.3.5   SELECTION OF STOPPING CRITERIA

The stop criteria is a fundamental aspect of training. The simple ideas of capping the number of iterations or of letting the system train until a predetermined error value is not obtained. The reason is that the ANN has to perform well in the test set data; i.e., ANN should perform well in data it never saw before (good *generalization*) [4]. The weights are changed during training, according to the optimization of some performance measure, which is a measure for the degree of fit (or difference) between the network estimates and the sample output values. The alteration of network parameters in the training phase is commonly stopped before the training optimum is found, because the network will start learning the noise in the training data and lose its generalization capability (overtraining). However, stopping too early means the ANN has not yet learnt all the information from the training data (undertraining). Both situations are likely to result in sub-optimal operational performance of an ANN model. It is for this reason that the available data are often split in three separate data sets (split-sampling method): (1) the training set, (2) the cross-validation set, and (3) the validation set. The first provides the data on which an ANN is trained. The second is used during the training phase to reduce the chance of overtraining of the network. The minimization of the training error is stopped as soon as the cross-validation error starts to increase. This point is considered to lie between undertraining and overtraining an ANN. The latter of the three data sets is used to validate the performance of a trained ANN. This is also called as early stopping approach. The use of early stopping reduced the training time four times and it provided better and more reliable generalization performance than the use of L-M algorithm alone [7]. Considering the advantages of early stopping approach, it is generally used in the research work.

2.3.6   INPUT PARAMETERS

In ANN modeling, choice of the input variables is an important issue. There is no general theory to solve this issue, and it is rather problem dependent. A trial and error procedure was generally adopted to finalize the input variables to the ANN model.

As the hydrological state (i.e., the amount and distribution of stored water in a catchment) for a great part determines the catchment's response to a rainfall event, it is critical as input to an ANN model. Previous discharge values are therefore often used as ANN inputs, since these are indirectly indicative for the hydrological state [2]. In addition to runoff values, current and previous day's rainfall are generally used as inputs to ANN models.

2.4 CONCLUSIONS

On the basis of results of this research, it is suggested to consider following points while developing ANN for R-R modeling using MLP-ANN technique, for precise and appropriate results.

    i.   A single hidden layer in network.

    ii. The trial and error method for determining optimum number of neurons in hidden layer.

    iii. Levenberg-Marquardt algorithm as a training algorithm.

    iv. Sigmoidal type transfer function in input hidden layer and linear transfer function in output layer.

    v. An early stopping criterion for stopping of training.

    vi. Previous runoff/inflow to reservoir along with rainfalls as input parameters.

2.5 SUMMARY

This chapter reviews the constraints in the formulation of MLP-ANN model for prediction of inflow to reservoir. Specifically the constraints like determination of number of hidden layers and hidden neurons in each layer, learning rule, activation function, training criteria and selection of input parameters are reviewed and discussed in detail. It is suggested that for rainfall-runoff modeling; use single hidden layer network as it yield good results than that with two hidden layer network; use trial and error method for determination of optimum number of neurons in hidden layer; for back propagation network – use Levenberg – Marquardt algorithm as it outperformed other algorithms in terms of accuracy and convergence speed; use sigmoidal type transfer function in input hidden layer and linear transfer function in output layer; use an early stopping criterion for training; and use previous runoff/inflow to reservoir along with rainfall as input parameters. From literature, it is also cleared that the climatic factors like temperature and wind have no significant effect on Rainfall-Runoff modeling using ANN technique.

KEYWORDS

- artificial neural network (ANN)
- hidden layer
- Levenberg – Marquardt algorithm

- modeling hidden layer
- multi layer
- rainfall-runoff
- reservoir flow
- reservoir inflow

REFERENCES

1. Abrahart, R. J. and L. See, 2000. Comparing neural network and autoregressive moving average techniques for provision of continuous river flow forecasts in two contrasting catchments. *Hydrol. Process.*, 14:2157–2172.
2. ASCE, Task committee on applications of ANNs in hydrology, 2000. Artificial Neural Networks in hydrology II: Hydrology application. *J. Hydrol Eng.*, 5(2): 124–37.
3. Bhola, P. K. and A. Singh, 2010. Rainfall-runoff modeling of river Kosi using SCS-CS method and ANN. Unpublished B. Tech. thesis, Department of Civil Engineering, National Institute of Technology, Rourkela. http://ethesis.nitrkl.ac.in/1883/1/RainfallRsunoff_modeling_using_SCS-CN_method_and_ANN.pdf
4. Bishop, C., 1995. *Neural Networks for Pattern Recognition. Oxford University Press, New York.*
5. Brath, A., Montanari, A., and Toth, E., 2002. Neural networks and non-parametric methods for improving real time flood forecasting through conceptual hydrological models. *J. Hydrology and Earth System Sciences,* 6(4):627–640.
6. Calvoa, I., and Portelab, M., 2007. Application of neural approaches to one step daily flow forecasting in Portuguese watersheds. *J. Hydrol.*, 332(1–2):1–15.
7. Coulibaly, P., Anctil, F. and Bobee, B., 2008. Daily reservoir inflow forecasting using artificial neural networks with stopped training approach. *J. Hydrology*, 230:244–257.
8. Dowson, C. W., and Wilby, R. L., 2001. Hydrological Modeling using artificial neural networks. *Prog. Phys. Geography,* 25(1):80–108.

9. De Villars, J., and Barnard, E., 1993. Back propagation neural nets with one or two hidden layers. *IEEE Trans. Neural Network,* 4(1):136–141.

10. El-Shafie, A., Mukhlisin, M., Najah, A, A., and Taha, M. R., 2011. Performance of artificial neural network and regression techniques for rainfall-runoff prediction. *Int. J. Physical Sciences,* 6(8):1997–2003.

11. Hornik, K., Stinchcombe, M., and White, H., 1989. Multilayer feed forward networks are universal approximators. *Neural Networks,* 2:359–366.

12. Hsu, K. L., Gupta, H., and Sorooshian, S., 1995. Artificial neural network modeling of rainfall runoff process. *Water Resour. Res.,* 31(10):2517–2530.

13. Kuligowski, R. J., and Barros, A. P., 1998. Experiments in short term precipitation forecasting using artificial neural networks. *Mon. Weather Rev.,* 126:470–482.

14. Maier, H. R., and Dandy, G. D., 2000. Neural Networks for prediction and forecasting of water resources variables: a review of modeling issues and applications. *Environ. Modeling Software,* 15:101–124

15. Minns, Anthony, W., and Hall, M. J., 1996. Artificial neural networks as rainfall-runoff models. *Hydrolog. Sci. J.,* 41:399–417.

16. Shamseldin, A. Y., 1997. Application of neural network technique to rainfall-runoff modeling. *J. Hydrology,* 199:272–294.

17. Solaimani, K., 2009. Rainfall-runoff prediction based on artificial neural network (A case study: Jarahi watershed). *American-Eurasian J. Agric. and Environ. Sci.,* 5(6): 856–865

18. Vos, N. J., de and Rientjes, T. H. M., 2005. Constraints of artificial neural networks for rainfall-runoff modeling: trade-offs in hydrological state representation and model evaluation. *Hydrology and Earth System Sciences,* 9:111–126

19. Wang Wen, Pieter H. A., Van, J. M., Gelder, J., Vrijling., K. and Ma Jun. 2006. Forecasting daily stream flow using hybrid ANN models. *J. Hydrology,* 324:383–399

20. Zealand, C. M., Burn, D. H., and Simonovic, S. P., 1999. Short term stream flow forecasting using artificial neural networks. *J. Hydrology,* 214(1–4):32–48.

CHAPTER 3

SOIL WATER BALANCE SIMULATION OF RICE USING HYDRUS-2D AND MASS BALANCE MODEL

LAXMI NARAYAN SETHI and SUDHINDRA NATH PANDA

Department of Agricultural Engineering, Assam University (A Central University), Silchar-788011, Assam, India, Phone: +91 9864372058, E-mail: lnsethi06@gmail.com

Department of Agricultural and Food Engineering, Indian Institute of Technology, Kharagpur-721302, India, Phone: +91 9434009156; E-mail: snp@iitkgp.ernet.in

CONTENTS

3.1   Introduction 42
3.2   Water Balance Simulation Models 44
    3.2.1   HYDRUS-2D Model 44
    3.2.2   Water Balance Model for Rice Field 45
        3.2.2.1   Formulation of Field Water Balance Model 45
        3.2.2.2   Ponding Depth 46
        3.2.2.3   Actual Evapotranspiration 46
        3.2.2.4   Surface Runoff 47
        3.2.2.5   Vertical Percolation and Lateral Seepage 45
        3.2.2.6   Bare Soil Evaporation During Germination
                Period of Rice 48

3.2.3   Simulation of Models 48

       3.2.3.1   Hydrus-2D Model Simulation 48

              3.2.3.1.1   Space and Time Discretization 48

              3.2.3.1.2   Boundary and Initial Conditions 49

       3.2.3.2   Field Water Balance Model Simulation 49

3.3   Materials and Methods 50

     3.3.1   Experimental Set-Up 50

3.4   Measurements and Analysis 51

3.5   Results and Discussions 54

     3.5.1   Water Balance Simulations 54

       3.5.1.1   Soil Water Content and Ponding Depth 55

       3.5.2.2   Supplemental Irrigation Requirement 58

       3.5.1.3   Actual Crop Evapotranspiration 59

       3.5.1.4   Rainfall and Surface Runoff 59

       3.5.1.5   Vertical Percolation and Lateral Seepage Flux 63

3.6   Conclusions 65

3.7   Summary 65

Acknowledgement 66

Keywords 66

References 67

3.1   INTRODUCTION

Rainfed agriculture in India extends over 94 M-ha that constitutes nearly 70% of the net sown area. The entire net sown area is under cultivation during monsoon season (June–September) but during non-monsoon season (October–January) most of the areas lying barren due to lack of supplemental irrigation facility. The variability of rainfall create situation like surface flooding that cause a lot of soil and nutrient erosion on one hand and water scarcity at the critical crop growth stages at the other hand [41].

     Rice (*Oryza sativa L.*) is the predominant cropping pattern adopted by most of the farmers in eastern India [8]. Farmers in the region have no well-defined irrigation and drainage systems and they advocate plot-to-plot method of irrigation and drainage. Thus, excess ponding water during initial and harvest stages of crop growth are drained out by cutting dikes of the rice fields. This causes inundation problem of rice and complete damage of seedlings and matured crop in low lands downstream. Thus, the study reveals that there is a need to introduce weir heights for effective water management in rice field and to conserve excess rainwater in water harvesting structure.

Conventional water management in the rice cultivation aims at keeping the fields continuously submerged (CS). But, excess ponding causes nutrient imbalance and reduction in yield. Rice plants performed better with respect to seedling establishment and grain yield in shallow water (i.e., <10 cm) than in deep water (i.e., >10 cm). In an experiment, 5 cm of ponding depth in the rainfed rice field was reported to give better yield and water use efficiency [25]. Depth of standing water in rice can be minimized with the proper dike/weir height which is the deciding parameter for conservation of rainwater, sediment, nutrient, control of declining ground water table and improvement in rice yield [16, 20, 34].

Since irrigation water is a scarce commodity in the rainfed farming system, water saving irrigation (WSI) is now gaining gradual attention. Experiments conducted with different WSI techniques in various regions of the world reveal that irrigation to rice at near saturation (SAT) gives comparable yield with continuous submergence and saves a lot of water [12, 32]. In another approach, rice is irrigated after standing water disappears from the field, which results in considerable reduction of the water requirement of rice and increase in water use efficiency [14, 19, 28, 36]. In one of the WSI techniques practiced in China, soil water is maintained at 60–100% SAT throughout the period following the start of the booting stage of rice [33]. Most sensitive period of water deficit in rice is the reproductive phase. Mitra [21] reported that under upland condition, optimum yield of dry seeded rainfed rice could be obtained by maintaining soil moisture at field capacity throughout the active growth stage. Hence in rainfed ecosystem, rice can be grown with maintaining soil water content in the effective root zone at field capacity with manageable allowable deficit (MAD) of 40% of SAT.

The concept of water balance is one of the greatest advances in understanding the response of crops in water-limited environments [4]. Often simple book keeping methods are used that specify gain and loss of water over specified depth, such as the root zone of the crop [5, 25]. The water balance components can be quantified theoretically as well as experimentally in the field. The principal loss is caused by vertical percolation and lateral seepage [11, 37]. Water balance cannot be achieved if evapotranspiration, surface runoff and vertical percolation below the root zone depth are taken to be the only sources of loss [5, 35, 38, 42]. The lateral seepage through the boundaries is the only other possible source of loss. So, there is need to use models to quantify water balance parameters such as vertical percolation, lateral seepage, surface runoff flux of the cropped field with different water management strategies.

Now a day, the mathematical models are used extensively in soil-water research. Once validated, these models allow quantitative estimation of the different water balance components under varying conditions. The models are also increasingly used to simulate the variably saturated flow and chemical transport process between the soil surface and the groundwater table [6, 18, 27, 39, 40]. Among the common software tools available, HYDRUS-2D [29] is a rather complex model involving different sub-models in Microsoft Windows based modeling environment for simulating the water and solute transport under variable saturated conditions [1, 9, 10, 13, 15, 22, 30, 31]. Thus, the main objective of this study was considered to quantify the water balance parameters such as runoff, actual evapotranspiration, vertical percolation and lateral seepage from the effective root zone of rice (45 cm) field with different weir heights (0, 50 and 100 mm) using HYDRUS-2D [29] and field water balance model.

3.2 WATER BALANCE SIMULATION MODELS

3.2.1 HYDRUS-2D MODEL

HYDRUS-2D [29] is a Microsoft Windows based modeling environment for simulating two-dimensional water, heat, and solute dynamics and root water uptake in variably saturated porous media. Assuming a homogeneous and isotropic soil, the flow equation (i.e., two-dimensional Richards' equation) in HYDRUS-2D model is solved numerically using a Galerkin-type linear finite element scheme. The soil hydraulic properties were modeled using the van Genuchten-Mualem (VGM) model constitutive relationships. The VGM model was also opted to optimize the shape parameters such as (α and n), residual water content and saturated water content with the observed data through inverse estimation process of the HYDRUS-2D model. The software includes a mesh generator and graphical user interface. The details of simulation for water balance parameter estimation are given in subsequent sections.

3.2.2   WATER BALANCE MODEL FOR RICE FIELD

3.2.2.1   Formulation of Field Water Balance Model

The maximum rooting depth of rice (45 cm) is considered as the soil reservoir. The water balance simulation model is developed for rainfed monsoon rice field with W1 (0 mm), W2 (50 mm) and W3 (100 mm) weir height. The water balance parameters considered are rainfall (P), runoff (SR) generated at different weir heights, actual evapotranspiration (AET), SI at critical growth stage, vertical percolation (VP) and lateral seepage (LS) through field boundary. The capillary rise from groundwater is ignored. The topography of fields is almost flat.

The generalized daily soil water balance model in the effective root zone of the crops (rice) ignoring upward flux because of capillary rise from groundwater is given as:

$$SMC_{jk} = SMC_{jk-1} + P_{jk} + SI_{jk} - AET_{jk} - SR_{jk} - VP_{jk} - LS_{jk} \quad (1)$$

where, SMC = soil moisture content in the effective root zone of rice (mm); P = rainfall (mm); AET = actual evapotranspiration (mm); VP = vertical deep percolation from effective root zone of the crop (mm); LS = lateral seepage from effective root zone of rice (mm); SI= Supplemental irrigation (mm); SR= surface runoff from rice field (mm); j = index for weir heights (0, 50 and 100 mm); and k = index for day (k = 1 to 105).

3.2.2.2   Ponding Depth

If soil moisture content is more than the saturation level then ponding will occur in the rice field. The ponding depth in the rice field on any day is given below:

$$H_{jk} = SMC_{jk} - SAT \qquad (2)$$

where, H = ponding depth in the rice field (mm); and SAT = saturation soil water content in the active root zone of rice (mm).

Under the ponding phase, the water balance in the rice field with different weir heights can be expressed as

$$H_{jk} = H_{jk-1} + P_{jk} + SI_{jk} - AET_{jk} - VP_{jk} - LS_{jk} - SR_{jk} \qquad (3)$$

3.2.2.3   Actual Evapotranspiration

Estimation of AET is most difficult in rainfed upland rice fields when most of the time soil remains under unsaturated/drying condition. While enough data are available for estimation of AET of rice in saturated or submerged case [7, 17] hardly any information is available for unsaturated condition. Crop coefficient (Kc) of rice is taken as 1.0, 1.15, 1.10 and 1.10 during crop establishment (25 days), crop development (20 days), reproductive (30 days) and maturity stages (26 days), respectively [7]. Reference crop evapotranspiration (ET0) was estimated on daily basis by the FAO Penman-Monteith method [3].

Under adequate soil moisture condition, evapotranspiration of plants occur at potential rate. However, as the ponding water from rice field decreases such that soil in the effective root zone of rice remains at moisture stress condition, evapotranspiration of plants decreases from the potential rate. The AET on any day under moisture stress condition is given as:

$$AET_{jk} = Kc \times Ksf_{jk} \times ET_{0jk} \qquad (4)$$

where, Kc = crop coefficient; Ksf = soil moisture stress factor; and ET_0 = reference crop evapotranspiration (mm).

The value of Ksf in Eq. (4) is assumed as 1.0 under no water stress condition. But as the ponding water disappears from the rice lands, soil moisture stress occurs that is usually provided by Ksf that consequently decreases AET. In the present study, Ksf is assumed to vary linearly with the relative available SMC in the field under unsaturated condition as [2, 23]:

$$Ksf_{jk} = \frac{SMC_{jk}}{SAT} \tag{5}$$

Under adequate soil moisture condition, potential evapotranspiration on any day k^{th} day (PET_{jk}) is given by:

$$PET_{jk} = Kc.ET_{0jk} \tag{6}$$

3.2.2.4   Surface Runoff

From sowing to first 10 days after germination of rice seed and last 10 days to the harvest, no standing water is allowed in the field. During these periods SR is given as

$$SR_{jk} = H_{jk-1} + P_{jk} - SAT \tag{7}$$

During rest of the periods except the fields with 0 mm weir height (W1= 0), 50, and 100 mm ponding depth are taken as the maximum limit (H_{jmax}) in the field (W2 = H_{jmax} = 50 mm and W3 = H_{jmax} = 100 mm) and any excess ponding above H_{jmax} is taken as the SR and is given as:

$$SR_{jk} = SMC_{jk-1} + P_{jk} + SI_{jk} - SAT - H_{jmax} \tag{8}$$

3.2.2.5   Vertical Percolation and Lateral Seepage

VP and LS under unsaturated condition of the cropped field were estimated using HYDRUS-2D model. Whereas under saturated condition the VP and LS were computed using Darcy and Dupuit's approach [26], respectively.

3.2.2.6 Bare Soil Evaporation During Germination Period of Rice

For computation of different parameters of water balance models during germination period (4 days) of rice, AET is replaced by bare soil evaporation (E) in Eq. (1). Initially when seed is sown in the field, SMC is assumed to be at permanent wilting point (PWP). Bare soil evaporation is estimated from ET_0 subjected to rainfall (P) condition of the day [14] as:

$$E_{jk} = 0.1\ ET_{0jk} \quad \text{if } P_{jk} = 0 \tag{9}$$

$$E_{jk} = ET_{0jk} \quad \text{if } P_{jk} > ET_{0jk} \tag{10}$$

$$E_{jk} = P_{jk} \quad \text{otherwise} \tag{11}$$

3.2.3 *SIMULATION OF MODELS*

In the present study, HYDRUS-2D model [29] was considered to simulate the vertical and lateral flux from the effective root zone of rice field with different W under variably saturated condition (up to saturation). However, under ponding situations (above saturation) of rice fields with weir W2 and W3, field water balance model was used to simulate water balance parameters. The detailed simulation and evaluation of water balance parameter using HYDRUS-2D as well field water balance model are explained in the following sections.

3.2.3.1 HYDRUS-2D model simulation

3.2.3.1.1 *Space and Time Discretization*

The graphical user interface (GUI) of HYDRUS-2D was used to define the domain geometry of the rice field cross section, generate the finite element method (FEM) mesh, define the initial and boundary conditions and observation nodes. The observation nodes, (corner points of the triangular elements in the FEM mesh) were assigned closest to the aqua-pro and tensiometer access tubes where observations were made. The size of model domain for rice field soil profile was considered as 3100 cm width including trapezoidal bund (90 cm wide, 30 cm height and 1:1 slope) and 45 cm depth. The soil profile of domain was divided into 3 layers of 15 cm intervals each for rice. Each layer was assumed to have homogeneous and uniform hydraulic properties. The finite element grid consisted of a total of 675 nodes and 1229 elements for rice field model domain.

The simulation period was started from the day after sawing to harvest day of rice, i.e., 105 days. Time discretization's were considered as: initial time step 1=0.0001 day, minimum time step 1=0.000001 day, and maximum time step 1 day.

3.2.3.1.2   *Boundary and Initial Conditions*

The daily rainfall/irrigation and potential evapotranspiration for the whole simulation period were used as a time-variable boundary for the soil surface of the application area. The amount of water irrigated was added with rainfall. The topsoil surface was assumed to have atmospheric for rice field model domain. The bottom boundary at the root zone depth was assumed to have free drainage boundary condition. The vertical boundaries were assumed to have seepage face conditions and rest of the domain was assumed no flow boundary conditions.

The measured and estimated values of hydraulic parameters for different layers of 15 cm interval from the cropped fields were applied to different layers of the model domain. Van Genuchten and Mualem hydraulic model was used for the hydraulic functions. The observed soil water content on day of sowing was used as initial condition. The soil water content measurements were considered to be representing the whole width of the cross section at a particular depth.

3.2.3.2   Field Water Balance Model Simulation

The developed simulation models for rice field with different W can meet both saturated and unsaturated condition in the field and quantify the effective root zone soil water status corresponding to different W. Based on the available meteorological parameters and field observations the field water balance model was used to simulate and test water balance parameters for the year 2002 to 2004. Water balance simulation for each year was started from the actual 1st day after sowing to harvest (105th) day of rice.

3.3   MATERIALS AND METHODS

3.3.1   EXPERIMENTAL SET-UP

The site selected for the present study is the experimental farm of the Department of Agricultural and Food Engineering, Indian Institute of Technology (IIT), Kharagpur, India. It is located at $22^0 19'$ N Latitude and $87^0 19'$ E Longitude with an altitude of 48 m above the mean sea level. It lies in the State of West Bengal of Eastern India. The dominant soil type in the study area is sandy loam (light textured), acidic (pH ranges from 4.8 to 5.6), and poor in organic matter. The study area receives about 1500 mm mean annual rainfall, about 80% of which is concentrated during the rainy season from June to September. The mean minimum and maximum air temperatures are 12°C in January and 40°C in May, respectively. In the corresponding months the mean relative humidity varies between 18–89% and 15.5–90.5%, respectively.

There are nine number of fields with dimensions of 40 m × 20 m each considered for field experiment. Based on topography and water management practices of the region, weir height (W) of 0 (W1), 50 (W2) and 100 (W3) mm were introduced in rice field to allow maximum standing water (ponding) in the field up to the depth of 0, 50 and 100 mm, respectively. Since, ponding of water in field during germination period, initial stage and harvest stage is harmful for seed germination and crop growth and yield, for W2 and W3 the maximum standing water was allowed in the field from 15 days after sowing (DAS) to 10 days to harvest and rest of the period no weir heights as W1 were maintained. Figure 3.1 shows the layout of field experiment.

Monsoon remains in the region for about 111 days [24]. Hence in the study, we have chosen rice (*Oryzasativa* L.) of variety MW-10 which is drought resistant and matures in 101 days. The present study is focused on rainfed ecosystem and irrigation to rice under such system is important. In the present study when the soil water content in effective root zone of rice

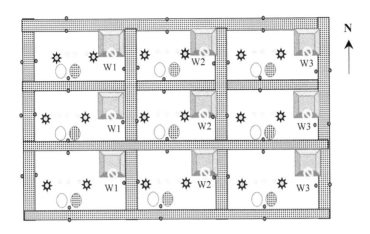

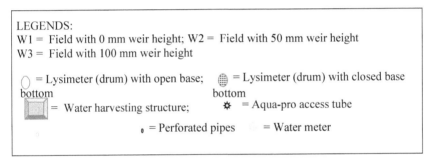

LEGENDS:
W1 = Field with 0 mm weir height; W2 = Field with 50 mm weir height
W3 = Field with 100 mm weir height

○ = Lysimeter (drum) with open base;   ⊕ = Lysimeter (drum) with closed base
bottom   bottom
□ = Water harvesting structure;   ✿ = Aqua-pro access tube
                         ₀ = Perforated pipes   = Water meter

FIGURE 3.1   Layout of experimental setup.

is depleted to 40% depletion from SAT (i.e., equivalent to 15% of FC in
the present soil) during critical growth stage (CGS), supplemental irrigation is provided.

3.4   MEASUREMENTS AND ANALYSIS

The soil profile of the cropped field was grouped into three depths of 15
cm intervals from the soil surface for in situ measurements of soil physical properties and soil water characteristics. Saturated hydraulic conductivities (K_s) in each depth were determined by falling head method. The
average of the soil texture and hydraulic properties measured at different
depths of the rice field are given in Table 3.1.

TABLE 3.1   Soil Texture and Hydraulic Properties Measured at Different Depths in Crop Field

Depth (cm)	Sand (%)	Silt (%)	Clay (%)	Bulk density (g cm^{-3})	K_S (cm day^{-1})	θ_s	FC	WP	θ_r
						(cm³/cm³)			
0–15	66.4	18.6	15.0	1.65	12.24	0.37	0.24	0.08	0.031
15–30	62.5	21.5	16.0	1.60	7.01	0.39	0.27	0.095	0.036
30–45	63.0	20.6	16.4	1.58	5.94	0.38	0.29	0.102	0.039

Note: θ_s = saturated moisture content, θ_r = residual moisture content, FC = field capacity and WP = wilting point.

Instruments such as aqua-pro soil water sensor and tensiometer were used to monitor the soil water content in different soil layers of the crop root zone depths (Figure 3.1). Soil water measurements were taken daily for each 15 cm intervals in soil profile using digital Aqua-pro moisture sensor, tensiometer as well as gravimetric measurements. In addition, perforated pipe on the both sides of boundary (dike), drum lysimeter and piezometers were installed in the root zone depth to measure the actual evapotranspiration, lateral seepage and vertical percolation using the methods followed by [26, 38]. Mechanical water meter and sloping gauges were used at the weir height to measure the excess runoff and ponding depth of the field with different weir heights.

The meteorological parameters namely rainfall, solar radiation, wind velocity, air temperature, relative humidity were collected from the meteorological center as well as Automatic Weather Station of Indian Institute of Technology, Kharagpur India which is located in the close vicinity of the experimental site.

A soil water balance model was used to simulate the soil moisture and water balance model parameters in the rice field. Various components of the model used in water balance are rainfall, supplemental irrigation, seepage and percolation, surface runoff, actual crop evapotranspiration and soil moisture/ponding depth in the rice field. A flow chart for simulation of soil water balance parameter in effective root zone of rice field with different W is also developed and shown in Figure 3.2.

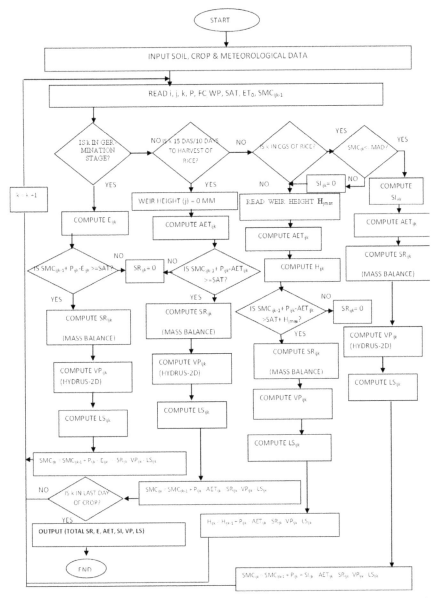

FIGURE 3.2   Flowchart for simulation of water balance parameters in the rice field with different weir heights.

3.5   RESULTS AND DISCUSSIONS

3.5.1   *WATER BALANCE SIMULATIONS*

In the rainfed rice field with different W, the water balance inflow components considered are P and SI if required from water harvesting structure, whereas outflow components include AET, RO, VP and LS. However, AET, RO, VP and LS and are the important parameters that influence the SMC/H of rainfed rice field with different W. The arbitrary assumption of these model parameters may result in some incredible estimates of SMC in the crop effective root zone. Hence, there is a need to provide a confirmation of the model parameters against the independently observed data.

These parameters were simulated using HYDRUS-2D and field water balance model as the method described in earlier sections. HYDRUS-2D model was calibrated using soil water content observation data (2002) for each layer of the rice field with W1 and validated for the year 2003 and 2004. By inverse modeling approach, calibration was carried out considering the measured soil hydraulic parameters and daily soil water content for each layer of soil depths. It was assumed that the data available for particular depth of soil layer to be uniform for the entire layer. The soil water content of each layer of soil depths (observation nodes in the model domain) to the corresponding field observation data points was considered for calibration and validation. The initial calibrated parameters such as coefficient (α) and exponent (n) in soil water retention function for each depth of soils were estimated using neural network prediction and Van Genuchten and Mualem hydraulic model. The inputs used for neural network predictions were measured soil texture (sand silt and clay), bulk density, field capacity and wilting point values of each layer of soils. If the simulated soil water content data fitted the observed data reasonably well and regression coefficient and water balance error were acceptable, then the model calibrated parameters used for further simulation of soil water content. The optimized model calibrated parameters (α and n) for each depth of soils, and these were estimated using inverse modeling with 95% confidence intervals (Table 3.2).

Similarly, water balance parameters predicted by field water balance models were also initially calibrated using observed parameters and

TABLE 3.2 Initial and Optimized Calibrated Parameters for Different Depth of Soils Predicted Using Inverse Modeling with 95% Confidence Intervals

Depth (cm)	α (cm^{-1})		n	
	Initial	Optimized	Initial	Optimized
0–15	0.0070	0.0072	1.4634	1.4500
15–30	0.0077	0.0057	1.4893	1.4749
30–45	0.0122	0.0039	1.5000	1.5619

validated for the rest two years. The model performance was also evaluated based on the statistical analysis of simulated and observed data. The statistical characteristics such as the root mean square error (RMSE) and prediction efficiency (PE) and coefficient of determination were used for model evaluation. The results presented below (Table 3.3) comprise both the daily as well as total comparisons of simulated and measured parameters.

3.5.1.1   Soil Water Content and Ponding Depth

Variation of soil water content (Figure 3.3) and ponding depth (Figure 3.4) in the different layers of crop root zone depths was found to have cyclic experience depending on rainfall and/or SI. However, during the period when there was no rainfall or no SI was applied, soil water content was found to decline gradually because of uptake of water by the plant roots, VP and LS from the soil profiles. Simulated and observed soil water content for 3 layers of soil of rice field with W1 for the year 2002 is found to be close to each other with high values of PE, R^2 and low values of RMSE (Table 3.3).

Similarly, field water balance model was used to simulate daily soil water content (unsaturated)/ponding (above saturation) as well as water balance parameters (RO, AET, VP and LS) in the rice field with different W for the year 2002. The predicted ponding water depth was estimated by mass balance as a residual of other water balance components. The graphical comparison of simulated and observed ponding depth (H) of rice field with W2 and W3, for the year 2003 and 2004 is shown in Figure 3.4. The statistic parameters such as R^2, PE and RMSE are also analyzed using

TABLE 3.3 Statistical Parameters for Prediction of Soil Water Content in Different Root Zone Depth and Ponding Depth in Rice Field

Parameter	2002			2003			2004		
	R^2	PE	RMSE	R^2	PE	RMSE	R^2	PE	RMSE
Rice field with 0 mm weir height (W1)									
SMC (0–15 cm)	0.9188	0.9326	0.023	0.8914	0.9132	0.0263	0.9426	0.95227	0.02069
SMC (15–30 cm)	0.9208	0.9475	0.0225	0.9161	0.9360	0.0209	0.9498	0.9623	0.02783
SMC (30–45 cm)	0.9388	0.9531	0.0221	0.9363	0.9525	0.02945	0.9529	0.9709	0.0191
Rice field with 50 mm weir height (W2)									
H (mm)	0.9679	0.9819	2.111	0.9270	0.9355	3.660	0.9517	0.9601	2.946
Rice field with 100 mm weir height (W3)									
H (mm)	0.9773	0.9805	3.597	0.9773	0.9454	5.451	0.9679	0.9786	4.582

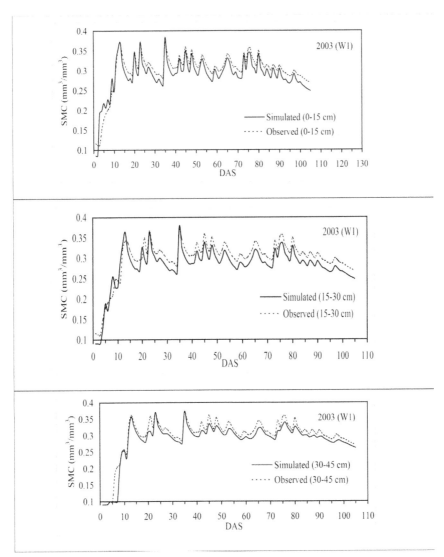

FIGURE 3.3   Variation of soil moisture content in rice field with weir height W1 in 2003.

simulated and observed H values for the year 2003 and 2004 (Table 3.3) and are found in acceptable range.

Hence, the calibrated water balance model to simulate the soil moisture content in rice field is justified and so it is used to simulate the soil moisture content in 2003 and 2004. Simulated and observed soil water content

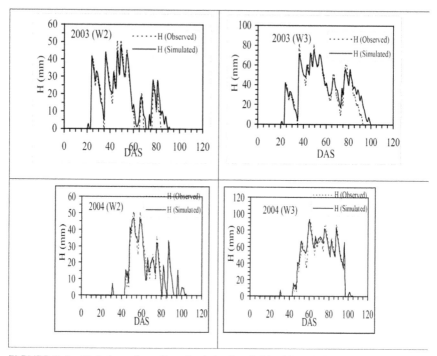

FIGURE 3.4   Variation of ponding depth in rice field with weir height W2 and W3 in 2003 and 2004.

for three layers of soil of rice field with W1 for the year 2003 and 2004 are studied and the variation for the year 2003 for weir height W1 is shown in Figure 3.3. The performance of the model was also evaluated using statistical parameters such as PE, RMSE and R^2 to predict soil water content in the different depths of the rice field for the year 2003 and 2004 also and the results are shown in Table 3.3.

3.5.2.2   Supplemental Irrigation Requirement

For rice, the critical growth stage is the reproductive phase of rice, which includes moisture sensitive periods of panicle differentiation and heading. For a short duration variety of rice of 101 days, it extends for 30 days (49–78 DAS). Soil water content in the effective root zone depth during this period should not decline below 15% soil FC/40% of SAT due to uniform rainfall distribution. So, supplemental irrigation was not required and/applied to the crop during CGS.

However, to get good yield of rice, supplemental irrigation requirement must be made available to rice during CGS. Since, there is less scope for either canal or groundwater supply in the region, rainwater management in the field as well as in harvesting structure is to be given importance.

3.5.1.3   Actual Crop Evapotranspiration

The parameter AET of rice crop grown with different water management strategies was tested using daily observations data of rice field with W for three years experiments. The comparison of simulated and observed values of AET for rice grown in the field with different W for the year 2002, 2003 and 2004 were studied. The temporal variation of observed and simulated AET for the year 2003 and 2004 are shown in Figure 3.5. The analysis reveals that the simulated values of AET are close to the observed values for all the years with high values of coefficient of determination, R^2 (Table 3.4). Total measured and simulated AET from the rice field with different weir heights for the year 2002, 2003 and 2004 are given in Table 3.5. It is observed that the average AET of rice grown in the fields with 0, 50 and 100 mm weir heights are 42.17, 47.87 and 48.87% of the total rainfall during the crop growth period, respectively.

3.5.1.4   Rainfall and Surface Runoff

The total rainfall measured during the crop growth period (105 days) for the year 2002, 2003 and 2004 are 1174.95, 692.95 and 847.80 mm, respectively with average of 905.23 mm. These values were considered as inflow component for the rice field with different weir heights.

The excess RO generated from the rice field with different W was also tested using daily measured data of three years experiments. Values of SR were maximum from the rice fields with 0 mm weir height and the values decreased with increasing weir heights. The variations are found only due to rainfall intensity, soil characteristics and water management strategies, i.e., weir height.

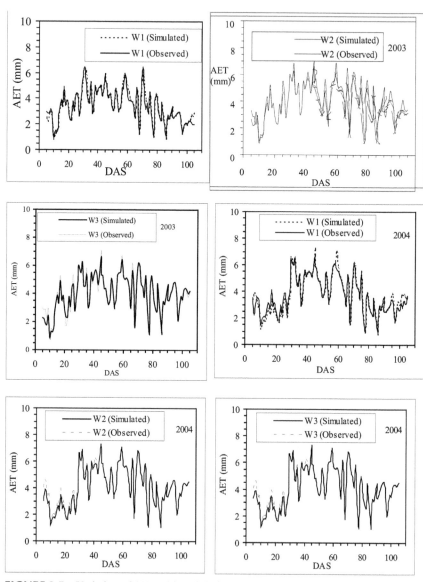

FIGURE 3.5   Variation of AET with weir height W2 and W3 in 2003 and 2004.

The analysis reveals that the simulated values of RO are close to the observed values for all the years with high values of coefficient of determination (R^2). Values of R^2 between the observed and simulated

TABLE 3.4 Statistical Parameters for Prediction of AET, VP and LS for Rice Field

Parameter	2002			2003			2004		
	R^2	PE	RMSE	R^2	PE	RMSE	R^2	PE	RMSE
Rice field with 0 mm weir height (W1)									
AET	0.8416	0.7781	0.414	0.9547	0.9542	0.243	0.8706	0.9420	0.306
RO	0.9804	0.9897	1.052	0.9847	0.9896	0.332	0.9177	0.989	1.545
VP	0.961	0.9515	0.6362	0.9843	0.9836	0.102	0.9847	0.9787	0.159
LS	0.9538	0.9482	0.365	0.8790	0.9584	0.274	0.9032	0.8960	0.297
Rice field with 50 mm weir height (W2)									
AET	0.891	0.8647	0.363	0.9587	0.9465	0.271	0.9815	0.9760	0.209
RO	0.9864	0.9867	0.865	0.9889	0.9892	0.342	0.9253	0.988	1.427
VP	0.9431	0.9475	0.856	0.9171	0.9075	0.266	0.9295	0.9488	0.582
LS	0.8954	0.9066	0.477	0.8646	0.8836	0.551	0.9148	0.9188	0.345
Rice field with 100 mm weir height (W3)									
AET	0.9100	0.9233	0.2883	0.9734	0.9718	0.201	0.9834	0.9783	0.1988
RO	0.9820	0.9876	0.856	0.989	0.99	0.592	0.9253	0.9627	2.587
VP	0.9049	0.9029	1.563	0.8928	0.8667	0.955	0.8875	0.8905	0.754
LS	0.8442	0.8765	0.925	0.8487	0.7632	0.468	0.7616	0.8679	0.565

values of RO for all the years are found to be more than 0.80. Since the values of R^2 for prediction of RO is high and that of RMSE is less (less than 5 mm), the models can be safely used to simulate the values of RO for rainfed rice field with different weir heights.

Total measured and simulated excess runoff from the rice field with different weir heights for the year 2002, 2003 and 2004 are given in Table 3.5. It is observed that the three years mean SR generated from the rice fields with 0, 50 and 100 mm weir heights are 17.60, 13.24 and 11.96% of the total rainfall during the crop growth period, respectively. This indicates that those percentages of rainfall can be conserved in the water harvesting structure. Thus there is a scope for harvesting of SR generated from the rice field with different W as well as rainfall in a water harvesting structure within-situ farming of fish and to re-utilize to meet SI need of a short duration winter crop that can be grown with residual moisture.

TABLE 3.5　Total Values of Simulated and Observed Water Balance Parameters of Rice Field with Different Weir Heights During the Study Period

Parameter	Year						% of total Rainfall
	2002		2003		2004		
	Simulated	Observed	Simulated	Observed	Simulated	Observed	
Rice Field with 0 mm weir height							
Rainfall (mm)		1174.95		692.95		847.80	
Runoff (mm)	340.03	324.00	95.33	91.14	106.25	81.84	17.60
AET (mm)	392.36	364.97	348.28	345.68	374.79	374.79	42.17
VP (mm)	241.15	277.67	122.82	119.40	149.95	156.26	19.20
LS (mm)	189.03	204.55	103.98	99.20	119.34	115.85	15.09
Rice Field with 50 mm weir height							
Runoff (mm)	321.68	298.79	45.27	40.86	76.05	44.21	13.24
AET (mm)	425.72	442.01	403.71	398.34	408.03	419.61	47.87
VP (mm)	303.72	330.10	157.66	141.25	194.34	225.47	24.43
LS (mm)	255.76	259.20	186.54	177.33	189.53	186.88	23.46
Rice Field with 100 mm weir height							
Runoff (mm)	321.68	285.97	26.02	15.46	76.05	42.96	11.96
AET (mm)	467.46	460.48	401.46	405.98	408.63	419.91	48.87
Vertical Percolation (mm)	374.625	389.78	204.06	184.01	253.85	270.50	30.48
Lateral Seepage (mm)	218.91	327.23	136.15	164.99	157.90	187.77	21.79

3.5.1.5 Vertical Percolation and Lateral Seepage Flux

The values of observed and simulated VP and LS in all the years were also varied with the input of rainfall and/or SI. The simulated and observed VP and LS from the effective root zone depth of rice field with different W for all the three years were studied.

The temporal variation of simulated and observed values of fluxes for the year 2003 and 2004 are presented in Figures 3.6 and 3.7, respectively.

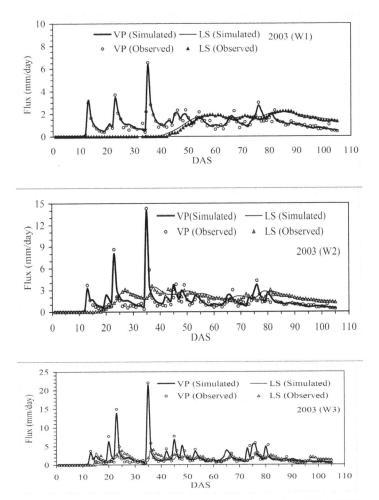

FIGURE 3.6   Variations of seepage and vertical percolation in rice field in 2003.

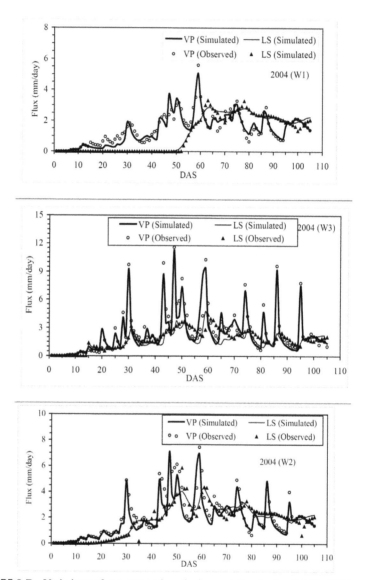

FIGURE 3.7   Variations of seepage and vertical percolation in rice field in 2004.

The predicted fluxes values were matched reasonably well with the observed values with acceptable R^2 (more than 0.80). Total measured and simulated excess runoff from the rice field with different weir heights for the year 2002, 2003 and 2004 are given in Table 3.4. It is observed that the mean VP from the rice fields with 0, 50 and 100 mm weir heights are 19.20, 24.43 and 30.48% of the total rainfall during the crop growth period, respectively. And the corresponding mean LS are 15.09, 23.46 and 21.79% of the total rainfall during the crop growth period, respectively.

Both the model and observed values reveals that VP and LS is function of depth of ponded water, but the variation of rate with depth of ponded water is small in model values as compared to observed value. Some of the discrepancy can be attributed to the fact that VP and LS depends on rainfall intensity, soil properties and surface storage capacity. Usually, only daily precipitation total not the intensity is known. Rainfall can be constant for 24 h or may come in short but heavy rainstorm. In this model it is assumed that all daily precipitation is received at the start of day. This assumption might have resulted in the discrepancy between observed and predicted values. Moreover fluctuations in fluxes are caused by several other factors such as chemical, biological properties of the soil that were not accounted for in the model.

3.6   CONCLUSIONS

HYDRUS-2D model and the water balance model have produced realistic results, which are supported by the field experiment. The predictions of water balance parameters were within the acceptable range during both the phases (ponded and unsaturated). The model predicts daily soil water content and ponding depth in the field at the end of each day. The models therefore, are useful for planning water management practices such as irrigation scheduling and better use of rainfall. At the same time, the model predicts all other water balance components on daily as well as in cropping season, which can be combined with any optimization model to achieve an effective use of rainwater with harvesting structure and its optimum design.

3.7   SUMMARY

A fundamental part of understanding and improving water management is quantitative estimates of the major components of field water balance in the rainfed agriculture. However, the proper estimation of different water balance components in the root zone of the cropped field can help to achieve the effective use of the rainwater to increase the productivity. So, in the present study, field water balance and HYDRUS-2D models was used to quantify the water balance parameters such as runoff, actual evapotranspiration, vertical percolation and lateral seepage from the effective root zone of rice (450 mm) field with different weir heights (0, 50 and 100 mm). Three years (2002–04) field observations from experimental Farm of IIT, Kharagpur, India were used for testing of model predictions. HYDRUS-2D was used to simulate soil water content, lateral seepage and vertical percolation in the effective root zone of rice field with different weir heights under variable saturated condition. However, field water balance model was used to simulate water balance parameters such as actual evapotranspiration, runoff, lateral seepage and vertical percolation under both ponding (above saturation) and unsaturated condition. Higher overall prediction efficiency and coefficient determination (more than 0.75) as well as lower RMSE values reveals that, the HYDRUS-2D and field water balance model can be used for to simulate the different water balance parameters in effective crop root zone of rice with different weir heights.

ACKNOWLEDGEMENT

The authors gratefully acknowledge the financial support from the National Agricultural Technology Project (Indian Council of Agricultural Research, New Delhi, India) and University Grants Commission, New Delhi to carryout the research work at the Department of Agricultural and Food Engineering, Indian Institute of Technology, Kharagpur, India.

KEYWORDS

- **coefficient of determination**
- **crop coefficient**
- **crop stress factor**

- effective root zone
- evapotranspiration
- field water balance
- **HYDRUS-2D**
- lysimeter
- percolation
- piezometer
- ponding depth
- prediction efficiency
- rainfall
- rainfed ecosystem
- reference crop evapotranspiration
- rice
- root mean square error
- saturation
- seepage
- simulation
- soil water
- supplemental irrigation
- surface drainage
- tensiometer
- water balance
- water saving irrigation
- weir height

REFERENCES

1. Abbasia, F., Feyena, J., and van Genuchten, M. T., 2004. Two-dimensional simulation of water flow and solute transport below furrows: model calibration and validation. *J. Hydrol.,* 290:63–79.
2. Agrawal, A., 2000. Optimal design of on-farm reservoir (OFR) for paddy-mustard cropping pattern using water balance approach. Unpublished MTech Thesis, Indian Institute of Technology, Kharagpur, India.

3. Allen, R. G., Pereira L. S., Raes D., and Smith, M., 1998. *Crop Evapotranspiration, Guidelines for Computing Crop Water Requirements*. Irrigation and Drainage Paper 56, Food and Agriculture Organization of the United Nations, Rome, Italy.

4. Angus, J. F., 1991. The evaluation of methods for quantifying risk in water limited environments. In: Muchow, R. C., Bellamy, J. A. (eds.), *Climatic risk in crop prediction models in management for the semi-arid tropics and sub-tropics*. CAB international, Wallingford, p. 39–53.

5. Bouman, B. A. M., Wopereis, M. C. S., Kropff, M. J., ten Berge, H. F. M., and Tuong, T. P., 1994. Water use efficiency of flooded rice fields: II. Percolation and seepage losses. *Agric. Water Manage.*, 26:291–304.

6. Clement, T. P., Wise, W. R., and Molz, F. J., 1994. A physically based, two dimensional, finite difference algorithms for modeling variably saturated flow. *J. Hydrol.*, 161:71–80.

7. Doorenbos, J., and Pruitt, W. O., 1977. *Guidelines for Predicting Crop Water Requirements*. FAO Irrigation and Drainage Paper 24, Food and Agriculture Organization of the United Nations, Rome, Italy.

8. Ghosh, B. C., Pande, H. K., and Mittra, B. N., 1978. Possibilities of growing a second crop on residual moisture of Aman paddy lands in Midnapore district. *Agronomy News Letter*, Agro Science Forum, Indian Institute of Technology, Kharagpur, India, 8:7–8.

9. Henry, E. J., Smith, J. E., and Warrick, A. W., 2001. Surfactant effects on unsaturated flow in porous media with hysteresis: horizontal column experiments and numerical modeling. *J. Hydrol.*, 245:73–88.

10. Hinz, C., 1998. Analysis of unsaturated/saturated water flow near a fluctuating water table. *J. Contam. Hydrol.*, 33:59–80.

11. Huang, M., Shao, M., and Li, Y., 2001. Comparison of a modified statistical dynamic water balance model with the numerical model WAVES and field measurements, *J. of Agric. Water Manage.*, 48:21–35.

12. Hukkeri, S. B., and Sharma, A. K., 1979. Water use efficiency of transplanted and direct-sown rice under different water management practices. *Indian J. Agril. Sci.*, 50:240–243.

13. Izbicki, J. A., Radyk, J., and Michel, R. L., 2000. Water movement through a thick unsaturated zone underlying an intermittent stream in the western Mojave Desert, southern California, USA. *J. Hydrol.*, 238:194–217.

14. Jensen, J. ., Mannan, S. M. A., and Uddin, S. M. N., 1993. Irrigation requirement of transplanted monsoon rice in Bangladesh. *Agric. Water Manage.*, 23:199–212.

15. Kao, C., Bouarfa, S., and Zimmer, D., 2001. Steady state analysis of unsaturated flow above a shallow water-table aquifer drained by ditches, *J. Hydrol.*, 250:122–133.

16. Khepar, S. D., Yadav, A. K., Sondhi, S. K., and Siag, M., 2000. Water balance model for rice fields under intermittent irrigation practices. *Irrig. Sci.*, 19:199–208.

17. Michael, A. M., 1978. *Irrigation Theory and Practices*. Vikash Publishing House Pvt. Ltd., New Delhi.

18. Miller G. R., Baldocchi, D. D., Beverly, E. L., and Meyers, T., 2007. An analysis of soil moisture dynamics using multi-year data from a network of micrometeorological observation sites. *Adv. Water Resour.*, 30:1065–1081.

19. Mishra, H. S., Rathore, T. R., and Pant, R. C., 1990. Effect of intermittent irrigation on groundwater table contribution, irrigation requirement and yield of rice in Mollisols of the Tarai region. *Agric. Water Manage.,* 18:231–241.

20. Mishra, P. K., Rao, C. A. R., and Prasad, S. S., 1998. Economic evaluation of farm pond in a micro-watershed in semi-arid alfisol Deccan plateau. *Indian J. Soil Conserv.,* 26:59–60.

21. Mitra, B. N., 1979. *Water management in rice fields: Process studies and problems in India.* Seminar on Research Technology and Rural Poor. IDS, Sussex, England.

22. Pang, L., Close, M. E., Watt, J. P. C., and Vincet, K. W., 2000. Simulation of picloram, atrazine and simazine leaching through two New Zealand soils and into groundwater using HYDRUS-2D, *J. Contam. Hydrol.,* 44:19–46.

23. Panigrahi, B., Panda, S. N., and Mull, R., 2001. Simulation of water harvesting potential in rainfed rice lands using water balance model. *Agricultural Systems,* 69: 165–182.

24. Panigrahi, B., Panda, S. N., and Mull, R., 2002. Prediction of hydrological events for planning rainfed rice. *Hydrol. Sci. J.,* 47:435–448.

25. Panigrahi, B., and Panda, S. N., 2003. Field test of a soil water balance simulation model. *J. Agric. Water Manage.,* 58:223–240.

26. Paulo, A. M., Pereira, L. A., Teixeira, J. L., and Pereira, L. S., 1995. Modeling paddy rice irrigation. In: *Crop-water-simulation models in practice* by L. S. Pereira, B. J. van Broek, P. Kabat, and R. G. Allen, eds., Wageningen Pers, Wageningen, The Netherlands, p. 287–302.

27. Rezzoug, A., Schumann, A., Chifflard, P., and Zepp, H., 2005. Field measurement of soil moisture dynamics and numerical simulation using the kinematic wave approximation. *Adv. Water Resour.,* 26:161–178.

28. Sandhu, B. S., Khera, K. L., Prihar, S. S., and Singh, B., 1980. Irrigation needs and yield of rice on a sandy loam soil as affected by continuous and intermittent submergence. *Ind. J. Agric. Sci.,* 50:492–496.

29. Simunek, J., Sejna, M., and Van Genuchten, M., 1999. *HYDRUS-2D: Simulating water flow, heat and solute transport in two-dimensional variably saturated media.* Version 2.0. U.S. Salinity Laboratory ARS/USDA, Riverside, California and IGWMC-TPS 53. Golden, Colorado School of Mines.

30. Simunek, J., Van Genuchten, M. T., Gribb, M. M., and Hopmans, J. W., 1998. Parameter estimation of unsaturated soil hydraulic properties from transient flow process. *J. Soil and Tillage Res.,* 47:27–36.

31. Simunek, J., Jarvis, N.J., van Genuchten, M. T., and Gardenas, A., 2003. Review and comparison of models for describing non-equilibrium and preferential flow and transport in the vadose zone. *J. Hydrol.,* 272:14–35

32. Singh, J. K., and Singh, P.M., 1989. Determination of different components of energy balance equation and estimation of evapotranspiration for winter paddy grown under saturated soil condition. *J. Agric. Eng.* (Indian Soc. of Agri. Engineers), 26:97–106.

33. SWIM Mission Report, 1997. *Water Saving Techniques in Rice Irrigation.* SWIM Mission to Guilin Prefecture, Guangxi Region, China.

34. Tuong, T. P., and Bhuiyan, S. I., 1999. Increasing water use efficiency in rice production: farm level perspectives. *Agric. Water Manage.,* 40:117–122.

35. Tuong, T. P., Wopereis, M. C. S., Marquez, J. A., and Kropff, M. J., 1994. Mechanisms and control of percolation losses in irrigated puddle rice fields. *Soil Sci. Soc. Am. J.,* 58: 1794–1803.
36. Tripathi, R. P., Kushwaha, H. S., and Mishra, R. K., 1986. Irrigation requirements of rice under shallow water table conditions. *Agric. Water Manage.,* 12:127–136.
37. Walker, S. H., and Rushton, K. R., 1986. Water losses through the dikes of irrigated rice fields interpreted through an analog model. *Agric. Water Manage.,* 11:59–73.
38. Walker, S. H., and Rushton, K. R., 1984. Verification of lateral percolation losses from irrigated rice fields by a numerical model. *J. Hydrol.,* 71: 335–351.
39. Warrick, A. W., Wierenga, P. J., and Pan, L., 1996. Downward water flow through sloping layers in the vadose zone: steady state analytical solutions. *J. Hydrol.,* 188/189:321–337.
40. Warrick, A. W., and Fennemore, G. G., 1995. Unsaturated water flow around obstructions simulated by two dimensional Rankine bodies. *Adv. Water Resour.,* 18:375–382.
41. Widawsky, D. A., and O' Toole, J. C., 1990. *Prioritizing the Rice Biotechnology Research Agenda for Eastern India.* The Rockefeller Foundation, New York, USA.
42. Wopereis, M. C. S., Bouman, B. A. M., Kroff, M. J., Ten Berge, H. F. M., and Maligaya, A. R., 1994. Water use efficiency of flooded rice fields: I. Validation of the soil-water balance model SAWAH. *Agric. Water Manage.,* 26:291–304.

CHAPTER 4

A MULTI-MODEL ENSEMBLE APPROACH FOR STREAM FLOW SIMULATION

DWARIKA MOHAN DAS,[1] R. SINGH,[1] A. KUMAR,[2]
D. R. MAILAPALLI,[1] A. MISHRA,[1] and C. CHATTERJEE[1]

[1]*Department of Agricultural and Food Engineering, Indian Institute of Technology, Kharagpur, 721302, West Bengal, India, E-mail: dwarika.dmd@gmail.com, Phone: 7548929957*

[2]*Ex-Director, ICAR-Indian Institute of Water Management, Opposite Rail Vihar, Chandrasekarpur, Bhubaneswar-751023, Odisha, India*

CONTENTS

4.1   Introduction 72
    4.1.1   Background 72
    4.1.2   The Bias/Variance Trade-Off and Ensemble Modeling 74
4.2   Materials and Methods... 77
    4.2.1   Description of the Study Area 77
    4.2.2   Data Availability.. 78
    4.2.3   Brief Descriptions About Models....................................... 83
        4.2.3.1   MIKE SHE 83
        4.2.3.2   Soil and Water Assessment Tool (SWAT) 84
        4.2.3.3   Hydrologic Modeling System (HEC-HMS)........ 84
        4.2.3.4   TANK 85

          4.2.3.5 Australian Water Balance Model (AWBM) 85
          4.2.3.6 Simplified Daily Conceptual
                   Rainfall-Runoff (SIMHYD) 85
          4.2.3.7 SACRAMENTO 86
          4.2.3.8 Soil Moisture and Accounting
                   Model (SMAR) 86
4.3   Ensemble Construction 87
      4.3.1 Linear Programming Technique of Ensemble Creation 87
      4.3.2 Ensemble Verification Criteria 89
             4.3.2.1 Nash-Sutcliff Efficiency (NSE) 89
             4.3.2.2 Brier Score (BS) 90
             4.3.2.3 Rank Probability Score (RPS) 91
             4.3.2.4 Continuous Rank Probability Score (CRPS) 92
4.4   Results and Discussion 92
      4.4.1 Ensemble Evaluation Based on NSE 92
      4.4.2 Selection of Optimum Ensemble Size Based on NSE 94
      4.4.3 Ensemble Evaluation based on Brier Score 94
      4.4.4 Selection of Optimal Ensemble Combinations
             Based on Brier Score 96
      4.4.5 Selection of Optimal Ensemble Combinations
             Based on NSE and Brier Score 96
4.5   Conclusions 97
4.6   Summary 98
Acknowledgements 99
Keywords 99
References 100

4.1 INTRODUCTION

4.1.1 BACKGROUND

Hydrological models are considered as an important tool for planning and management of the complex hydrologic system. Despite the improvement in actual level of knowledge of the physical processes of the hydrological system and the computational advances in the recent years, the models, still can provide an approximation of reality. With advancement in modeling methodology, several hydrological models of varying complexity have been developed. Applications of these models, however, are associated with several kinds of uncertainties dealing with model structure, model parameters and input data requirements. Hence, uncertainty in model predictions arises from several sources like natural randomness, errors in the system input and output data, model parameter error and model structural error. Natural randomness arises from the random temporal and spatial fluctuations of natural processes and induces a large amount of randomness also in the physical processes that produce the responses of the hydrological system. Error in data affects both the magnitude and the timing of the measurements of metrological and hydrological observations.

For calibrated models, these errors accounts for the fact that the parameter set adjustment can compensate for the other types of errors. Model structural error can have several origins, such as incorrect representation of the processes, errors in numerical algorithms and computer codes. These errors propagate through the different processes of the model and finally sum up; which leads to a considerable uncertainty in model predictions. This means that we can never create a perfect prediction because it is very difficult to observe every detail of the natural system. Different methods have been proposed to handle these uncertainties independently but ensemble modeling is considered the best as it can address the combined uncertainty in input data, model parameters and model structure [22].

In general, different model perform differently, some provides more accurate predictions than others. It is therefore useful to develop a number of different models to ensure that at least one model provides the good representation of different parts of the predictions. One can select the best performing model and discard others less successful models. However, selecting the 'best' model is not necessarily the ideal choice, because potentially valuable information may be wasted by discarding the results of less-successful models [14, 21]. This leads to the concept of 'combining', where the outputs of several models are pooled before a decision (collective prediction) is made [21].

The word "ensemble" is a French word, meaning "together" or "at the same time" and usually refers to a unit or group of complementary parts that contribute to a single effect. Ensemble modeling is being used for climate forecasting from several decades but it is overlooked in many other areas of environmental modeling. In the field of hydrology, ensemble modeling has been applied only over last decade through cooperative initiatives such as, Distributed Model Inter-comparison Project (DMIP) [18], Hydrological Ensemble Prediction Experiment (HEPEX) project [17], Ensemble Streamflow Prediction (EPS) [24] and Land Use Change on Hydrology by Ensemble Modeling (LUCHEM) [22].

Ensemble offers greatly improved result over using single models even when number of members is as few as two [3]. There are two types of ensemble: Single model ensemble and multi model ensemble. Single model ensemble involves use of realizations of a single deterministic model where as a multi-model ensemble involves realizations from different models having different structural complexities. Multi-model ensemble, however, outperforms both single models and single-model ensemble [7, 10, 13]. Error in ensemble is related to the prediction error of constituent members. Mean square error of an ensemble can be expressed as an average of individual model error. Theoretically, it states that by increasing the number of members in the population, the error of an ensemble can be made to be arbitrarily small when compared to the error of the individual models [14].

4.1.2 THE BIAS/VARIANCE TRADE-OFF AND ENSEMBLE MODELING

The effect of combining models to reduce errors may be expressed in terms of bias and variance. Mean square error (MSE) can be expressed as the sum of model variance and square of bias. Model bias and model variance are decreasing and increasing functions, respectively with model complexity. A model to perform well requires small MSE, which corresponds to small variance and small bias. However, an attempt to decrease the bias is likely to increase variance and vice-versa as both are opposite in nature. Hence, optimal fitting needs a tread-off between bias and variance.

A model that performs well (an optimal compromise between low variance and low bias) must take sufficient account of the data to avoid bias towards a particular sub-set of the data, but also should avoid over-fitting (i.e., due to increase in variance).

In ensemble modeling, imperfect predictions are combined which creates a way of managing individual limitations i.e., each component model is known to make errors (bias, variance and noise), but they are combined in such a way as to minimize the effect of these errors. This is considered as one of the major advantage of ensemble modeling. The bias measures the extent to which the ensemble output averaged over all the ensemble members differs from the target function, whereas the variance is a measure of the extent to which the ensemble members disagree [11]. The improvement of predictions that can arise from ensemble combinations is usually the result of a reduction (if not complete elimination) in variance, whilst leaving the bias unaltered [21, 23].

Main reason for combining models in ensembles is to improve predictive ability and to guard against the failure of individual member models. Here, the term 'fail' refers to the fact that, individual models will make predictions that will not usually be identical to the target function, and will usually under or over-estimate the observed value(s). There is clearly no other advantage to be gained from an ensemble that is composed of a set of identical models because the bias of the individual models will be identical, and the variance of the individual predictions/simulation equal to zero. Thus, the aim is to create models that predict differently or models of different structural complexity can be used (i.e., multi-model ensemble) because error cancelation and non-linearity in diagnosis can only be achieved by combining different models. Due to this reasons in many cases multi-model ensemble performs better than that of single-model ensemble.

There are generally four parameters that may be changed to produce different realizations. These are (1) initial conditions, (2) model topology, (3) training algorithm, and (4) data. When considering the initial conditions, what we are primarily concerned with is the initial boundary conditions of the model, which can be perturbed to produce different realizations. Differing the model architecture (topology) allows each model to produce different predictions. Moreover, multi-model ensembles fall into this category, where we combine the outputs of different models to predict a common target. Modeling methods that employ the use of training (calibration) algorithms usually allow the choice of different algorithms to be used for training. Each of these algorithms provides a different means for traversing the error surface, and thus, using different training algorithms yields models that generalize differently. Since the error surface is fixed for a given dataset and by varying the data from one ensemble member to the other, the error surface will be changed which produces a new realization.

There are several methods available to combine the outputs of different models, i.e. mean, median, trimmed mean, constrained multiple linear regression, unconstrained linear regression, weighted average method, linear programming technique, multi-model super ensemble technique, Supra Bayesian method, etc. [1, 6]. A major challenge in ensemble modeling is to determine the optimal size of an ensemble because the best ensemble may not contain only the best individual models [5]. Increasing the ensemble size improves performance but it is strongly dependent on the objective function used to assess performance [5].

In this chapter, *Linear Programming* (LP) *technique* was selected to develop multi-model ensembles. This technique determines the optimal weights by minimizing the deviations between observed and simulated discharge of individual models by considering all events together in the time series. Different commonly used techniques like Nash Sutcliffe Efficiency (NSE), Brier Score (BS), Rank Probability Score (RPS), Rank Probability Skill Score (RPSS), and Continuous Rank Probability Score (CRPS) are selected to evaluate different possible ensembles in order to find out the optimal model combination for the study area. These criteria include different aspects of ensemble performance viz., overall performance for NSE, Uncertainty and Reliability for BS, and relative skillfulness for RPSS and CRPS.

The specific objectives of this study are as follows:

i. To develop multi model ensembles using linear programming technique.
ii. To evaluate the developed ensembles using: Nash Sutcliffe efficiency, Brier score, Rank probability score, Rank probability skill score and Continuous rank probability score.

4.2 MATERIALS AND METHODS

4.2.1 DESCRIPTION OF THE STUDY AREA

Salebhata catchment, a catchment of Ong River, situated in the middle reach of Mahanadi River Basin of Odisha was selected for this study (Figure 4.1). The catchment covers 4515 km2 area and lies between 20°40'12" and 21°25'08"N latitude, and 82°33'24" and 83°34' 11"E longitude. The elevation of the area ranges from 100 to 1000 m. An extensive part of the study area is under forest. Paddy is the main crop grown on the cultivable land. The annual rainfall ranges from about 1200 to 1700 mm.

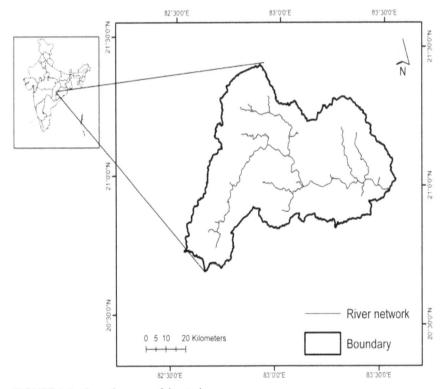

FIGURE 4.1 Location map of the study area.

4.2.2 DATA AVAILABILITY

The simulation results of eight hydrological models from the previous studies [2, 9, 15, 16] were used for this study. The models include:

i. Six lumped conceptual models: HEC-HMS, TANK, AWBM, SIMHYD, SACRAMENTO, and SMAR; and

ii. Two physically based distributed models: SWAT and MIKE SHE.

The eight models were calibrated for the period 1st June 2004 to 31st May 2007 and validated for the period 1st June 2007 to 31st May 2009. The results of these models were used to develop multi-model ensembles (MME). The simulation results of these models are shown in Figures 4.2–4.9 for calibration period and Figures 4.10–4.17 for validation periods, respectively.

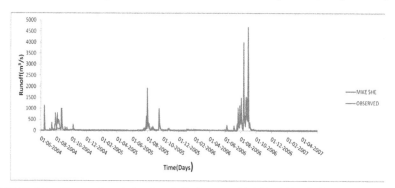

FIGURE 4.2 Observed and simulated discharges of MIKE SHE during the calibration period.

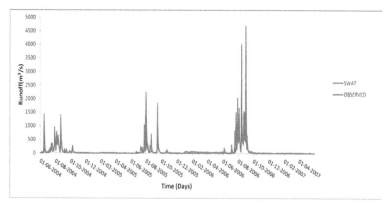

FIGURE 4.3 Observed and simulated discharges of SWAT during the calibration period.

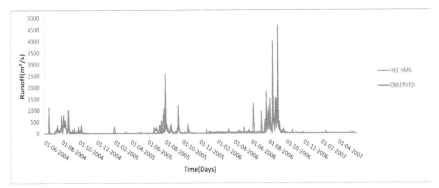

FIGURE 4.4 Observed and simulated discharges of HEC HMS during the calibration period.

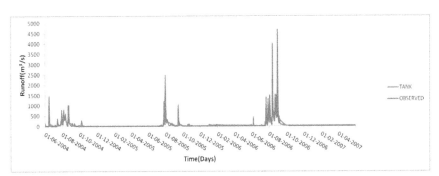

FIGURE 4.5 Observed and simulated discharges of TANK during the calibration period.

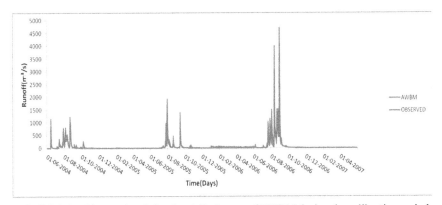

FIGURE 4.6 Observed and simulated discharges of AWBM during the calibration period.

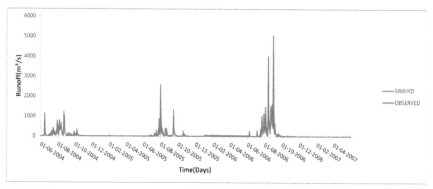

FIGURE 4.7   Observed and simulated discharges of SIMHYD during the calibration period.

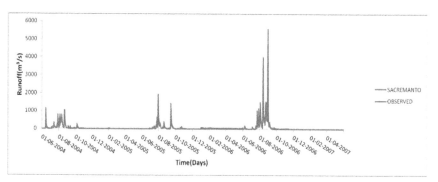

FIGURE 4.8   Observed and simulated discharges of SACREMANTO during the calibration period.

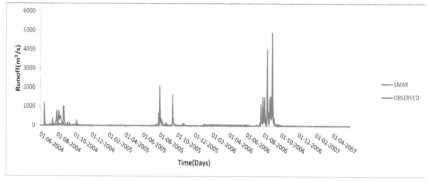

FIGURE 4.9   Observed and simulated discharges of SMAR during the calibration period.

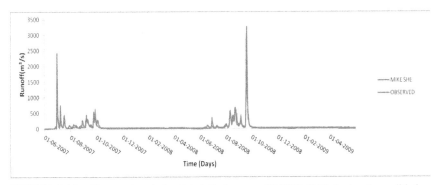

FIGURE 4.10   Observed and simulated discharges of MIKE SHE during the validation period.

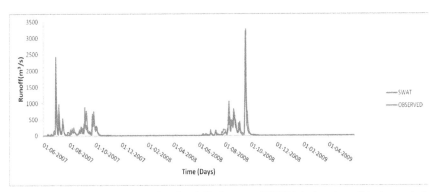

FIGURE 4.11   Observed and simulated discharges of SWAT during the validation period.

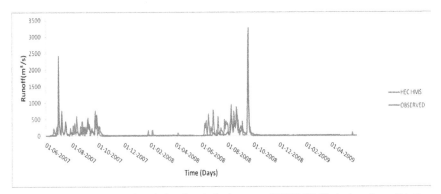

FIGURE 4.12   Observed and simulated discharges of HEC HMS during the validation period.

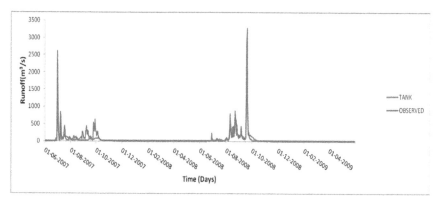

FIGURE 4.13   Observed and simulated discharges of TANK during the validation period.

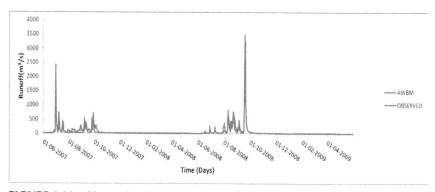

FIGURE 4.14   Observed and simulated discharges of AWBM during the validation period.

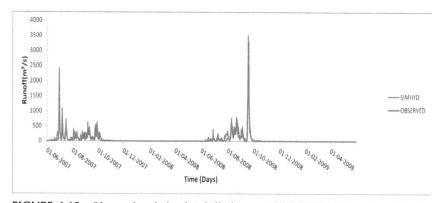

FIGURE 4.15   Observed and simulated discharges of SIMHYD during the validation period.

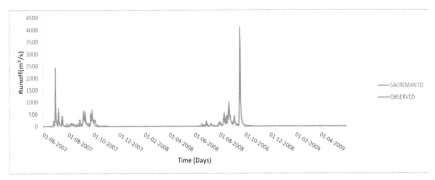

FIGURE 4.16 Observed and simulated discharges of SACREMANTO during the validation period.

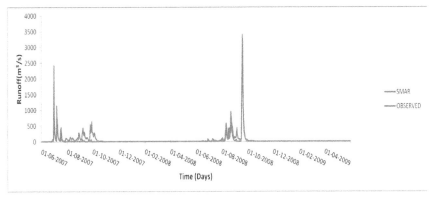

FIGURE 4.17 Observed and simulated discharges of SMAR during the validation period.

4.2.3 BRIEF DESCRIPTIONS ABOUT MODELS

In this section, brief description is given for each of eight hydrological models (MIKE SHE and SWAT, IIEC HMS, AWBM, SIM HYD, SECREMENTO, SMAR and TANK) are considered for this study.

4.2.3.1 MIKE SHE

It is an integrated, physically based, distributed model that simulates the hydrological and water quality process on a basin scale. MIKE SHE modeling system simulates most major hydrological process, including canopy and land surface interpretation after precipitation, snowmelt, evapotranspiration, overland flow, channel flow, unsaturated subsurface flow and saturated groundwater flow using physically based methods. The system has no limitations regarding watershed size. In this model area is discretized by horizontal as well as vertical square grid networks for surface and groundwater flow components.

4.2.3.2 Soil and Water Assessment Tool (SWAT)

SWAT is a physically-based, continuous-time watershed hydrological model that operates on a daily time step. In this model, watershed is divided into multiple sub-watersheds, which are further subdivided into HRUs that consist of homogeneous land use, management, and soil characteristics. The HRUs represent percentages of the sub-watershed area and are not identified spatially within a SWAT simulation. Alternatively, a watershed can only be subdivided into sub watersheds that are characterized by dominant land use, soil type, and management. The model has been adopted as part of the USEPA's Better Assessment Science Integrating Point and Nonpoint Sources (BASINS) system and AGWA software systems because it is a widely accepted continuous model suitable for agricultural and forest land uses.

4.2.3.3 Hydrologic Modeling System (HEC-HMS)

The HEC-HMS is designed to simulate the rainfall-runoff processes of dendritic watershed systems. It is designed for a wide range of geographic areas for solving the widest possible range of problems. This includes large river basin water supply and flood hydrology, and small urban or natural watershed runoff. Hydrographs produced by this model may be used directly or in conjunction with other software for studies of water availability, urban drainage, flow forecasting, future urbanization impact, reservoir spillway design, flood damage reduction, floodplain regulation, and systems operation. HEC-HMS is a product of the Hydrologic Engineering Center within the U.S. Army Corps of Engineers. The program was developed in 1992 as a replacement for HEC-1, which has long been considered as a standard for hydrologic simulation. The new HEC-HMS provides almost all of the same simulation capabilities, but has modernized them with advances in numerical analysis that take advantage of the significantly faster desktop computers available today. It also includes a number of features that were not included in HEC-1, such as continuous simulation and grid cell surface hydrology. It also provides a graphical user interface to make it easier to use the software.

4.2.3.4 TANK

TANK models were developed by Sugawara et al. [19, 20] and have been popular in Japan, Korea, and many other countries of Asia for flood forecasting, watershed modeling, reservoir operation, etc. The TANK model is a simple model, composed of four tanks laid vertically in series. Precipitation is put into the top tank, and evaporation is subtracted sequentially from the top tank downwards. As each tank is emptied the evaporation shortfall is taken from the next tank down until all tanks are empty. The outputs from the side outlets are the calculated runoffs. The output from the top tank is considered as surface runoff, output from the second tank as intermediate runoff, from the third tank as sub-base runoff and output from the fourth tank as base flow.

4.2.3.5   Australian Water Balance Model (AWBM)

The AWBM was developed in the early 1990s and is now one of the most widely used rainfall-runoff models in Australia. There are two main versions: one for daily water yield and low flow studies; the other for continuous simulation of flood runoff at hourly time steps. The AWBM is a conceptual model, and was developed from concepts of saturation overland flow generation of runoff. The model can relate runoff to rainfall on daily time step basis, and calculates losses from rainfall for flood hydrograph modeling.

4.2.3.6 Simplified Daily Conceptual Rainfall-Runoff (SIMHYD)

SIMHYD is a simplified version of the daily conceptual rainfall-runoff model. It simulates daily runoff (surface runoff and base flow) using daily precipitation and potential evapotranspiration (PET) as input data. In this model, the rainfall first fills the interception store, which is depleted by evaporation subject to potential evapotranspiration rate. The excess rainfall is then subjected to an infiltration function that determines the infiltration capacity. The excess rainfall that exceeds the infiltration capacity becomes infiltration excess runoff. Moisture that infiltrates is subjected to a soil moisture function that diverts the water to the interflow, groundwater recharge and soil moisture store recharge.

4.2.3.7 SACRAMENTO

The SACRAMENTO model is a conceptual rainfall-runoff model with spatially lumped parameters. The model represents the moisture distribution in a physically realistic manner within hypothetical zones of a soil column. The model attempts to maintain percolation characteristics to simulate stream flow contributions from a basin. The components of SACRAMENTO are: tension water, free water, surface flow, lateral drainage, ET and vertical drainage (percolation). The model uses comprehensive runoff analysis in water balance accounting system.

4.2.3.8 Soil Moisture and Accounting Model (SMAR)

The SMAR Model is a development of the 'Layers' conceptual rainfall-runoff model introduced by O'Connell et al. [12]. Its water balance component was in 1969. After several modifications/improvements, the present form of SMAR model was obtained. The SMAR is a lumped conceptual rainfall-runoff water balance model with soil moisture as a central theme. The model provides daily estimates of surface runoff, groundwater discharge, evapotranspiration and leakage from the soil profile for the catchment as a whole. The surface runoff component comprises overland flow, saturation excess runoff and saturated through-flow from perched groundwater conditions with a quick response time. The model consists of two components in sequence, a water balance component and a routing component. The model utilizes time series of rainfall and pan evaporation data to simulate stream flow at the catchment outlet.

4.3 ENSEMBLE CONSTRUCTION

In ensemble modeling, different model predictions are pooled in a statistical procedure to improve the prediction accuracy. Multi-model ensemble approach generally helps in reducing prediction uncertainty by sampling models with a range of structural uncertainties. Different models have different strengths and weaknesses. In ensemble, the deficiencies in one model may be masked by the strengths in other and thus ensemble modeling provides an estimate of the most probable system.

In this present study, an attempt has made to create multi-model ensembles using the simulation results of eight hydrological models, i.e., MIKE SHE, SWAT, HEC-HMS, TANK, AWBM, SIMHYD, SACRAMENTO AND SMAR. Linear Programming technique (LP) has been used to combine model simulations. Using this technique, weights for constituent members of an ensemble are calculated for calibration periods and used directly to create ensembles for calibration and validation periods.

4.3.1 LINEAR PROGRAMMING TECHNIQUE OF ENSEMBLE CREATION

The basic principle involved in formulation of linear programming problem is to formulate an objective function in order to minimize the sum of absolute deviations of simulation and observation. The weights are constrained to sum to unity and the deviations of the predictions are calculated by subtracting the observed prediction from estimated prediction for each day. The formulation to solve the linear programming is given below:

Objective function

$$\text{Minimize: } z = \sum_{j=1}^{n} (U_j + V_j) \tag{1}$$

$$\text{Subject to: } \sum_{i=1}^{m}(S_{ij}W_i) + U_j - V_j = O_j \tag{2}$$

$$\sum_{i=1}^{m}W_i = 1 \tag{3}$$

where, S = simulated discharge, O = observed discharge, W = weight of different models, U and V = +ve and –ve deviational variables, j = index for number of days in the time series i = model index number, n = number of days in the time series and m = number of models.

The selected models are mainly categorized into two classes, i.e., physically based distributed models (MIKE SHE and SWAT) and lumped conceptual models (HEC-HMS, TANK, AWBM, SIMHYD, SACRAMENTO AND SMAR). Different multi-model ensembles have been developed from these models by considering the fact that each ensemble combination must possess at least one model from each class. Table 4.1 represents 189 possible combinations of selected models. Index numbers of 1, 2, 3, 4, 5, 6,

TABLE 4.1   Possible Ensembles and Their Combinations

Ensemble size	No. of physically based models	No. of conceptual models	Ensemble combination
2	1	1	13, 14, 15, 16, 17, 18, 23, 24, 25, 26, 27, 28
3	1	2	134, 135, 136, 137, 138, 145, 146, 147, 148, 156, 157, 158, 167, 168, 178, 234, 235, 236, 237, 238, 245, 246, 247, 248, 256, 257, 258, 267, 268, 278
3	2	1	123, 124, 125, 126, 127, 128
4	1	3	1345, 1346, 1347, 1348, 1356, 1357, 1358, 1367, 1368, 1378, 1456, 1457, 1458, 1467, 1468, 1478, 1567, 1568, 1578, 1678, 2345, 2346, 2347, 2348, 2356, 2357, 2358, 2367, 2368, 2378, 2456, 2457, 2458, 2467, 2468, 2478, 2567, 2568, 2578, 2678
4	2	2	1234, 1235, 1236, 1237, 1238, 1245, 1246, 1247, 1248, 1256, 1257, 1258, 1267, 1268, 1278

TABLE 4.1 Continued

Ensemble size	No. of physically based models	No. of conceptual models	Ensemble combination
5	1	4	13456, 13457, 13458, 13467, 13468, 13478, 13567, 13568, 13578, 13678, 14567, 14568, 1457814678, 15678, 23456, 23457, 23458, 23467, 23468, 23478, 23567, 23568, 23578, 23678, 24567, 24568, 24578, 24678, 25678
5	2	3	12345, 12346, 12347, 12348, 12356, 12357, 1235812367, 12368, 12378, 12456, 12457, 12458, 12467, 12468, 12478, 12567, 12568, 12578, 12678
6	1	5	134567, 134568, 134578, 134678, 135678, 145678, 234567, 234568, 234578, 234678, 235678, 245678
6	2	4	123456, 123457, 123458, 123467, 123468, 123478, 123567, 123568, 123578, 123678, 124567, 124568, 124578, 124678, 125678
7	1	6	1345678, 2345678
7	2	5	1234567, 1234568, 1234578, 1234678, 1235678, 1245678
8	2	6	12345678

7, and 8 are assigned to MIKE SHE, SWAT, HEC-HMS, TANK, AWBM, SIMHYD, SACRAMENTO and SMAR, respectively. Each combination is denoted by the concentration of its corresponding models index numbers. For example, combination "247" denotes the combination of SWAT (2), TANK (4) and SACRAMENTO (7). Last column in Table 4.1 represents the ensemble members with their respective notations.

4.3.2 ENSEMBLE VERIFICATION CRITERIA

Ensemble verification is used to select the best ensemble out of different possible ensembles. Following are the four evaluation criteria, which are considered for ensemble verification.

4.3.2.1 Nash-Sutcliff Efficiency (NSE)

The Nash–Sutcliffe efficiency, NSE, which may be estimated by Eq. 4, is traditionally used in many hydrological applications. A perfect simulation would have a NSE value of one. A value of zero indicates a simulation consisting of the mean of the observations. A NSE value of below 0.6 commonly indicates a non-skillful prediction [6]. It is often considered that a score higher than 0.7 characterizes a very good simulation of the discharges. It measures how much of the variability in observations is explained by the simulations. It is given by:

$$NSE = 1 - \frac{\sum(Obs - Sim)^2}{\sum(Obs - Obs_{Avg})^2} \tag{4}$$

where, Obs = the measured runoff, Sim = simulated runoff, Obs_{Avg} = average of measured runoff. NSE is a score of between $-\infty$ and 1. Table 4.2 gives some suggested ranges for evaluation of the efficiency.

4.3.2.2 Brier Score (BS)

One of the most common measures of accuracy for verifying two-category probability forecasts is the Brier score [4]. The Brier score can be computed from Eq. (5) [8].

$$BS = \bar{O}(1 - \bar{O}) + \sum_{j=1}^{N} \frac{N_j}{N} \left[(P_j - \bar{O}_j)^2 - (\bar{O}_j - \bar{O})^2 \right] \tag{5}$$

TABLE 4.2   Ranges for NSE

Nash Sutcliffe Correlation Efficiency (NSE)	Fit
<0.2	Insufficient
0.2–0.4	Satisfactory
0.4–0.6	Good
0.6–0.8	Very good
>0.8	Excellent

where, P_j = the central forecast probability of the jth probability class; $P_j = 0.05 + (j-1) \times 0.2), \overline{O}_j = \dfrac{b_j}{a_j + b_j}, O_j$ = Relative frequency of the event, a_j = Sum of non events, b_j = Sum of events, $\dfrac{N_j}{N} = \dfrac{a_j}{A}, \dfrac{N_j}{N}$ = Relative population of forecasts in the jth class, $A = \Sigma_j \, a_j, \overline{O} = \dfrac{B}{A+B}, \overline{O}$ = Sample Simulation, $B = \Sigma_j \, b_j$.

The Brier Score is the mean-square error of probability forecasts. It is negatively orientated, with a perfect score of BS=0. As observations and probability forecasts are bounded by 0 and 1, the Brier Score equally ranges between 0 and 1. A score of greater than 0.30 in most cases represents a poor prediction. For better prediction they tend to lie between 0.1–0.25. Some rare predictions whose score lies below 0.1 are considered as the best prediction. The Brier Score is the most important score to verify prediction models, because it accounts for reliability, sharpness and uncertainty of the prediction. However, as the score depends on the verification dataset, a model comparison should be based on that same dataset.

4.3.2.3   Rank Probability Score (RPS)

The RPS is intended for verifying continuous multi-category probability predictions. For verifying prediction, j number of categories is defined, which covers all possible outcomes. For all categories the squared differences between the cumulative forecast probability and the corresponding cumulative observation of each category are averaged to gain the RPS. It can be expressed as

$$RPS = 1 - \frac{1}{j-1}\sum_{j=1}^{j}\left[\left(\sum_{i=1}^{j}P_i\right)-\left(\sum_{i=1}^{j}d_i\right)\right]^2 \qquad (6)$$

where, $P = (p_1, p_2,....p_j)$ simulation probability, j = No. of classes or categories, the vector d = $(d_1, d_2 ...dj)$ represents the observation vector such that d_i equals 1 if the event occurs, and zero otherwise.

The RPS has a range of zero to one and is positively oriented (higher the value, better will be the prediction). A perfect categorical prediction always receives a score of one. The worst possible categorical prediction receives a score of zero. Ranked Probability Skill Score (RPSS) is an extension of RPS, which may be estimated from Eq. (6).

$$RPSS = 1 - \frac{RPS_{forecast}}{RPS_{reference}} \qquad (7)$$

The *RPSS* measures skill with respect to a reference standard forecast, and ranges from +1 (perfect forecast) to −∞. Negative RPSS values indicate that the forecast has less accuracy than the standard. It is unstable when applied to small datasets, which may result in large changes in the score's value when one dataset is compared to another. Interpretation of the RPSS is similar to the interpretation of the BSS.

4.3.2.4 Continuous Rank Probability Score (CRPS)

The CRPS (Eq. (8)) measures the integral square difference between the cumulative distribution function (CDF) of the prediction $F_X(q)$, and the corresponding CDF of the observed variable $F_Y(q)$,

$$CRPS = \int_{-\infty}^{\infty} \{F_X(q) - F_Y(q)\}.dq \qquad (8)$$

where $F_Y(q) = \begin{cases} 0, < observed\ value \\ 1, \geq observed\ value \end{cases}$

The CRPS has a negative orientation; smaller the CRPS better will be result. The minimal value 0 is achieved $F_X = F_Y$.

In this study eight hydrological models of varying complexity are used to develop 189 ensemble combinations using linear programming technique. Resulting ensembles are evaluated using NSE and Brier score, which are two oppositely oriented scores, i.e. higher NSE gives better performance and lower Brier score indicates better performance. The aim of the whole exercise is to find out optimal model combination, i.e., how many minimum numbers of models are required to ensure high accuracy out of all possible multi model ensembles. This chapter deals with analysis of ensembles and their results.

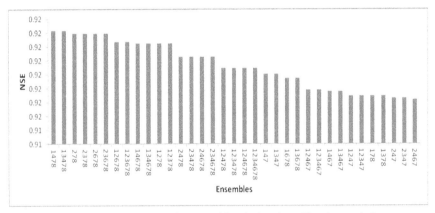

FIGURE 4.18　NSE values of top 35 ensembles during the calibration period.

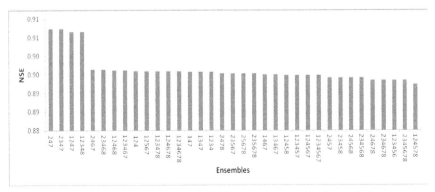

FIGURE 4.19　NSE values of top 35 ensembles during the validation period.

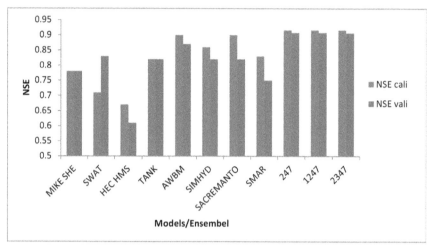

FIGURE 4.20 NSE of individual models and better performing ensemble during calibration and validation.

4.4   RESULTS AND DISCUSSION

4.4.1   *ENSEMBLE EVALUATION BASED ON NSE*

Nash Sutcliff Efficiency is, used to evaluate all the ensemble combinations. The combinations are arranged according to decreasing order of their NSE values, which are shown in Figures 4.18 and 4.19 for calibration and validation periods, respectively. However, top 35 combinations are plotted.

NSE value for all combinations ranges from 0.75 to 0.92 for the calibration period and 0.79–0.91 for the validation period. No combination has NSE value below 0.7 either during calibration or validation. Out of 189 Ensemble combinations 179 combinations have NSE above 0.85 for calibration period and 157 combinations have NSE above 0.85 for validation period. The analysis shows that the best ensemble combinations have higher NSE than individual models both for calibration and validation periods (Figure 4.20).

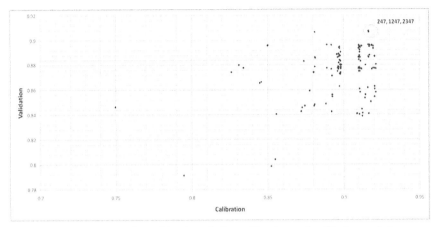

FIGURE 4.21   NSE of all combinations during calibration and validation periods.

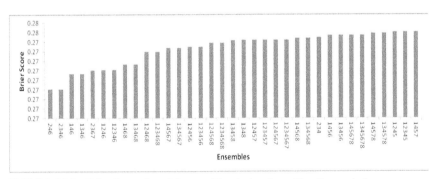

FIGURE 4.22   BS values of top 35 ensembles during the calibration period.

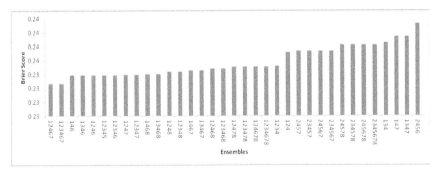

FIGURE 4.23   BS values of top 35 ensembles during the validation period.

4.4.2 SELECTION OF OPTIMUM ENSEMBLE SIZE BASED ON NSE

Respective NSE values of all ensemble combinations for calibration and validation periods are plotted in a scatter diagram to find out better performing combinations for both the periods. Three ensemble combinations 247, 1247 and 2347 falling in the top right corner (i.e., having higher NSE during calibration and validation periods) of the scatter diagram are selected as better performing combinations (Figure 4.21).

4.4.3 ENSEMBLE EVALUATION BASED ON BRIER SCORE

Brier score, which is the most commonly used, probabilistic ensemble verification score is taken to verify the ensemble combinations. The combinations are arranged with increasing order of their Brier score values (as BS is a negatively oriented score), which are shown in Figures 4.22 and 4.23 for calibration and validation periods respectively. However, also only top 35 combinations are plotted.

The range of Brier score varies from 0.27–0.32 during the calibration period and 0.23–0.31 during the validation period. 121 ensemble combinations have Brier score below 0.3 for the calibration period and 128 combinations have Brier score below 0.3 during the validation periods. The upper range is also not very far from 0.3 for both calibration and validation periods, hence; almost all ensembles have a better prediction.

4.4.4 SELECTION OF OPTIMAL ENSEMBLE COMBINATIONS BASED ON BRIER SCORE

The scattered plot of BS for calibration and validation period is also prepared in similar way as that of NSE and is presented in Figure 4.24. The ensemble combinations falling in the bottom left corner of the scatter plot are considered as the better performing ensemble as it corresponds to lowest BS during both the calibration and validation periods.

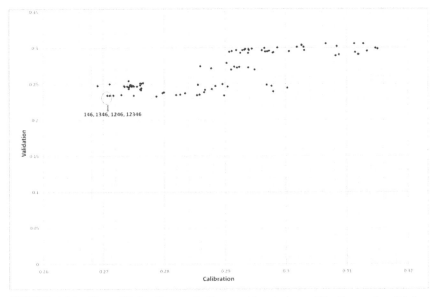

FIGURE 4.24 Plot of Brier Score of all ensembles during calibration and validation periods.

TABLE 4.3 Better Performing Combinations Bases on NSE and BS

	Combinations	NSE Cal	NSE Val	BS Val	BS Val
Better performing models based on NSE	247	0.9163	0.9073	0.2903	0.2460
	1247	0.9164	0.9073	0.2898	0.2338
	2347	0.9163	0.9065	0.2903	0.2460
Better performing modes based on BS	146	0.8814	0.8856	0.2706	0.2338
	1346	0.8814	0.8856	0.2706	0.2338
	1246	0.8813	0.8862	0.2710	0.2338
	12346	0.8813	0.8862	0.2710	0.2338

It is evident from Figure 4.24 that the ensemble combinations 146, 1346, 1246, and 12346 are better performing combinations during calibration and validation periods.

4.4.5 SELECTION OF OPTIMAL ENSEMBLE COMBINATIONS BASED ON NSE AND BRIER SCORE

Sections 4.4.2 and 4.4.3 indicate that the better performing combinations are different for different criteria, i.e., NSE and BS. Hence, both criteria are considered together for further analysis. Table 4.3 presents the better performing models on the basis of both the criteria.

Table 4.3 indicates that the better performing ensemble combinations based on NSE are not performing well, when evaluated using Brier Score. The ensemble combinations "247, 1247 and 2347," which are the better performing ensembles on the basis of NSE are also not giving a consistent Brier Score during calibration and validation periods.

The better performing combinations based on Brier score are not giving best NSE value during calibration and validation periods. However, they are showing NSE value near 0.88 both for calibration and validation periods and 0.88 can be considered as a better NSE value. Hence, for the time being before calculation of other evaluation scores ensemble combinations 146, 1346, 1246 and 12346 can be considered as better performing combinations.

4.5 CONCLUSIONS

In the present study, linear programming technique was used to develop 189 possible ensemble combinations using results of eight hydrological models. The ensembles were then evaluated using Nash Sutcliff Efficiency and Brier Score. Following conclusions were drawn on the basis of analysis made:

    i. NSE value for all combinations ranges from 0.75 to 0.92 for the calibration period and 0.79–0.91 for the validation period.

    ii. Three ensemble combinations 247, 1247 and 2347 are selected as better performing combinations based on NSE.

    iii. The range of Brier score varies from 0.27–0.32 during the calibration period and 0.23–0.31 during the validation period.

    iv. Four ensemble combinations 146, 1346, 1246, and 12346 are selected as better performing combinations based on Brier Score.

v. The better performing ensemble combinations based on Brier Score though do not result in higher NSE; NSE of these ensembles is around 0.88 during both the calibration and validation periods. Hence, for the time being (before calculation of other evaluation scores), ensemble combinations 146, 1346, 1246 and 12346 may be considered as better performing combinations.

4.6 SUMMARY

Now-a-days, hydrological models play a crucial rule for planning and management of the complex hydrologic system. Although several improvements have been made in the field of modeling so that accuracy can be maintained in forecasting the variates in hydrology, the models, still can provide an approximation of reality. With advancement in modeling methodology, several hydrological models of varying complexity have been developed. Applications of these models, however, are associated with several kinds of uncertainties. In this present study, an attempt has made to create multi-model ensembles using the simulation results of eight hydrological models: MIKE SHE, SWAT, HEC-HMS, TANK, AWBM, SIMHYD, SACRAMENTO and SMAR.

Linear Programming (LP) technique is selected to develop multi-model ensembles. Different commonly used techniques like Nash Sutcliffe Efficiency (NSE), Brier Score (BS), Rank Probability Score (RPS), Rank Probability Skill Score (RPSS), and Continuous Rank Probability Score (CRPS) are selected to evaluate different possible ensembles in order to find out the optimal model combination for the study area. NSE value for all combinations are found to range from 0.75 to 0.92 for the calibration period and 0.79–0.91 for the validation period and the range of Brier score varies from 0.27–0.32 during the calibration period and 0.23–0.31 during the validation period. The better performing ensemble combinations based on Brier Score though is found not to result in higher NSE. Hence, for the time being (before calculation of other evaluation scores), ensemble combinations 146, 1346, 1246 and 12346 may be considered as better performing combinations.

ACKNOWLEDGEMENTS

The authors gratefully acknowledged the financial support from Ministry of Water Resources, Govt. of India for funding the Project entitled "*Ensemble modeling of rainfall-runoff transformation process*" at Indian Institute of Technology, Kharagpur. This chapter is an outcome of this project.

KEYWORDS

- AWBM
- Brier score
- Calibration
- Ensemble
- HEC-HMS
- Hydrology
- Linear Programming
- MIKE SHE
- Model
- Nash Sutcliffe Efficiency
- Prediction
- Randomness
- SACRAMENTO
- Scatter diagram
- SIMHYD

- Simulation
- SMAR
- SWAT
- TANK
- Validation

REFERENCES

1. Ajami, N. K., Duan, Q., Gao, X., and Sorooshian S., 2006. Multimodel combination techniques for analysis of hydrological simulations: Application to distributed model inter comparison project results. *Journal of Hydrometeorology*, 7:755–768.
2. Akbari, S., 2011. *Hydrological modeling of a watershed using MIKE SHE.* Unpublished MTech Thesis, Indian Institute of Technology, Kharagpur.
3. Baker, L., and Ellison, D., 2008. Optimization of pedo-transfer functions using an artificial neural network ensemble method. *Geoderma*, 144(1–2):212–224.
4. Brier, G. W., 1950. Verification of forecasts expressed in terms of probability. *Monthly Weather Review,* 78:1–3.
5. Buizza, R., and Palmer, T. N., 1997. Impact of ensemble size on ensemble prediction. *Monthly Weather Review,* 126:2503–2518.
6. Freer, J. E., McMillan, H., McDonnell, J. J., and Beven, K. J., 2004. Constraining dynamic TOPMODEL responses for imprecise water table information using fuzzy rule based performance measures. *Journal of Hydrology*, 9:571–582.
7. Georgakakos, K. P., Seo, D. J., Gupta, H., Schaake, J., and Butts, M. B., 2004. Towards the characterization of stream flow simulation uncertainty through multi-model ensembles. *Journal of Hydrology*, 298:222–241.
8. Hsu, W., and Murphy, A. H., 1986. The attributes diagram: A geometrical framework for assessing the quality of probabilistic forecasts. *International Journal of Forecasting*, 2:285–293.
9. Kadam, S. R., 2011. *Hydrological modeling of catchments in Mahanadi river basin using ArcSWAT and Multi-Model ensemble.* Unpublished MTech Thesis, Indian Institute of Technology, Kharagpur.
10. Krishnamurti, T. N., Kishtawal, C. M., Larow, T. E., Bachiochi, D. R., Zhang, Z., Williford, C. E., Gadgil, S., and Surendran, S., 1999. Improved weather and seasonal climate forecasts from multimodel super ensemble. *Science,* 285:1548–1550.
11. Krogh, A., and Vedelsby, J., 1995. Neural network ensembles cross validation and active learning. In: *Advances in Neural Information Processing Systems.* MIT Press, Cambridge, 7:231–238.
12. O'Connell, P. E., Nash, J. E., and Farrell, J. P., 1970. River flow forecasting through conceptual models, Part 2, The Brosna catchment at Ferbane. *Journal of Hydrology,* 10:317–329.
13. Palmer, T. N., Doblas-Reyes, F. J., Hagedorn, R., and Weisheimer, A., 2005. Probabilistic prediction of climate using multi-model ensembles: from basics to applications. *Philosophical Transactions of the Royal Society,* 360:1991–1998.
14. Perrone, M. P., and Cooper, L. N., 1993. *When networks disagree: Ensemble method for neural networks. Neural networks for speech and image processing.* Chapman-Hall, London, p. 126–142.
15. Sahni, M., 2011. *Hydrological modeling using Rainfall Runoff Library (RLL) and Multi-model ensemble.* Unpublished MTech Thesis, Indian Institute of Technology, Kharagpur.
16. Samal, A., 2011. *Runoff estimation using HEC-HMS model and bootstrap based artificial network (BANN).* Unpublished M. Tech Thesis, Indian Institute of Technology, Kharagpur.

17. Schaake, J., Franz, K., Bradley, A., and Buizza, R., 2007. The hydrologic ensemble prediction experiment (HEPEX). *Hydrology and Earth System Sciences Discussions*, 3: 3321–3332.

18. Smith, M. B., Seo, D. J., Koren, V. I., Reed, S., Zhang, Z., Duan, Q. Y., Cong, S., Moreda, F., and Anderson, R., 2004. Runoff response to spatial variability in precipitation: an analysis of observed data. *Journal of Hydrology*, 298:267–286.

19. Sugawara, M., 1979. Automatic calibration of the tank model. *Hydrological Sciences Bulletin*, 24:375–388.

20. Sugawara, M., 1995. Tank model. In: V. P. Singh, (ed.), *Computer Models of Watershed Hydrology*. Water Resources Publications, Littleton, Colorado.

21. Tumer, K., and Ghosh, J., 1996. Error correlation and error reduction in ensemble classifiers. *Connection Science*, 8:385–404.

22. Viney, N. R., Bormann, H., Breuer, L., Bronstert, A., Croke, B. F. W., Frede, H., Graff, T., Hubrechts, L., Huisman, J. A., Jakeman, A. J., Kite, G. W., Lanini, J., Leavesley, G., Lettenmaier, D. P., Lindstrom, G., Seibert, J., Sivapalan, M., and Willems, P., 2009. Assessing the impact of land use change on hydrology by ensemble modeling (LUCHEM) II: Ensemble combinations and predictions. *Advances in Water Resources*, 32:147–158.

23. Wang, Y., 1998. *Active learning based on a hybrid neural network modeler*. PhD Thesis, University of Abertay Dundee, p. 42–45.

24. Wood, A. W., and Lettenmaier, D. P., 2008. An ensemble approach for attribution of hydrologic prediction uncertainty. *Geophysical Research Letters*, Vol. 35, L-14401, 10.1029/2008GL034648.

25. Reddy, B. R. S., 1999. Development of a decision support system for estimating evapotranspiration. MTech Thesis, Department of Agricultural and Food Engineering, Indian Institute of Technology Kharagpur, India.

26. Swarnakar, R. K. and Raghuwanshi, N. S., 2000. *DSS_ET user Manual*. Department of Agricultural and Food Engineering, Indian Institute of Technology, Kharagpur, India.

27. Tabari, H. and Talaee, P. H., 2011. Local calibration of the Hargreaves and Priestley-Taylor equations for estimating reference evapotranspiration in arid and cold climates of Iran based on the Penman-Monteith Model. *Journal of Irrigation and Drainage Engineering, ASCE*, 138(2):111–119.

28. Xu, C., Gong, L., Jiang, T., Chen, D., and Singh V. P., 2004. Analysis of spatial distribution and temporal trend of reference evapotranspiration and pan evaporation in Changjiang (Yangtze River) Catchment. *Journal of Hydrology*, 327:81–93.

CHAPTER 5

MULTI-CRITERIA ANALYSIS FOR GROUNDWATER STRUCTURES IN HARD ROCK

RANU RANI SETHI, ASHWANI KUMAR, MADHUMITA DAS, and S. K. JENA

ICAR-Indian Institute of Water Management, Opposite Rail Vihar, Chandrasekarpur, Bhubaneswar-751023, Odisha, India. Phone: +91 8763667446. E-mail: ranurani@yahoo.com; skjena_icar@yahoo.co.in

CONTENTS

5.1 Introduction.. 104
5.2 Materials and Methods.. 107
    5.2.1 Study Area.. 107
    5.2.2 Development of an Optimization Model......................... 107
    5.2.3 Formulation of Optimization Equations......................... 109
        5.2.3.1 Area Constraints 109
        5.2.3.2 Rainfall Recharge Constraints...........................110
        5.2.3.3 Recharge and Groundwater
                Draft Constraints 110
        5.2.3.4 Area of Influence Constraints...........................111
    5.2.4 Multi Criteria Analysis...112
5.3 Results and Discussion 113

5.3.1   Determination of Optimal Number of Groundwater
        Recharge Structures...113

5.3.2   Development of an Integrated Spatial Decision
        Support System (SDSS)...114

5.3.3   Thematic Mapping on
        Groundwater Table Fluctuation...114

5.3.4   Thematic Mapping on Slope of the Watershed.................116

5.3.5   Drainage and Classified drainage Density Map...............117

5.3.6   Remote Sensing and GIS to Classify Land Use,
        Land Cover of Study Area..118

5.3.7   Prioritization of Suitable Sites for
        Construction of Recharge Structures..................................119

        5.3.7.1   Multi-Criteria Analysis for
                  Delineation of Sites..119

5.4   Conclusions... 125

5.5   Summary.. 126

Keywords 127

References... 128

5.1   INTRODUCTION

The practice of artificial recharge is increasingly emerging as a powerful tool in water resources management [22]. Surface and subsurface hydrological features (such as lithology, geological structure, drainage density, groundwater flow and boundary conditions of aquifer system) play an important role in a groundwater system. With increasing demand of water over the period of time, there is need to develop groundwater resources on sustainable basis, so that there is balance between demand and supply of water. It is not an easy task to study the hydro-geological basin parameters through conventional methods to identify suitable areas for construction of groundwater recharge structures. For this purpose, different controlling parameters must be independently derived and integrated, which involves additional cost, time and manpower. Integration of remote sensing information and field survey data on GIS platform provides convergent analysis of diverse data sets for decision making in groundwater management.

Many assessments of groundwater conditions made with remote sensing techniques have been reported [2, 5, 15, 43]. Geographic Information System (GIS) techniques have many advantages over older, improved geo referenced thematic map analysis and interpretations [8, 23, 40]. Cowen [6] defined *GIS as a decision support system involving the integration of spatially referenced data in a problem solving environment.* In addition, unlike conventional methods, GIS methods for demarcation of suitable areas for ground water replenishment are able to take into account the diversity of factors that control groundwater recharge. Thematic map integrated various features derived from data in a GIS environment [1, 4, 14, 19, 24, 35, 39]. Determining groundwater potentiality, movement, storage and other parameters in an area are only possible when the characteristics of the rock formation are known [16, 20, 45]. In this focus area, attempts have been already made in different parts of India by scholars [12, 13, 18, 27, 31, 33, 36]. Some other scholars [7, 21, 29, 38, 41, 42] have arrived at the groundwater prospects by considering the geo-morphological aspects that have been derived out of satellite images.

Dominance of hard rock terrain is more than 80% of geographical area of India, and is considered as one of the major limitations in exploitation of groundwater resources in few areas. Most of the area remains fallow during *rabi* season (non monsoon) due to non availability of required amount of water. During this period, all the groundwater sources like dug wells get dried off and yield of tube-wells get reduced due to lowering of water table depth. Knowing the status of groundwater availability, trend of groundwater level fluctuations and geological features of the aquifer, groundwater recharge structures can be planned in order to maintain the water budget within the watershed. GIS based hydro-geomorphic approach has been used to identify the site-specific artificial groundwater recharge techniques in *Deccan Volcanic Province* of India [32].

Well-planned optimal combination of recharge measures, to maximize area of influence due to groundwater recharge structures, can be an instrument of equitable water access to the users and a means of providing 'on demand' delivery. In shallow aquifers and hard rock areas with very limited aquifer storage capability, the improvement in the irrigation service can alone justify the investment and efforts on artificial recharge. For this purpose it is necessary to use a combination of distributed small storages, recharge wells and ponds. Integrated development and management of surface storages and recharge through a combination of wells, ponds or channels emerges as a cost effective option to facilitate optimization of the farming system with user initiative.

Optimization techniques have been proved to be one of the powerful set of tools that are important in efficiently managing the resources and thereby maximizing its impact. Models, which solve governing flow and transport equations in conjunction with optimization techniques, have become powerful aquifer management tools. The need to develop management models and evolve policy guidelines for optimal utilization of surface and groundwater is stressed [26].

Selection of suitable sites for construction of appropriate recharge structures is critical for effective recharge and is dependent upon several parameters with has been analyzed together in GIS environment. A study on percolation tanks showed that if the site of a percolation tank is properly selected and the tank designed appropriately, the groundwater recharge through the tanks could go up to 70% [30]. Selection of potential zones for construction of groundwater recharge structures are being assessed by remote sensing techniques. Few studies have been carried out to identify these potential zones by multi criteria analysis. Suja Rose and Krishnan [38] identified groundwater potential zones in the Kanyakumari and Nambiyar basins of Tamil Nadu in India based on multi criteria analysis. In this study, multivariate statistical technique was used to find out the relationship between rainfall and groundwater resource characteristics, which resulted to identify the groundwater potential zones.

In this chapter, optimization model has been developed in order to determine the number of groundwater recharge structures in different geological formations within the watershed in Odisha – India. The aim was to:

- identify the prioritized areas, where groundwater recharge structures would have maximum impact on area of influence so that availability of water during non monsoon season will meet the crop water demand.
- Further, remote sensing technique was used to categorize land use land pattern area of the watershed.

• Multi-criteria analysis was carried out to prepare preference index for construction of groundwater recharge structures within Munijhara watershed located in Odisha, India.

5.2   MATERIALS AND METHODS

5.2.1   STUDY AREA

The research study was carried out in Munijhara micro-watershed, which lies between 20° 05' and 20° 09'N latitude and 85° 05' to 85° 09'E longitude (Figure 5.1). This watershed is located in Nayagarh block of Nayagarh district of Odisha (India) and falls under fissured formations (hard rock) with low to moderate yield potential areas underlain by weathered and fracture granite gneiss. Altitude of the watershed varies from 80 to 100 m above mean sea level (msl). Apart from forest cover, the topography is comparatively flat and has gently sloping terrain with an average gradient of 0.5 m/km in southwest to northwest direction. The area falls under sub humid agro-ecosystems with variation of average annual rainfall of 1037–1483 mm.

The texture of surface soil varies from red loamy soil to red and yellow soil, whereas texture of subsurface horizons is mainly dominated by clay-to-clay loam. The saturated hydraulic conductivity varies from 0.9 to 2 cm/h for surface soils and from 0.7 to 0.9 cm/h for subsurface soils. The soils are non-saline and non-sodic in nature. Principal agriculture seasons are *kharif* (June–October) and *rabi* (November–February). Paddy is the major crop grown in more than 90% of the area in *kharif* season followed by pulses on residual moisture in *rabi* season. More than 58% area remains under fallow during *rabi* season due to non availability of required amount of water. Groundwater is the only source for domestic, industrial and crop demand throughout the year. It is being abstracted from 166 tube wells and 450 open wells. Due to increase in groundwater use, drying up of wells and decline in water table depth has become major issue in this area.

5.2.2   DEVELOPMENT OF AN OPTIMIZATION MODEL

Linear Programming (LP) model was used to work out number of recharge structures including water harvesting structures and dug well based on

FIGURE 5.1   Location map of the study area.

hydraulic properties, geology, draft, rainfall, run-off and feasibility of its installation. In formulating the linear program factors like existing groundwater structures, area of each structure, their respective area of influence, and water table depth during pre and post monsoon season, capacity, groundwater draft from each structures and rainfall-recharge relationship in the study area were taken as the basis. Linear Programming model was formulated for both upland and flood plain area. This model was formulated with the objective of maximizing area of influence of recharge structures to ensure availability of ground water to suffice crop water requirement in *rabi* season. Following few factors were kept in view in deciding the variables and constraints to optimize area of influence of different recharge structures.

i.   Groundwater balance: for environmental reasons, abstraction rates from wells and boreholes will not be allowed to exceed the allowable aquifer capacity and well supply needs.

ii. Ground water draft: The needs of households, industry and irrigation must be satisfied by the optimization procedures.

iii. Water quality: Groundwater quality should be within the permissible limit of the requirement.

5.2.3   FORMULATION OF OPTIMIZATION EQUATIONS

The objective of the optimization was to maximize area of influence of each groundwater recharge structures so that water withdrawals from wells and boreholes for crop production during *rabi* season could be met. The governing equations were derived with the dependent variables like:

i. x = Number of dug wells that could be used for agricultural and related use in the area.

ii. y = Number of small water harvesting structures used for both domestic, industrial and agricultural uses, which also act as means to recharge groundwater.

iii. z = Number of large water harvesting structures, which needs generally for household use in the region but due to its large storage capacity it recharge to groundwater.

The objective is to Maximize

$$P = I_d x + I_s y + I_l z \qquad (1)$$

where, P = total area of influence, m²; I_s = area of influence of small water harvesting structures, m²; I_d = area of influence of dug-well, m²; and I_1 = area of influence of large water harvesting structures, m².
   This objective function was subjected to following constraints.

5.2.3.1   Area Constraints

Area of the watershed that can be exploited for constructing the groundwater recharge structures to ensure sufficient area of influence that was limited to 10% of total area on each geological condition [3].

$$[A_d x + A_s y + A_l z] \leq 0.1 A_g \tag{2}$$

where, A_d = cross-sectional area of dug well, m²; A_s = cross sectional area of small water harvesting structures, m²; A_l = cross sectional area of large water harvesting structures, m²; and A_g = Total area of particular geologic conditions (flood plain/ upland), ha.

In the Munijhara watershed, commonly used dug wells, small water harvesting structures and large water harvesting structures has dimensions of 4 m diameter, 30 × 30 m², 40 × 40 m², respectively. Out of the total watershed area, area under flood plain and upland was 2300 and 1600 ha, respectively.

5.2.3.2   Rainfall Recharge Constraints

$$V_d x + V_s y + V_l z \geq 0.09 V_r \tag{3}$$

where, V_d = capacity of water to be retained in dug well, m³; V_s = capacity of water to retain in small water harvesting structures, m³; V_l = capacity of water to retain in large water harvesting structures, m³; V_r = Rainfall volume that could recharge the groundwater under hard rock areas.

In this case, the 9% of the total rainfall was assumed to get recharge to aquifer. Depth of dug well was taken as 6–10 m. For calculating volume of small and large water harvesting structures, area at the bottom surface was considered as 26×26 m² and 36×36 m², respectively. Side slope for both the structures was assumed as 1:1, depth as 2 m and seepage rate as 10 mm/day for 150 days.

5.2.3.3   Recharge and Groundwater Draft Constraints

$$R_d x + R_s y + R_l z \geq D_g \tag{4}$$

where, R_d=amount of recharge from dug well, m; R_s=amount of recharge from small water harvesting structures, m; R_l=amount of recharge from large water harvesting structures, m. Groundwater recharge from each of the structures was calculated as follows:

$$R = S_y \times \Delta S \times I \tag{5}$$

where, R = volume of recharge from each structures, m³; S_y = specific yield; ΔS = change in water table depth during pre- and post-monsoon period, m; I = area of influence of each structures determined from piezometric studies in different geologic conditions, m²; and D_g = annual groundwater draft from all the structures

Area of influence was assumed as square in shape around the structures, considering structures are located based on appropriate groundwater flow direction. It was observed from the piezometric study that 20×20 m², 125×125 m² and 500×500 m² areas were being influenced by dug well, small and large water harvesting structures, respectively. Average water table fluctuations in all the structures and specific yield of the watershed were considered as 2 m and 0.01, respectively. It was also considered that total recharge from different structures within the watershed area should be more than the groundwater draft. Hence for calculating the groundwater draft, field survey was carried out and questionnaires were prepared to monitor the amount of groundwater draft for both domestic and agricultural purposes from each structures in both *kharif* and *rabi* season. Total annual draft from all structures was estimated as 91,200 m³.

5.2.3.4 Area of Influence Constraints

It has been considered that impact of each recharge structures should influence at least 20% of the total area under each geologic condition.

$$I_d x + I_s y + I_l z \geq 0.2 A_g \tag{6}$$

where, I_d = area of influence of dug well, m²; I_s = area of influence of small water harvesting structures, m²; I_l = area of influence of large water harvesting structures, m²; and A_g = area under different geologic conditions within the watershed, ha.

Area of influence of small and large water harvesting structures was calculated based on the piezometric observations. For upland, A_g was taken as 1600 ha. As the recharge structures and recharge volume of water in the upland would influence the areas in the flood plain, A_g in the flood plain had been calculated accordingly.

5.2.4 MULTI CRITERIA ANALYSIS

Multi-criteria decision analysis is a sub-discipline of operational research that explicitly considers multiple criteria in decision-making environments. Multi-criterion technique allows map layers to be weighted to reflect their relative importance [9, 10, 25, 44]. Saaty's [34] analytic hierarchy process is the most widely accepted method for scaling the weight of parameters whose entries indicate the strength with which one element dominates over the other in relation to the relative criterion. The basic input is the pair-wise comparison matrix of n parameters constructed based on the Saaty's scaling ratios, which could be of the order ($n \times n$), and is defined as:

$$GWS_{site} = F (G_{wtf}, Geom, D_d, Sl, Lu, R_{rech}) \qquad (7)$$

where, GWS_{site} = sites for groundwater structures; G_{wtf} = groundwater table fluctuations; Geom = geomorphology; D_d = drainage density; Sl = slope; L_u = land use; and R_{rech} = recharge. Suitable zones for groundwater structures can be expressed as:

$$GWS_{site} = \Sigma W_i \times Cv_i \qquad (8)$$

where, W_i = map weight; and Cv_i = capability value. The algorithm used in the derivation of suitable zones for groundwater structures was:

$$GWS_{site} = 0.2 \times [Cv\ G_{wtf}] + 0.17 \times [CvGeom] + 0.16 \times [CvD_d]$$
$$+ 0.15 \times [CvSl] + 0.12 \times [CvL_u] + 0.2 \times [CvR_{rech}] \qquad (9)$$

where, CvG_{wtf} = groundwater table fluctuations with capability value; CvD_d = drainage density layer with capability value; CvGeom = geomorphology with capability value; $CvSl$ = slope layer with capability value; CvR_{rech} = recharge with capability value; and CvL_u = land use layer with capability value.

5.3   RESULTS AND DISCUSSION

5.3.1   DETERMINATION OF OPTIMAL NUMBER OF GROUNDWATER RECHARGE STRUCTURES

Based on the optimization model, a combination of number of different types of structures has been worked out to optimize the area of influence in upland and flood plain areas. For upland area, by solving these above inequalities by simplex method, the optimal combination was worked out as:

x (number of dug wells) = 70,
y (number of small water harvesting structures) = 130, and
z (number of large water harvesting) = 10.

With these combination around 451.24 ha of area can be influenced, which is 25% of the total upland areas. The upland area in the watershed is located with an elevation of more than 90 m and flood plain is located towards the outlet of the watershed with an elevation 70 to 90 m. Considering these geological conditions, it was considered that the number of recharge structures, volume of water recharged and influence of the structures thereafter in the upland area will have a similar impact on the flood plain. Hence, to determine the number structures required in the flood plain, combined effect of structures in upland and flood plain was taken into account in the L.P. model. For flood plain, by solving these above inequalities by simplex method, the optimal combination was worked out as:

x (number of dug wells) = 278,
y (number of small water harvesting structures) = 252, and
z (number of large water harvesting) = 10.

With these combinations around 654.87 ha area can be influenced, which is 25% of the total flood plain areas. There are a number of dug wells, small and large water harvesting structures existing in the watershed, but annual recharge within the watershed was estimated to be nearly 43–69 mm, which accounts to only 4–5% of the annual rainfall. This lower rate of recharge has attributed to drying up of most of the structures during non-monsoon period (*rabi*). In Indian conditions, it has been estimated that the groundwater recharge in the hard rock areas could be upto 9–12% of rainfall [11]. Hence there is a potential to increase the groundwater recharge by construction of different recharge structures in appropriate locations within the watershed. It is estimated that only 8–10% of runoff is harvested through natural storage structures, thus there is good scope for harvesting rest of the runoff by various means [28]. With population growth and increasing demand for water, construction of more number of similar structures in the watershed to suffice the water requirement have become need of the hour.

5.3.2 DEVELOPMENT OF AN INTEGRATED SPATIAL DECISION SUPPORT SYSTEM (SDSS)

Soil water balance and groundwater model was used to quantify the spatial distribution of recharge within Munijhara watershed. Spatial change in groundwater table depth, slope, and drainage density was used to develop a spatial decision support system for delineating suitable locations for construction of groundwater recharge structures within the study area. Thereafter, remote sensing image analysis was carried out to generate land use, land pattern map during *kharif* and *rabi* season. The information generated from the study was used in GIS to develop an integrated groundwater recharge model to identify and prioritize locations for construction of recharge structures within the watershed.

5.3.3 THEMATIC MAPPING ON GROUNDWATER TABLE FLUCTUATION

Munijhara watershed lies within the topographical elevation within 70–120 m. Depth to water table was monitored from 64 numbers of spatially distributed monitoring wells. Water table elevation was calculated

for each of the wells during January 2006–June 2009. Groundwater elevation above mean sea level was calculated based on the following equation:

$$E_w = E - D \tag{10}$$

where, E_w = elevation of water above mean sea level (m) or local datum;
E = elevation above sea level or local datum at point of measurement (m);
and D = depth to water (m).

Ground water level data were collected observing wells and sum-
marized on hydrographs of each well. Hydrograph of each well showed
the fluctuation of ground water levels during a given period of time and
allows for comparison of ground water levels from year to year. It was
observed that groundwater table depth was lowest during the month of
May, which was considered for pre monsoon water table depth, Similarly
November month was considered for post monsoon period as the water
table depth reaches to the highest level. Contour map on water table elevation was prepared for both the season by using TNT MIPS GIS software. Spatial distribution of water table elevation analysis showed that
water table depth varied between 5.4–7.8 m, 5.3–7.5 m and 3.9–6.8 m
during May 2006, 2007 and 2008, respectively. Similarly it varied within
2.5–5.8 m, 3.4–5.9 m and 1.2–3.3 m during November 2006, 2007 and
2008, respectively.

It was observed that rainwater influences depth of water table and net
change in water table depth during pre and post monsoon season was
around 2.38, 1.63 and 3.28 m during 2006, 2007 and 2008, respectively.
The water table fluctuation is also influenced by amount of draft during the
period of monitoring. Practically at all places within the watershed, water
table fluctuates up and down in response to additions to or withdrawals
from the nearby structures within the area of influence of different structures. It was observed that during dry seasons of reported periods, there
was not much change in water table elevation in the watershed. However
changes were marked in wet season for three consecutive years. Spatial
variation of water table elevation showed that areas covered under forest
cover always falls under the contours of >100 m in most of the season. But
areas near to the outlet of the watershed and low-lying areas the fluctuation
in water table depth is high due heavy pumping.

5.3.4   THEMATIC MAPPING ON SLOPE OF THE WATERSHED

The response of a particular watershed to different hydrological processes and its behavior depends upon various physiographic, hydrological and geomorphological parameters. Slope of the area has a direct
control on runoff and infiltration to the ground surface. A digital elevation model (DEM) derived slope model was generated for the watershed.
The topographic maps from Survey of India maps were scanned and first
geo-referenced to specific coordinates and used to generate a DEM with
32-bit resolution using TNT (GIS) modules. The slope maps were then
generated from DEM and reclassified into four slope groups as 0–1%,
1–2%, 2–3% and >3% slopes (Figure 5.2). Steep slopes were found in
north and south western part of the watershed covered by forest areas.
Most of the area under settlements have low slope of 1–2%. These are
the determinant factors for selecting the suitable areas for construction
of recharge structures in order to conserve maximum runoff during the
monsoon periods.

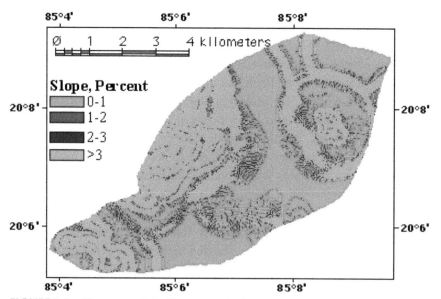

FIGURE 5.2   Slope map of Munijhara watershed.

5.3.5   DRAINAGE AND CLASSIFIED DRAINAGE DENSITY MAP

Certain physical properties of watersheds significantly affect the characteristics of runoff and infiltration process in hydrologic analyzes. Morphological characteristics like stream order, drainage density, aerial extent, watershed length and width, channel length, channel slope and relief aspects of watershed are important in understanding the hydrology of the watershed. A detailed analysis of the drainage network in a watershed can provide valuable information about watershed behavior, which is also useful for detail hydrological analysis. The drainage density, expressed in terms of length of channels per unit area (km/km^2) indicates an expression of the closeness of spacing of channels. It is expressed as:

$$D = Lu/A \qquad\qquad (11)$$

where, D = Drainage density; A = Area of the watershed, km^2; and Lu = Total stream length of all orders, km.

It thus provides a quantitative measure of the average length of stream channels within different portions of the whole basin. Drainage density indirectly indicates its permeability and porosity due to its relationship with surface run-off. Areas with high drainage density values indicate high surface run-off and higher permeability; hence such areas are to be considered favorable for arresting excessive run-off. The major drainage system in the study area comprises of SW-NE flowing Munijhara Rivers and their distributaries. The dominant drainage pattern was observed in upland granitic zones. The lithological control results in the evolution of dentratic and subdendratic pattern due to low infiltration. Strahler's system of stream ordering [37] is used, in which smallest finger-tip tributaries are designed as order 1 and when two 1st order channels join, a channel segment of order 2 is formed (and so on). The highest-order stream in the basin defines the order of that basin and here the watershed is of 3rd order basin.

The study area was divided into square grids of 0.5 sq. km and the total lengths of all streams in each grid were calculated in order to determine the drainage density values in km/km^2. These values were regrouped to produce a drainage density map that was classified into four categories, i.e., very low 0–1, low (1–2), medium (2–3) and high (3–4) for the watershed (Figure 5.3). A major portion (>50%) of the region has low drainage density (<1 km/km^2).

5.3.6 REMOTE SENSING AND GIS TO CLASSIFY LAND USE, LAND COVER OF STUDY AREA

Land-use planning is defined as a systematic assessment of land and water potential, alternatives for land use, and the economic and social conditions required to select and adopt the best land-use options. Land-use planning aims at achieving a balance among these goals through the use of information on appropriate technology, and consensus-based decision-making. Effective land-use planning often involves local communities, scientific

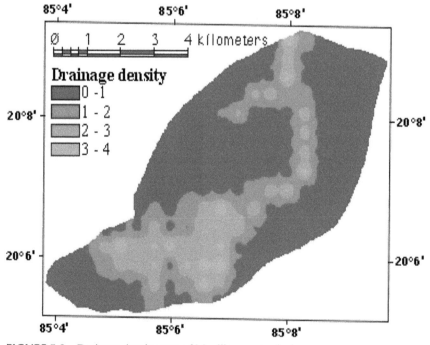

FIGURE 5.3 Drainage density map of Munijhara watershed.

information on land resources, appropriate technologies, and integrated evaluation of resource use. The Remote Sensing technology along with GIS is a perfect device to identify, locate and map various types of lands associated with different landform units. In this study, land use map shows the different types of land cover pattern present in the study area. Vegetation cover is an important factor, which influences the occurrence and movement of the rainfall.

The watershed area is characterized by the dense forest, agricultural land, forest plantation, grass land; land with or without scrub, open forest and villages during both pre monsoon and post monsoon season. In this study, land use land cover image for the study area was prepared by using supervised classification. Data were collected from survey of India topo-sheet 73H/4, Satellite images of IRS-P6; LISS-III of November 2008 and March 2009. Geology/geomorphology map showing the faults, upland and flood plain areas were extracted from Geological Survey of India (GSI) map of Nayagarh district. TNT MIPS software was used to prepare slope map, drainage density, spatial variation in water table depth within the watershed. Net groundwater recharge obtained from MODFLOW was used in GIS to prepare the distribution of net recharge for the watershed.

The multi-spectral satellite imageries of 2 overpass dates were classified to generate the land use land cover map. For this purpose the supervised classification technique was used. The error matrix of land use and land cover classification was presented in Table 5.1. In order to quantify the components of water budget, 8 classes were identified out of which 2473 ha of land (>58%) is fallow or barren where as only 239 ha (<6%) is cultivated during summer or *rabi* season. Total area under forest and settlements were 311 ha and 757 ha respectively. Total area under water bodies was only 46 ha area is under the water bodies during pre-monsoon season and during the monsoon season it was increased to 213 ha.

TABLE 5.1   Error Matrix of Land Use Land Cover Classification of Satellite Image of March 2009

Verified	Water Bodies	Current Fallow	Barren Land	Trees/ Shrubs	Rabi Crop land	Fallow Land	Forest Cover	Settlements	Total
Water Bodies	35 (99.8%)	0	0	0	0	0	0	0	35
Current fallow	0	1332 (93.9%)	7	0	0	0	0	80	1419
Barren Land	0	49	170 (72.6%)	0	0	0	0	15	234
Trees/Shrubs	0	0	0	17 (99.8%)	0	0	0	0	17
***Rabi* Crop**	0	1	0	0	15 (93.7%)	0	0	0	16
Fallow Land	0	1	0	0	0	24 (96%)	0	0	25
Forest Cover	0	0	0	0	0	0	367 (99.9%)	0	367
Settlements	1	141	3	0	0	0	0	423 (74.4%)	568
Total	36	1525	180	17	15	24	367	518	2681

Overall accuracy = 88.92%.

5.3.7 PRIORITIZATION OF SUITABLE SITES FOR CONSTRUCTION OF RECHARGE STRUCTURES

5.3.7.1 Multi-Criteria Analysis for Delineation of Sites

In this study, all six parameters namely the geo-morphological unit, water table fluctuation, slope, drainage density, land use, recharge were subjected to multi criterion evaluation by TNT MIPS GIS software that uses the weighted aggregation method. The summation of the product of all maps weights of thematic layers, each with related capability values of corresponding categories, finally depicts the suitable area for groundwater structures.

In this method, the total weights of the integrated polygons that were ultimately formed were derived as a sum or product of the weights that had been assigned to the different layers according to their suitability. The sequence followed to create the final integrated layers was: In the first step, each two layers were integrated with one another. In this study multi criteria analysis was carried out and the decision rules were derived by scoring hydro-geomorphic parameters corresponding to the medium to low groundwater potential zones where adoption of recharge structures could be most effective as shown in Table 5.2. For Munijhara watershed, analysis on pre- and post-monsoon groundwater fluctuation maps indicated that poor groundwater condition resulted from high water table fluctuation (greater than 5 m) are found mainly in the high slope regions, which are not suitable from any groundwater structures. On the other hand, in the

TABLE 5.2 Decision Rules for Selecting Favorable Sites for Recharge Structures

	Parameter	Value
Water table depth	Water table fluctuation	>2.5 m
Geomorphologic data	Drainage density	<3 km/km^2
	Land forms	Plains
	Land use/land cover	Barren, fallow, cultivated land
	Slope	<3%
Geological data	Depth to bed rock	8 m
	Soil cover	>0.9 m
Potential recharge zone	Net recharge	40–50 mm

most suitable zones, which have gentle slopes, the water table fluctuation was low (<3 m, implying that these regions essentially do not require groundwater augmentation. As most wells are shallow, a fluctuation in water table higher than average results in inadequate supply of water to meet the demands of the people. Hence, a water table fluctuation greater than 3 m, which is the average water table fluctuation in both upland and flood plain areas, were taken as one of the criteria for consideration of sites for recharge. These adopted criteria encompass both the recharge and storage zones where appropriate techniques could be implemented.

Another equally important criteria is that availability of storage space in subsurface for augmenting the groundwater supply and storing it, so that water is available in the event of any eventuality like bore wells predominantly in the plains as well as field studies, indicate that the top soil and highly weathered zone has an average thickness of 5 m while underlying moderate rely weathered zone has an average thickness of 10 m. Hence soil thickness and depth to bedrock both satisfy the general criteria adopted. In this watershed faults and lineament density was low and the soil and weathered zone is sufficiently thick to be considered suitable for recharge structures. Gently slopes (<3%) serve to build up the hydraulic gradient and are thus considered most suitable. Areas of steep slopes, hills were considered unfavorable. Moderate drainage density (<3 km/km^2) was taken as an optimum balance between runoff and infiltration. The areas where the area is favorable for implementation of recharge structures was first demarcated using hydro geomorphologic parameters. Details of thematic layers, their categories, weights are given in Table 5.3.

Among six thematic layers; layer rank has been decided based on the importance for deciding the location for groundwater structures. Highest map weight of 0.2 has been assigned to the geo morphological layer, water table fluctuation and recharge map. Then from thematic layers categories has been delineated by mapping in GIS. In geo morphology layer, fault zone, flood plain Charnokite, flood plain upland, Flood plain granite gneiss, upland granite gneiss, upland Charnokite rank has been assigned as 1–6 respectively. Water table fluctuation has been categorized as 0–1, 1–2, 2–3 and > 3 m and highest rank was assigned to the category of >3 m. Drainage density was categorized as 0–1, 1–2, 2–3 and 3–4 km/km^2. The highest rank was assigned to the area having more drainage density. Construction

TABLE 5.3 Thematic Layers, Their Categories, Weight

Thematic layer	Layer rank	Map weight Wi	Category	Category rank	Capability value, Cvi
Geo morphology	1	0.2	Upland Charnokite	6	0.12
			Upland granite gneiss	5	0.13
			Flood plain granite gneiss	2	0.15
			Flood plain upland	3	0.15
			Flood plain Charnokite	4	0.20
			Fault zone	1	0.25
Water table fluctuation, m	2	0.2	0–1	4	0.05
			1–2	3	0.2
			2–3	2	0.25
			>3	1	0.5
Drainage density, km/km^2	3	0.16	0–1	4	0.05
			1–2	3	0.2
			2–3	2	0.25
			3–4	1	0.5
Slope	4	0.15	0–1%	1	0.75
			1–2%	2	0.20
			2–3%	3	0.04
			>3%	4	0.01

TABLE 5.3 Continued

Thematic layer	Layer rank	Map weight Wi	Category	Category rank	Capability value, Cvi
Land use land cover	5	0.12	Barren lands	3	0.15
			Fallow land	1	0.33
			Forest cover	5	0.02
			Rabi crop land	4	0.18
			Settlements	5	0.05
			Trees/shrubs	5	0.01
			Water bodies	5	0.01
			Fallow crop land	2	0.25
Recharge, mm	6	0.2	0–10	5	0.05
			10–20	4	0.1
			20–30	3	0.2
			30–40	2	0.25
			40–50	1	0.4

of structures should be almost in flat land so that it can retain water for a longer period. Hence, areas with slopes of 0–1% was assigned as highest rank, among different categories of land use highest rank was given to the fallow land, followed by fallow crop and barren lands. Areas having the capacity to recharge more, i.e., 40–50 mm were assigned as highest rank followed by other areas.

Category rank and value was assigned to each layer based on their influence to the groundwater recharge structures within the watershed. Capability value for different categories has been decided from scale 0.5 and assigned to each based on the influence over the recharge structures. Finally preference index has been prepared for suitable locations for appropriate recharge structures (Figure 5.4). From the figure it was clear that most preferred zones were identified in different locations within the watershed and mainly land with fallow/barren lands were found out for these structures. Land use, land pattern maps indicated the suitable spatially distributed within the watershed. Areas with forest cover and having very steep slopes were demarcated as least preferred zone for construction of suitable recharge structures.

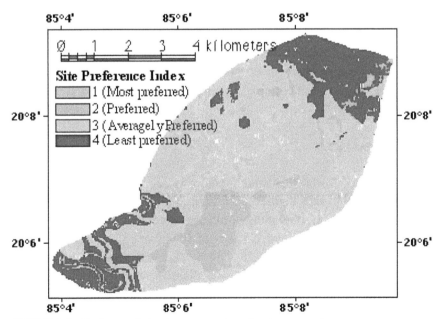

FIGURE 5.4   Preference index for construction of recharge structures.

5.4   CONCLUSIONS

A regional groundwater recharge model was developed for 4200 ha hard rock area in Nayagarh district of Odisha State. Out of the total geographical area of the watershed, 311 ha and 757 ha is under forest cover and settlement area. The model resulted that optimal combination of 20 number of large water harvesting structures of 40 × 40 m² size with carrying capacity of 30,000 m³, 382 number of small water harvesting structures of 30 × 30 m² size with capacity of 1570 m³ and 340 numbers of dug wells of 4 m diameter could influence around 3132 ha area of the watershed. These structures would recharge on average of 12–18% of rainfall of water annually. Farmers can utilize additional 91 ha (4%) area in *kharif* and 2473 ha (>80%) area in *rabi* season, which could enhance income generation and ultimately increase irrigation and cropping intensity of the region. The developed integrated model could be used to define the location to construct the different recharge structures to ensure sustainable yield from the aquifer and make the fallow lands suitable for cropping in *rabi* season. This model can be utilized to improve groundwater development status of the region scientifically integrated with balanced effort to ensure sufficient recharge to the aquifer to maintain sustainable yield and availability of water in *rabi* season. This provides an optimal combination of structures that can be constructed in specific locations in different geological conditions of the watershed that can be used to address the necessity of agricultural development of the region without losing sight of the pertinent question of ensuring sustainability for future generation.

5.5   SUMMARY

A regional groundwater recharge model was developed for 4200 ha hard rock area in Nayagarh district of Odisha, India. The model resulted that optimal combination of 20 number of large water harvesting structures of 40 × 40 m² size with carrying capacity of 30,000 m³, 382 number of small water harvesting structures of 30 × 30 m² size with capacity of 1570 m³ and 340 numbers of dug wells of 4 m diameter could influence around 3132 ha area of the watershed. These structures would recharge on average of 12–18% of rainfall of water annually. Farmers can utilize additional 91 ha (4%) area in *kharif* and 2473 ha (more than 80%) area in *rabi* season, which could enhance income generation and ultimately increase irrigation and cropping intensity of the region. The developed integrated model could be used to define the location to construct the different recharge structures to ensure sustainable yield from the aquifer and make the fallow lands suitable for cropping in *rabi* season.

KEYWORDS

- aquifer
- cropping intensity
- Decision Support System
- delineation
- drainage density
- dug well
- flood plain
- geomorphology
- GIS
- groundwater
- hydrograph
- kharif
- land capability
- land use
- lithology
- model
- multi-criteria analysis
- optimization
- rabi
- recharge structure
- remote sensing
- software
- tube-well
- water harvesting structure
- watershed

REFERENCES

1. Baker, P., Please, P., Coram, J., Dawes, W., Bond, W., Stauffacher, M., Gilfedder, M., Probert, M., Huth, N., Gaydon, D., Keating, B., Moore, A., Simpson, R., Salmon, L., and Stefanski, A. 2001. Assessment of salinity management options for upper Billabong Creek catchment, NSW: Groundwater and farming systems water balance modeling. *Bureau of Rural Sciences*, Canberra.
2. Bastiaansen, W., Menenti, R, Feddes, R., and Holtslag, A., 1998. A remote sensing surface energy balance algorithm for lands (SEBAL). *J Hydrol*, 212:198–229.
3. Central Groundwater Board (CGWB), 2004. Report of Central Groundwater Board, Ministry of Water Resources, Govt. of India, Bhubaneswar.
4. Chowdary, V. M., Ramakrishnan, D., Srivastava, Y. K., Chandran, V., and Jeyaram, A., 2009. Integrated water resource development plan for sustainable management of Mayurakshi Watershed, India using remote sensing and GIS. *Water Resour Manag.*, 23:1581–1602.
5. Chowdary, V. M., Ramakrishnan, D., Srivastava, Y. K., Chandran, V., and Jeyaram, A., 2009. Integrated water resource development plan for sustainable management of Mayurakshi Watershed, India using remote sensing and GIS. *Water Resour Manag.*, 23:1581–1602.
6. Cowen, D. J., 1988. GIS versus CAD versus DBMS: what are the difference? *Photogramm Engg Remote Sensing,* 54(11):1551–1555.
7. Das, S., Behera, S.C., Kar, A., Narendra, P., and Guha, S., 1997. Hydrogeomorphological mapping in groundwater exploration using remotely sensed data – A case study in Keonjhar district in Orissa. *J Indian Soc. Remote Sen.,* 25(4):247–259.
8. Dixon B., 2005. Applicability of neuro-fuzzy techniques in predicting ground-water vulnerability: a GIS-based sensitivity analysis. *J Hydrol.,* 309:17–38.
9. Eastman, J. R., Jin, W., Kyemi, P. A. K., and Toledano, J., 1995. Raster produce for multicriteria/multi objective decisions. *Photogramtric Engineering and remote sensing,* 61:539–547.
10. Eastman, J. R., 1996. Multi criteria evaluation and graphical Information systems Eds. Longley P. A., Godchild M. F., Magurie D. J., Rhind D. W., second edition, I. John Wiley and Sons, New York, 493–502.
11. GEC, 1997. *Report of the Groundwater Resource Estimation Committee. 1997. Ground Water Resource Estimation Committee.* Groundwater Resource Estimation Methodology. Ministry of Water Resource, Govt. of India.
12. Goyal, R., and Arora, A. N., 2004. Use of remote sensing in groundwater modeling. Map Asia 2004. Beijing International Convention Center, Beijing, China.
13. Goyal, S., Bhardwaj, R. S., and Jugran, D. K., 1999. *Multicriteria analysis using GIS for groundwater resources evaluation in Rawasen and Pili watershed.* UP Proc. Map India 99, New Delhi, India.
14. Henry, N. N. B., James, W. M., David, B. M., John, C. H., Aris, A. H., 2007. A GIS-based approach to watershed classification for Nebraska reservoirs. *J Am. Water Resour Assoc.,* 43(3):605–621.
15. Jasrotia, A. S., Majhi, A., and Singh, S., 2009. Water balance approach for rainwater harvesting using remote sensing and GIS techniques, Jammu Himalaya, India. *Water Resour Manag.* doi: 10.1007/ s11269-009-9422-5.

16. Karanth, K. R., 1987. *Groundwater assessment, development and management.* McGraw Hill Publishing Company Limited. New Delhi.

17. Krishnamurthy, J., and Srinivas, G., 1995. Role of geological and geomorphological factors in groundwater exploration: a study using IRS LISS data. *International Journal of Remote Sensing,* 16 (4):2595–2618.

18. Krishnamurthy, J., Manavalan, P., and Saivasan, V., .1992. Application of digital enhancement techniques for groundwater exploration in a hard rock terrain. *International Journal of Remote Sensing,* 13:2925–2942.

19. Krishnamurthy, J., Kumar, N. V., Jayaraman, V., and Manivel, M., 1996. An approach to demarcate groundwater potential zones through remote sensing and a geographic information system. *International Journal of Remote Sensing,* 17(10):1867–1884.

20. Kruseman, G. P., de Ridder, N. A., (2nd Edition). 1970. *Analysis and Evaluation of Pumping Test Data.* ILRI publications. Netherlands.

21. Kumar, A., and Tomas S., 1998. Groundwater assessment through hydro geomorphological and geophysical survey – A case study in Godaveri sub-watershed, Giridih, Bihar. *J Indian Soc. Remote Sen.,* 26(4):177– 184.

22. Ma, L., and Spalding, R. F., 1997. Effects of artificial recharge on ground water quality and aquifer storage recovery. *J Am Water Resour Assoc.,* 33(3) : 561–572.

23. Martin, P. H., LeBoeuf, E. J., Dobbins, J. P., Daniel, E. B., and Abkowitz, M. D., 2005. Interfacing GIS with water resource models: a state-of-the-art review. *J Am Water Resour Assoc.,* 41(6):1471–1487.

24. Murthy, K. S. R., 2000. Ground water potential in a semi-arid region of Andhra Pradesh—a geographical information system approaches. *Int J Remote Sens.,* 21(9): 1867–1884.

25. Navalgund, R. R., 1997. *Revised Development plan of Ahmedabad Urban Development Authority Area-2011.* Ahmedabad Urban Development Authority and space applications Center, Ahmedabad, India, Technical report 1 SAC/RSAG/TR/12/AUG/1997:142.

26. NIH report, http://www.indiawaterportal.org/node/10264 (NIH report).

27. Obi Reddy, G. P., Suresh, B. R., and Rao Sambasiva, M., 1994. Hydrogeology and hydrogeomorphological conditions of Anantapur district using remote sensing data. *Indian Geographical Journal,* 69(2):128–235.

28. Pradhan, K., 2004. Artificial recharge to groundwater-the need and scope in hill regions of Orissa. In Proceedings of workshop on Groundwater –The need and scope in hill regions of Orissa. p. 50–54.

29. Pratap K., Ravindran, K. V., and Prabakaran, B., 2000. Groundwater prospect zoning using remote sensing and geographical information system – A case study in dala-Renukoot area, Sonbhadra district, Uttarpradesh. *J Indian Soc. Remote Sen.,* 28(4): 249–263.

30. Raju, K. C. B., 1998. Importance of recharging depleted aquifers: State-of-the-art artificial-recharge in India, *J. Geol. Soc., India,* 51:429–454.

31. Rao, D. P., Bhattacharya, A., and Reddy, P. R., 1996. Use of IRS 1C data for geological and geographical studies. *Current Science,* IRS 1C, 70(7):619–623.

32. Ravi Shankar, and G. Mohan., 2005. A GIS based hydrogeomorphic approach for identification of site-specific artificial-recharge techniques in the Deccan Volcanic Province. *J. Earth Syst. Sci.,* 114(5):505–514.

33. Reddy, P.R., Vinod Kumar, K., and Seshadri, K., 1996. Use of IRS-1C data in ground water studies. *Current Science.* Special session, IRS 1C 70(7): 600–605.
34. Saaty, T. L., 1980. *The Analytic Hierarchy Process.* McGraw Hill, New York.
35. Saraf, A. K., and Choudury, P. R., 1998. Integrated remote sensing and GIS for groundwater exploration and identification of artificial recharge sites. *Int J Remote Sens.*, 19(10): 1825–1841.
36. Sikdar, P. K., Chakraborty, S., Enakshi, A., and Paul, P. K., 2004. Land use/Land cover changes and groundwater potential zoning in and around Raniganj coal mining area, Bardhaman District, West Bengal: A GIS and Remote Sensing Approach. *Journal of Spatial Hydrology,* 4(2):1–24.
37. Strahler, A. N., 1957. Quantitative analysis of watershed geomorphology. *Trans. Am. Geophys Union,* 38:913–920.
38. Suja Rose, R. S., and Krishnan, N., 2009. Spatial Analysis of Groundwater Potential using Remote Sensing and GIS in the Kanyakumari and Nambiyar Basins, India. *J. Indian Soc. Remote Sens.*, 37:681–692.
39. Tabesh, M., Asadiyani Yekta, A.H., and Burrows, R., 2009. An integrated model to evaluate losses in water distribution systems. *Water Resour Manag.*, 23:477–492.
40. Thapinta, A., and Hudak, P. F., 2003. Use of geographic information systems for assessing groundwater pollution potential by pesticides in Central Thailand. *Environ Int.*, 29:87–93.
41. Tomas, A., Sharma, A. K., Manoj, K. S., and Sood, A., 1999. Hydrogeomorphological mapping in assessing groundwater by using remote sensing data – A case study in Lehra Ganga block, Sangrur district, Punjab. *J. Soc. Remote Sens.*, 27(1):31–42.
42. Vaidyanathan, R., 1964. Geomorphology of Cuddapah basin. *Journal of Indian Geosciences Association,* 4:29–36.
43. Venkata, R. K., Eldho, T. I., Rao, E. P., and Chithra, N. R., 2008. A distributed kinematic wave–Philip infiltration watershed model using FEM, GIS and remotely sensed data. *Water Resour Manag.*, 22:737–755.
44. Voogd, H., 1983. *Multi-Criteria Evaluation for Urban and Regional Planning.* Pion, London.
45. Walton, W. C., (2nd Edition), 1989. *Numerical Groundwater Modeling.* Lewis Publishers.

CHAPTER 6

HYDROLOGIC MODELING OF WATERSHEDS USING REMOTE SENSING, GIS AND AGNPS

SUSANTA KUMAR JENA[1] and KAMLESH NARAYAN TIWARI[2]

[1]*Indian Institute of Water Management (ICAR), Opposite Rail Vihar, Chandrasekarpur, Bhubaneswar-751023, Odisha, India, Phone: +91 9437221616, E-mail: skjena_icar@yahoo.co.in*

[2]*Department of Agricultural and Food Engineering, Indian Institute of Technology, Kharagpur, West Bengal-721302, India, Email: kamlesh@agfe.iitkgp.ernet.in*

CONTENTS

6.1   Introduction.. 132
6.2   Materials and Methods... 134
    6.2.1   AGNPS (Agricultural Non-Point Source Pollution) Model.. 136
        6.2.1.1   Model Structure 136
        6.2.1.2   Hydrology 138
        6.2.1.3   Erosion and sediment transport processes........ 141
    6.2.2   AGNPS Model Use 143
6.3   Results and Discussions.. 145
    6.3.1   Model Calibration 145
    6.3.2   Model Validation 150
    6.3.3   Sensitivity Analysis 156

6.3.4   Assessment of Critical Area and
            Best Management Practices (BMP) 158
6.4   Conclusions.. 161
6.5   Summary... 162
Keywords... 162
References.. 163

6.1   INTRODUCTION

Land and water are two most vital natural resources of the world and these resources must be conserved and maintained carefully for environmental protection and ecological balance. Prime soil resources of the world are finite, non-renewable over the human time frame, and prone to degradation through misuse and mismanagement. Total global land degradation is estimated at 1964.4 M-ha, of which 38% is classified as light, 46% as moderate, 15% as strong and the remaining 0.5% as extremely degraded, whereas present arable land is only 1463 M-ha, which is less than the land under degradation [7]. The annual rate of loss of productive land in the whole world is 5–7 M-ha, which is alarming. In India, out of 328 M-ha of geographical area, 182.03 M-ha is affected by various degradation problems out of which 68 M-ha are critically degraded and 114.03 M-ha are severely eroded whereas total arable land is only 156.15 M-ha [21]. In India, Singh et al. [16] reported that 0.97% of total geographical area is under very severe erosion (>80 t.ha^{-1}.yr^{-1}), 2.53% area under severe erosion (40–80 t.ha^{-1}.yr^{-1}), 4.86% area under very high erosion (20–40 t.ha^{-1} yr^{-1}), 24.42% area under high erosion (10–20 t.ha^{-1}.yr^{-1}), 42.64% area under moderate erosion (5–10 t.ha^{-1}.yr^{-1}) and rest 24.58% area under slight erosion (0–5 t.ha^{-1}.yr^{-1}). Therefore, the problem of land degradation due to soil erosion is very serious, and it will further aggravate with increasing population pressure, exploitation of natural resources, faulty land and water management practices. Land degradation also reduces the world's fresh water reserves.

Water resources degradation is an issue of significant societal and environmental concern. Water pollution originates either from point or non-point source or from both. Point source pollution enters water resources directly through a discrete pipe, ditch or other conveyance. Industrial and municipal discharges fall into this category. Non-point source pollution enters water diffusely through runoff or leachate from rain or melting snow and is often too costly to observe and to measure. In addition non-point emissions are typically stochastic (random) due to impact of weather related and other environmental processes. Consequently, non-point pollution has been identified as a major reason for water quality problems. Perrone and Madramootoo [14] stated that contribution of agriculture to non-point source pollution in the United States is 64% of total suspended sediment and 76% of total phosphorous. In India, the quality of water flowing in the streams is getting polluted due to improper and over dose application of fertilizers, pesticides, and herbicides. Also point source pollutions, such as effluent from industries, feedlots and erosion from gully, are also getting mixed with stream water causing pollution of water resources.

Proper watershed management, which is a comprehensive term meaning the rational utilization of land and water resources for optimal production and minimum-hazard to natural resources, may be the solutions to all these problems. Accurate estimation/prediction of runoff, sediment yield, nutrient loss, its amount and rate are the key parameters for watershed management. In many instances, non-availability of these data is a major handicap to start with watershed development program.

Use of mathematical models for hydrologic evaluation of watersheds is the current trend and extraction of watershed parameters using remote sensing and geographical information system (GIS) with high-speed computers are the aiding tools and techniques for it. Agricultural Nonpoint Source Pollution (AGNPS) model [26, 27] is one such mathematical, hydrologic, distributed, event based model that is used as a watershed analysis tool for evaluating non-point source pollution from agricultural watersheds.

The AGNPS model has been validated in different environments worldwide mainly for prediction of runoff, peak discharge, sediment load and nutrient discharge. Chahor et al. [2] applied this model for prediction of runoff and sediment yield in small sized Latxaga watershed, in Spain and found its performance satisfactory. Taguas et al. [18] carried out a study in an olive orchard micro catchment in Spain under different soil management to model the contribution of ephemeral gully erosion. Momm et al. [12] used AGNPS for mapping and identification of cropland potential ephemeral gullies. Due to integration and applicability with GIS, the popularity of AGNPS model has increased recently. Emili and Greene [4] developed GIS protocol and showed the useful and practical application of GIS in AGNPS. This model has been validated using field data from agricultural watersheds in Minnesota, Iowa, and Nebraska [26], Illinois [9] and watersheds in Northeast Kansas [6]. Mitchell et al. [11] evaluated the AGNPS model for predicting runoff and sediment yield from small watersheds of mild topography in Illinois. Lo [10] integrated AGNPS and ARC/INFO to quantify erosion problems at the Bajun river basin and the Tsengwen reservoir watershed in Taiwan.

While reviewing the works on comparison of different models performance and its use to combat NPS pollution, it is also seen that AGNPS predicted well in flat land area of uniform slope. Since for design of any hydraulic structure or to evaluate best management practices (BMPs), it is desirable to simulate under severe conditions of rainfall event, AGNPS is a better model for the above purpose as it is event based. Also the distributed nature of AGNPS model has the other advantages of applying to medium watersheds, where variability in land use/ land cover (LU/LC) is there spatially as well as temporally. When topographic, LU/LC and other inputs is required grid wise, remote sensing and GIS techniques are best tools for obtaining them. The application of remote sensing and GIS for hydrologic studies is numerous.

Hence realizing the success of these techniques it is also thought that these methodologies can be used in this research study using IRS satellite data. Considering all above points, AGNPS model was selected for this study for hydrologic evaluation of watersheds.

6.2 MATERIALS AND METHODS

This study was undertaken in two agricultural watersheds namely Tarafeni and Bhairabbanki and its sub watersheds located respectively in Midnapore and Bankura district of West Bengal state in India (Figure 6.1). Both the watersheds lie within 22°37' to 22°51' north latitude and 86°38'

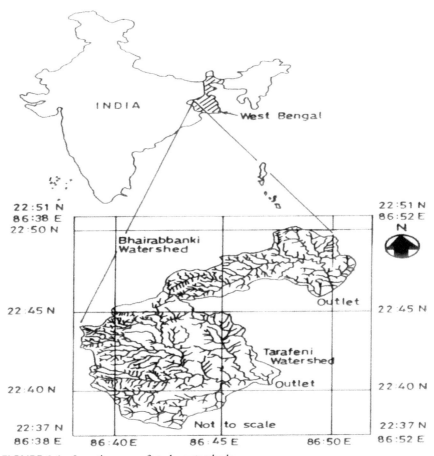

FIGURE 6.1   Location map of study watersheds.

to 86°52′ east longitude. The area of Tarafeni watershed is 158.06 km^2 and elevation varies between 110 m and 290 m above mean sea level (msl). The area has a sub-tropical sub-humid climate with occasional high intensity of rainfall during monsoon (May–October). Area of Bhairabbanki watershed is 69.15 km^2 and its elevation varies between 130 m to 270 m above msl. Both the watersheds lie on the SOI toposheet number: 73/J-9, J-10, J-13, and J-14 (1:50,000 scale). The outlet of river Tarafeni is considered at Tarafeni barrage and for Bhairabbanki watershed it is considered at Bhairabbanki barrage. The two major landscapes of the watersheds are undulating plane interspersed with mounds and valleys in which red and lateritic and old alluvial soils and Dharwar landscape in which lateritic soils are present.

The major soil textural classes present are sandy loam, silt loam and loam. The soil map of study area was prepared by using *All India Soil and Land Use Survey map*. Indian remote sensing satellite images for the year 1989 (IRS-1A-LISS-II-B1-20-52, date of pass (DOP) 21–02–1989 and 13–11–1989 and IRS-1D-LISS-III-107–56, DOP 22–02–2000 and 14–11–2000) were used for land use land cover classification and updating drainage network. The different image processing and GIS software used in this study are ERDAS/IMAGINE and ARC/INFO. The toposheets were scanned using A0 size scanner. The rainfall amount, intensity, runoff amount, sediment and nutrient loss were observed at the outlet of watersheds and data were analyzed storm wise.

The Survey of India toposheets covering the study area were scanned, rectified and digitized for elevation contours, drainage network, and prominent land cover using ARC/INFO GIS software. Then GIS analysis was made to convert the whole watersheds into grids of 400×400 m^2 and grid wise slope, aspect, slope shape factor, soil characteristics, etc. were found out. The IRS satellite images for 1989 and 2000 (both monsoon (*kharif*) and winter (*rabi*) seasons) were classified using supervised classification (after several ground truth verifications) with maximum likelihood classification algorithm in ERDAS/IMAGINE software. The classification accuracy was found to be acceptable ranging from 88.7 to 91.2% for all the classifications.

6.2.1 AGNPS (AGRICULTURAL NON-POINT SOURCE POLLUTION) MODEL

6.2.1.1 Model Structure

The USDA Agricultural Research Service (USDA-ARS) in cooperation with the Minnesota Pollution Control Agency and the Soil Conservation Service (now the Natural Resources Conservation Service) developed the Agricultural Non-Point Source Pollution (AGNPS) model [25–27].

AGNPS is an event based distributed model, which simulates runoff, sediment, and nutrient transport from agricultural watersheds. The nutrients considered include nitrogen (N) and phosphorous (P), both essential plant nutrients and major contributors to surface water pollution. Basic model components were hydrology, erosion, sediment and chemical transport. In addition, the model considers point sources of sediment from gullies and inputs of water, sediment, nutrients, and chemical oxygen demand (COD) from animal feedlots, springs and other point sources. COD is a measure of the oxygen required to oxidize organic and oxidizable inorganic compounds in water. As such it is used as an indicator of the degree of pollution. Water impoundments, such as tile outlet terraces, also are considered as depositional areas of sediment associated nutrients.

The model operates on cell basis with cells of uniformly square size subdividing the watershed, allowing analysis at any point within the watershed. Hydrology, erosion or sediment transport, and chemical transport are calculated for each cell within the watershed and routed to the watershed outlet.

The calculations made by AGNPS occur in three stages, or loops:

Initial calculations for all cells in the watershed are made in the first loop. These calculations include estimates for upland erosion, overland runoff volume, time until overland flow becomes concentrated, level of soluble pollutants leaving the watershed via overland runoff, sediment and runoff leaving impoundment terrace systems, and pollutants coming from point source input such as tile lines or feedlots.

The second loop calculates the runoff volume leaving the cells containing impoundments and the sediment yield for primary cells. A primary cell is that no other cell drains into. The sediment from these cells and other cells is broken down into five particle-size classes: clay, silt, small aggregates, large aggregates and sand.

In third loop, the sediment and nutrients are routed through the rest of the watershed. Calculations are made to establish the concentrated flow rates, to derive the channel transport capacity, and to calculate the actual sediment and nutrient flow rates.

6.2.1.2   Hydrology

(a) Runoff
The model requires daily or total storm rainfall amount per event. Runoff
volume and peak flow rate are calculated in hydrology component. The
runoff volume from each cell is calculated by using SCS runoff curve
number (CN) equation [19]. The overland runoff duration, or the time
needed for concentrated flow to occur, is calculated using the runoff velocity as determined [19]. The equation used for computation is:

$$OFT = \frac{L_s}{V_o} \tag{1}$$

where, OFT = overland flow time (s), L_s = field slope length (m), and
V_o = overland flow velocity (m s^{-1}). The velocity is calculated as:

$$V_o = 10^{(0.5 \times \log_{10}(S_L \times 100) - SCC)} \tag{2}$$

where, S_L = land slope (m m^{-1}), and SCC = overland surface condition
constant, which is a cell characteristic that accounts for the effects of land
use and vegetation.

(b) Peak runoff rate
The model uses two different ways for calculating peak runoff rate for
each cell. An empirical relationship [17] and used in the CREAMS model
can be selected. This method assumes a triangular shaped channel and
uses the following equation:

$$Q_p = 3.79(A)^{0.7}(CS)^{0.16}(RO/25.4)^{0.903A^{0.017}}(LW)^{-0.187} \tag{3}$$

where, Q_p = peak runoff rate (m^3 s^{-1}), A = drainage area (km^2), CS = channel slope (%), RO = daily runoff volume (mm), LW = length–width ratio
of the watershed = L^2/A, and L = maximum flow path (km).
      The second option uses a method based on SCS TR55 [20], which is
a simplified procedure for estimating runoff and peak discharges in small
watersheds. The peak runoff rate is dependent on the rainfall distribution
and amount, runoff curve number, and the time of concentration. In this

method, a rectangular shaped channel is assumed and the peak flow is based on the time of concentration (T_c). The total travel time for any cell is the time required for runoff to travel from the hydraulically most distant point of the watershed to the outlet of that cell and the T_c is computed by summing all the travel times for consecutive cells in a specific flow path in the watershed. Time of concentration is estimated with the equation

$$T_c = T_{cc} + T_{cs} + T_{sf}$$ (4)

where, T_C = watershed time of concentration (h), T_{CC}, T_{CS}, and T_{SF} are components of T_C attributed to channel flow, surface flow and shallow channel flow. The channel component is computed with the equation:

$$T_{CC} = \frac{(L - \lambda - L_{SF})^n}{3.6 d^{0.67} \sigma^{0.5}}$$ (5)

where, T_{cc} = channel time of concentration (h), L = channel length from the most distant point to the watershed outlet (km), λ = surface slope length (km), L_{SF} = shallow flow length (km), n = Manning's roughness coefficient, d = average channel flow depth (m), and σ = channel slope (m m^{-1}). The shallow flow component of T_C is estimated with the equation:

$$T_{SF} = \frac{L_{SF}}{V_{SF}}$$ (6)

where, V_{SF} = average shallow flow velocity (km h^{-1}), and is estimated with equation:

$$V_{SF} = 17.7 S^{0.5} \leq 2.19 \text{ km h}^{-1}$$ (7)

where, S = surface flow slope (m.m^{-1}). The length of shallow flow, L_{SF} is estimated with the equations

LSF = 0.05   (If, L > 0.1 km)

$$L_{SF} = L - 0.05 \quad \text{(If, } 0.05 < L < 0.1 \text{ km)}$$

$$L_{SF} = 0.0 \quad \text{(If, } L < 0.05 \text{ km)}$$ (8)

The surface flow component of T_{CS} is estimated with equation:

$$T_{CS} = \frac{0.0913(\lambda - n)^{0.8}}{S^{0.4}R^{0.5}} \tag{9}$$

where, T_{CS} = surface flow component of time of concentration (h), and R = storm rainfall (mm).

The peak flow is calculated from T_C by:

$$Q_p = 10^{\log[C_0 + C_1(\log T_C) + C_2(\log T_C)]^2}[0.000000672]AQ \tag{10}$$

where, Q_p = peak flow rate in ($m^3 \ s^{-1}$), A = drainage area (km^2), Q = run-off volume (mm), C_0, C_1, and C_2 = coefficients based on 24-hour precipitation and initial abstraction as determined from the curve number.

With either method of calculating peak flow, model has the option of entering known channel characteristics, i.e., channel length, width, and depth or using hydro-geomorphic relationships to determine channel geometry. The hydraulic geometry predicted by geomorphic calculation allows the user to estimate the channel dimensions as a function of total drainage area into the cell.

(c) Hydrograph generation

The peak flow rates calculated from either method are used to generate a triangular hydrograph for each cell. The triangular hydrograph is partitioned into uniform increments with at least three increments on the rising limb for sediment routing. Flow rates are calculated for each increment using the average time to peak, the flow duration and the increment duration. The peak flow rate in SCS TR55 method is derived from the SCS triangular hydrograph resulting from a rainfall excess of duration D. The peak discharge is given by:

$$q_p = \frac{0.028AQ}{t_p} = \frac{0.028AQ}{0.5D + 0.6T_C} \tag{11}$$

where, q_p = peak discharge ($m^3 \ s^{-1}$), A = basin size (km^2), Q = runoff depth (mm), t_p = time of rise to the peak of hydrograph (h), and T_C = time of concentration (h).

6.2.1.3   Erosion and sediment transport processes

(a) Erosion

AGNPS model uses modified form of the universal soil loss equation (USLE) to estimate upland erosion for single storm [23] as follows:

$$SL = (EI) \times K \times L \times S \times C \times P \times (SSF) \tag{12}$$

where, SL = soil loss (t ha^{-1}), EI = rainfall energy intensity (N h^{-1}), K = soil erodibility factor ((th^{1}) (ha N^{-1})), L = slope length factor, S = slope steepness factor, C= cover and management factor, P = support practice factor and SSF = a factor to adjust for slope shape within the cell (concave, convex, uniform).

Soil loss is calculated for each cell of the watershed. Eroded soil and sediment yield are sub divided into five particle size classes: clay, silt, sand, small aggregates and large aggregates.

(b) Sediment transport

After runoff and upland erosion are calculated for each cell, detached sediment is routed from cell to cell through the watershed to the outlet. The sediment routing through the watershed is done in loops 2 and 3 of the computer programs. The primary cells are routed in loop 2 and rest of cells in loop 3. The routing is done as per cell and per particle size basis proceeding from the headwaters of the watershed to its outlet.

The method used for sediment routing involves equations for sediment transport and deposition [5, 8]. The basic routing equation is derived from the steady state continuity equation as follows:

$$Q_s(x) = Q_s(0) + Q_{sl}(x/L_r) - \int_o^x D(x)wdx \tag{13}$$

where:

$Q_s(x)$ = sediment discharge at the downstream end of the channel reach (kg s^{-1}),

$Q_s(0)$ = sediment discharge into the upstream end of the channel reach (kg s^{-1}),

Q_{sl} = lateral sediment inflow rate (kg s^{-1}),

x = down slope distance (m),

L_r = reach length (m),

D(x) = sediment deposition rate (kg s^{-1} m^{-2}) at the point X, and

w = channel width (m).

The deposition rate is estimated by using following equation:

$$D(x) = [V_{ss}/q(x)][q_s(x)-g'_s(x)]$$ (14)

where, V_{ss} = particle fall velocity (m s^{-1}), q(x) = discharge per unit width (m^2 s^{-1}), $q_s(x)$ = sediment load per unit width (kg s^{-1} m^{-1}), and $g'_s(x)$ = effective sediment transport capacity per unit width (kg s^{-1} m^{-1}). The effective transport capacity is computed using a modified Bagnold stream power equation as follows [1]:

$$g'_s = \frac{\eta k \tau V_c^2}{V_{ss}}$$ (15)

where, η = an effective transport factor, k = transport capacity factor, τ = shear stress (kg m^{-2}), and V_c = average channel flow velocity (m s^{-1}). The transport capacity factor is calculated as:

$$k = (1-e_b).e_s\left[\frac{\gamma_w}{\gamma_s - \gamma_w}\right]$$ (16)

where, e_b = bed load transport efficiency, e_s = suspended load transport efficiency, γ_s = sediment specific weight (kg m^{-3}), and γ_w = specific weight of water (kg m^{-3}).

From flume studies, the combined efficiency term, $(1-e_b)e_s$, has been found to be about 0.01 [15]. Since the actual value of $(1-e_b)e_s$ would vary with the size of the particle being transported, the combined efficiency term was adjusted by an effective transport factor, η [25]. The value of η can be estimated by:

$$\eta = 0.74 Ef^{-1.98}$$ (17)

where, Ef = an entrainment function [15] and calculated as

$$Ef = \frac{\tau}{(\gamma_s - \gamma_w)P_d}$$ (18)

where, P_d = particle diameter (m). Sediment load for each of the five particle size classes leaving a cell is calculated as follows:

$$Q_s(x) = \left[\frac{2q(x)}{2q(x) + "xV_{ss}}\right] \left[Q_s(o) + Q_{sl}\frac{x}{L_r} - \frac{w"x}{2}\left[\frac{V_{ss}}{q(o)}[q_s(o) - g_s'(o)] - \frac{V_{ss}}{q(x)}g_s'(x)\right]\right]$$

(19)

where, Δx = change in channel length across the cell (m), and other symbols are as defined above. Equation 19 is the basic routing equation that derives the sediment transport model.

The sediment discharge is calculated in two periods: the first period during which the eroded sediment from the upland portions of the cell enter the channel and the remaining period during which upland erosion has stopped but channel flow continues. During both periods, the sediment flow at point O remains constant. The sediment transport calculations allow for deposition, and/or scouring of all particle sizes during channel flow based on transport capacity and sediment availability. Model has the option of not allowing any scouring, as in the case of non-erodible channel bed, or allowing only specific particle sizes to scour.

6.2.2   AGNPS MODEL USE

AGNPS version 5.00 [28] was used to simulate runoff and sediment yield from different storms for the watershed. The energy intensity (EI) values for the storms were computed using maximum 30-minute rainfall intensity and break point or time intensity rainfall approach. The curve numbers coverage for the watershed land use was prepared based on hydrologi-cal soil groups, hydrologic condition and antecedent moisture conditions (AMC). The CN coverage was generated by converting the classified image in raster form to polygons and intersecting with the watershed grids. If more than one land use was found in a grid weighted CN was calculated for individual grid, similarly for other parameters like surface condition constant, overland Manning's coefficient, crop management factor, etc. their weighted values were found out.

Soil erodibility (K) factor for watershed soils were computed using the nomograph developed by Wischmeier et al. [22] based on particle size parameter, percentage organic matter, soil texture and soil permeability class. In AGNPS simulations, slope shape for different grids were found from GIS analysis and different slope length was considered with maximum value not exceeding 100 m. The values of the agricultural management parameters [such as crop and management factor (C), conservation practice factor (P) and Manning's roughness coefficient (n), watershed condition constant, fertilizer availability factor, COD (chemical oxygen demand) factor for watershed land uses] were taken from the available literatures [AGNPS Users Guide, USDA Agricultural Handbook, 537, USDA ARS Conservation Research report 35, and [3]].

The parameters required for the model calibration were extracted from the analysis of digital elevation model (DEM), soil map and land use/land cover based on satellite imagery and the available literature. For model calibration 31 rainfall-runoff events were used. The calibration was made in three steps. Initially the values from standard tables and measured values were used to generate a base input file and all errors were removed and it was asserted that the model is running and giving some output without any error message. In the second stage of model calibration, different hydrologic calculation options available in AGNPS model for peak flow calculation and channel parameter calculation, etc. were tried. Both the methods of peak flow computation (SCS-TR55 and AGNPS) with geomorphic option for known storm EI (energy intensity) values were attempted. Also both the methods for hydrograph shape factor, k-coefficient values and percent runoff prior to peak (default values) as well as average of the observed percentage of runoff prior to peak were adjudicated. Also EI calculated by break point or time intensity rainfall method and based on maximum 30-minute intensity were tried separately to find out which method of EI calculation gives better prediction. In this stage the best combination of hydrologic computation options for predicting runoff volume, peak flow rate and sediment close to the observed value was determined. In the final stage the model was calibrated for different input parameters such as CN, surface condition constant, Manning's *n,* C factor, etc. The model was calibrated using trial and error procedure of parameter adjustment and optimization.

After each parameter adjustment, the simulated and observed runoff volume, peak flow and sediment yield were compared to determine the accuracy of predictions. The USLE C factor and Manning's *n* for over-land flow for watershed land uses were also varied according to growth stages of vegetation. AMC prior to each storm was estimated using the limits of five day antecedent rainfall (FDAR) amount as suggested by USDA-SCS and antecedent precipitation index (API) values. The calibration was done for Tarafeni watershed for continuous 31 rainfall runoff events spreading over two years, whereas validation was done for both Tarafeni and Bhairabbanki watersheds for continuous 18 rainfall- runoff events of the next year. The calibrated values of model input parameters were used for model validation. In order to determine the relative sensitivity of model input parameters on model output values, sensitivity analysis was performed for the calibrated parameters of the model for a representative rainfall amount in an average condition of AMC and fair hydrologic condition of watershed.

6.3 RESULTS AND DISCUSSIONS

6.3.1 MODEL CALIBRATION

For calibration of different hydrological and channel parameter computation options, different calibration trials (trial – 1 to 5) were carried out. The trial-1 calculates peak flow using TR55 method, geomorphic option was used for channel variable calculations, runoff percentage prior to peak was input as average value of the observed events. In trial-2 same TR55 method, geomorphic option was used. But the only difference was either model default value of k coefficient = 484 or 37.5% of runoff prior to peak was entered. Either of the input was used as it was found during model initial runs that they yielded same results as explained in previous section. For trial 3 through 5 instead of TR55, AGNPS option was used for peak flow computation method. Rainfall energy intensity was calculated using two methods (EI_{30} or EI_{BR}) as explained in chapter IV, the EI_{30} method was used in trial-4 and in all other trials the EI_{BR} method was used.

For the selected watershed, runoff predictions do not vary with the use of different available options of hydrological computations in AGNPS model version 5.00, rather all the trials predict same values of runoff depth. This is due to the fact that model computes runoff using SCS–CN method and runoff computations are not affected with those available hydrological computation options. During calibration of peak rate of runoff it is found that TR55 and AGNPS method of calculation give two different sets of results and peak flow calculation is insensitive to other factors like method of calculation of rainfall energy intensity or model default value of percentage of runoff prior to peak, k coefficient or average value of observed runoff hydrographs, etc. AGNPS method of calculation give more closer result with the observed than TR55 method of calculation as evident from statistical parameters. Peak rate of runoff predicted using AGNPS method was accepted, as it gave model efficiency as 0.979, coefficient of determination (R^2) as 0.993, RMSE as 5.58 m^3 s^{-1}, and in students t-test calculated value was less than table value.

Among all the trials made for sediment yield prediction the option, which includes AGNPS option of peak flow calculation, geomorphic option of channel parameter calculation, default values of k coefficient or percentage of runoff prior to peak used for hydrograph shape factor, and rainfall energy intensity calculated using break point or time intensity rainfall method [23] gave better prediction as evident from statistical parameters. The model efficiency was found to be 0.965, R^2 as 0.9959, RMSE 173.9 tons (11.02 kg ha^{-1}) at the outlet of watershed, students' t-test also confirmed that there is no significant difference between the mean of observed and predicted value at 95% level of confidence.

In the final stage of calibration, only the above option was used. The AGNPS input parameters like CN, Manning's roughness coefficient (N), C (crop management factor) and surface condition constants (SCC) are varied within their specific limit and optimized for calibration. The calibrated CN, N, SCC and C values were found out for Tarafeni watershed in this process. Using these calibrated parameters simulations were performed for all 31 events and statistical parameter were found out to verify the adequacy of calibration and evaluation of model performance. For runoff, peak flow and sediment prediction most of the deviation was within 20% from observed value (Tables 6.1 and 6.2). Runoff and peak flow values were found to be under predicted for most of the small rainfall events (<50 mm) and under AMC I condition and over predicted for medium and

TABLE 6.1 Simulated Runoff and Peak Flow Deviation and Other Model Evaluation Parameters Obtained During AGNPS Model Parameters Calibration

Storm event number	Rain (mm)	Storm size	AMC	HYC	Runoff (mm)			Peak flow rate		
					Observed	Simulated	Dv (%)	Observed	Simulated	Dv (%)
1	48.6	25–50	II	Poor	10.5	10.3	1.9	50.1	53.4	−6.6
2	77.6	>75	I	Poor	26.2	28.3	−8.0	125.0	138.8	−11.0
3	70.0	>50–75	I	Fair	21.3	22.8	−7.0	123.0	119.1	3.2
4	30.0	>25–50	III	Fair	9.8	9.7	1.0	30.5	33.1	−8.5
5	73.6	>50–75	III	Fair	41.0	42.7	−4.1	121.0	125.6	−3.8
6	80.0	>75	III	Fair	28.5	30.2	−6.1	135.0	148.5	−10.0
7	31.0	>25–50	II	Fair	3.1	3.3	−6.5	18.0	16.9	6.1
8	23.0	0–25	I	Fair	2.0	1.8	10.0	7.1	7.6	−7.0
9	36.0	>25–50	II	Good	6.2	5.7	8.1	24.5	26.2	−6.9
10	29.4	>25–50	III	Good	2.1	2.3	−8.9	21.5	22.2	−3.3
11	65.2	>50–75	III	Good	28.2	29.6	−5.0	92.0	100.8	−9.6
12	21.4	0–25	III	Good	3.8	4.0	−5.3	16.2	14.6	9.9
13	32.6	>25–50	II	Good	4.0	4.1	−1.6	23.0	21.1	8.3
14	63.0	>50–75	I	Good	18.5	19.6	−5.9	92.0	102.7	−11.6
15	28.0	>25–50	I	Poor	3.1	2.7	12.9	14.1	13	7.8
16	27.0	>25–50	III	Poor	2.6	2.4	7.7	14.5	13.2	9.0
17	40.0	>25–50	II	Poor	7.0	6.9	2.0	33.9	35.8	−5.6
18	28.0	>25–50	III	Poor	6.5	7.1	−9.2	20.1	19.9	1.0
19	37.0	>25–50	I	Poor	6.1	5.6	8.4	31.0	29.6	4.5

TABLE 6.1 Continued

Storm event number	Rain (mm)	Storm size	AMC	HYC	Runoff (mm)			Peak flow rate		
					Observed	Simulated	Dv (%)	Observed	Simulated	Dv (%)
20	51.0	>50–75	I	Fair	11.2	11.7	-4.5	56.0	61.1	-9.1
21	27.0	>25–50	I	Fair	2.7	2.6	3.7	14.1	13.1	7.1
22	21.0	0–25	I	Fair	2.0	1.7	15.0	6.1	5.6	8.2
23	38.0	>25–50	II	Fair	6.1	5.9	3.3	28.5	30.3	-6.3
24	23.0	0–25	I	Fair	1.1	1.2	-9.1	7.1	6.6	7.0
25	34.0	>25–50	I	Fair	3.9	4.4	-12.8	22.4	23.1	-3.1
26	20.0	0–25	I	Good	1.1	0.9	18.2	5.1	4.4	13.7
27	24.0	0–25	I	Good	2.2	1.9	13.6	9.0	7.9	12.2
28	25.0	0–25	II	Good	2.3	1.9	17.4	9.1	9.6	-5.5
29	30.0	>25–50	I	Good	3.8	3.4	10.5	17.5	16.4	6.3
30	34.4	>25–50	III	Good	9.7	10.1	-4.1	39.0	36.8	5.6
31	38.0	>25–50	I	Good	6.2	5.9	4.8	32.0	30.4	5.0
Statistical parameters										
Model efficiency (E) [13]						0.9938			0.9857	
Coefficient of residual mass (CRM)						-0.0277			-0.0396	
Coefficient of determination (R²)						0.9988			0.9947	
Root mean square error (RMSE)						0.77 mm			4.68 m³ s⁻¹	
t-test for difference						-1.9019			-1.9638	
t-table value (t$_{0.975,30}$) for two-tailed t distribution						2.0423				

N.B. AMC: antecedent moisture condition; HYC: hydrologic condition; D$_v$: deviation.

TABLE 6.2 Simulated Sediment Yields and Model Evaluation Parameters Obtained During AGNPS Model Parameters Calibration

Storm event number	Rain (mm)	Storm size	Rainfall intensity (mm h^{-1})	AMC	HYC	Sediment yield (tons)		
						Observed	Simulated	Dv(%)
1	48.6	25–50	30.2	II	Poor	908	1020	−12.3
2	77.6	>75	21.5	I	Poor	2750	3055	−11.1
3	70.0	>50–75	52.6	I	Fair	2568	3010	−17.2
4	30.0	>25–50	20.8	III	Fair	345	340	1.4
5	73.6	>50–75	45.5	III	Fair	2689	3020	−12.3
6	80.0	>75	55.2	III	Fair	3400	3655	−7.5
7	31.0	>25–50	22.5	II	Fair	360	372	−3.3
8	23.0	0–25	15.3	I	Fair	172	170	1.2
9	36.0	>25–50	16.9	II	Good	515	460	10.7
10	29.4	>25–50	21.0	III	Good	405	385	4.9
11	65.2	>50–75	38.5	III	Good	2168	2320	−7.0
12	21.4	0–25	15.2	III	Good	245	240	2.0
13	32.6	>25–50	18.9	II	Good	395	400	−1.3
14	63.0	>50–75	42.5	I	Good	2285	2370	−3.7
15	28.0	>25–50	20.5	I	Poor	265	215	18.9
16	27.0	>25–50	15.9	III	Poor	267	230	13.9
17	40.0	>25–50	28.7	II	Poor	685	690	−0.7
18	28.0	>25–50	20.5	III	Poor	389	315	19.0
19	37.0	>25–50	35.1	I	Poor	595	560	5.9
20	51.0	>50–75	45.2	I	Fair	1109	1210	−9.1
21	27.0	>25–50	20.0	I	Fair	285	235	17.5
22	21.0	0–25	15.8	I	Fair	145	140	3.4
23	38.0	>25–50	21.4	II	Fair	610	520	14.8
24	23.0	0–25	19.8	I	Fair	160	155	3.1
25	34.0	>25–50	22.6	I	Fair	468	410	12.4
26	20.0	0–25	16.8	I	Good	113	110	2.7
27	24.0	0–25	20.2	I	Good	190	180	5.3
28	25.0	0–25	18.7	II	Good	205	210	−2.4
29	30.0	>25–50	22.8	I	Good	320	325	−1.6
30	34.4	>25–50	26.4	III	Good	567	630	−11.1
31	38.0	>25–50	32.2	I	Good	550	560	−1.8

TABLE 6.2   Continued

Storm event number	Rain (mm)	Storm size	Rainfall intensity (mm h^{-1})	AMC	HYC	Sediment yield (tons)		
						Observed	Simulated	Dv(%)
Model efficiency (E) [13]						0.9793		
Coefficient of residual mass (CRM)						–0.0530		
Coefficient of determination (R^2)						0.9966		
Root mean square error (RMSE)						133.0 ton (8.41 kg ha^{-1})		
t-test for difference						–1.9515		
t-table value (t$_{0.975,30}$) for two-tailed t distribution						2.0423		

N. B. AMC: antecedent moisture condition; HYC: hydrologic condition; D$_v$: deviation.

large events and also in AMC III condition. The model efficiency found to be 0.994, 0.986 and 0.979; R^2 as 0.999, 0.995 and 0.997; RMSE as 0.77 mm and 4.68 m^3 s^{-1} and 133 tons (8.41 kg ha^{-1}), respectively for runoff, peak flow and sediment yield at outlet. Very small (close to zero) and negative value of CRM for all the parameters indicate that AGNPS model has a slight tendency of over prediction but the model is acceptable as the value is very low. Student's t-test for all parameters showed there is no significant difference between the observed and simulated value and the calculated t value is less than the table value, hence the calibrated parameters are acceptable.

The graph plotted between the observed and simulated showed the data points are close to 1:1 line except for few medium and large events (Figures 6.2–6.4). For all the above results, it can be concluded that AGNPS model after calibration can be used for simulation of runoff, peak flow and sediment.

6.3.2   MODEL VALIDATION

After the model was adequately calibrated, it was validated for 18 rainfall events for both Tarafeni and Bhairabbanki watersheds (Tables 6.3 and 6.4). The calibrated parameters and the best combination of options, which had given best result, were used for validation. The deviations of different parameters from the observed was found within 20% for most

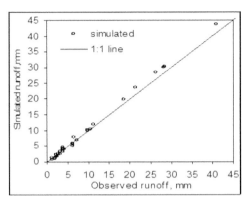

FIGURE 6.2 Observed and simulated runoff value during AGNPS calibration.

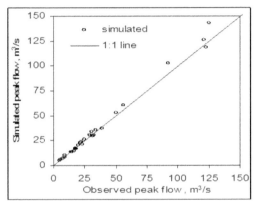

FIGURE 6.3 Observed and simulated peak flow rate during AGNPS calibration.

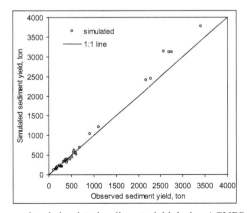

FIGURE 6.4 Observed and simulated sediment yield during AGNPS calibration.

TABLE 6.3　Simulated Runoff, Peak Flow Rate, and Sediment Yield During Validation of AGNPS Model for Tarafeni Watershed

Storm event number	Rainfall (mm)	Storm size	Rainfall intensity (mm h⁻¹)	AMC	HYC	Runoff		Peak rate of runoff		Sediment yield	
						Observed	Dv (%)	Observed	Dv (%)	Observed	Dv (%)
1	45.5	Small	30.5	I	Poor	7.8	2.3	37.1	3.4	915	-4.0
2	44.2	Small	32.0	II	Poor	8.4	-2.8	49.4	9.3	890	-4.0
3	37.8	Small	30.0	III	Poor	7.1	3.4	37.2	4.0	656	4.1
4	43.0	Small	28.0	I	Poor	7.2	-2.3	33.1	-1.0	835	-4.8
5	33.0	Small	22.0	III	Poor	10.8	-5.8	53.2	-10.6	705	4.5
6	32.8	Small	15.0	I	Poor	4.0	11.1	21.3	9.5	475	11.6
7	37.2	Small	23.0	I	Poor	5.1	-4.6	30.1	8.3	590	-2.0
8	32.0	Small	14.0	II	Fair	3.6	-5.8	19.8	0.5	425	7.8
9	28.0	Small	20.0	III	Fair	8.0	-4.8	39.4	-8.4	466	-3.0
10	41.0	Small	33.0	III	Fair	19.6	-10.2	95.7	-13.7	1085	-5.0
11	44.2	Small	28.0	I	Fair	7.1	-7.3	38.1	6.6	996	5.1
12	42.0	Small	34.0	III	Fair	24.0	7.9	98.7	-12.2	1097	-6.4
13	48.0	Small	38.0	I	Fair	9.8	-3.7	51.4	5.2	1058	-3.0
14	31.0	Small	26.0	III	Good	10.1	-3.1	46.8	-13.3	645	8.7
15	62.0	Medium	45.0	III	Good	30.5	-11.6	155.0	-9.5	2100	-12.8
16	33.2	Small	29.0	I	Good	3.8	-0.3	23.4	16.4	486	12.6
17	40.0	Small	30.0	II	Good	6.7	-2.4	36.2	1.3	601	-3.3
18	48.0	Small	26.0	I	Good	9.5	-1.6	45.0	-5.5	992	-9.8

TABLE 6.3 Continued

Storm event number	Rainfall (mm)	Storm size	Rainfall intensity (mm h⁻¹)	AMC	HYC	Runoff		Peak rate of runoff		Sediment yield	
						Observed	Dv (%)	Observed	Dv (%)	Observed	Dv (%)
Statistical parameters											
Model efficiency (E) [13]						0.9758		0.9647		0.9573	
Coefficient of residual mass (CRM)						−0.033		−0.04		−0.024	
Coefficient of determination (R²)						0.9853		0.9935		0.9932	
Root mean square error (RMSE), ton or, kg ha⁻¹ of sediment at outlet						1.10		6.18		78.5 ton (4.97 kg ha−1)	
t-test for difference						−1.3472		−1.4322		−1.1056	

t-table value ($t_{0.975,17}$) for two-tailed t distribution= 2.1098

TABLE 6.4 Simulated Runoff, Peak Flow Rate, and Sediment Yield During Validation of AGNPS Model for Bhairabbanki Watershed

Storm event number	Rainfall (mm)	Storm size	Rainfall intensity (mm.h⁻¹)	AMC	HYC	Runoff		Peak rate of runoff		Sediment yield	
						Observed	Dv (%)	Observed	Dv (%)	Observed	Dv (%)
1	45.5	Small	30.5	I	Poor	7.6	-10.3	25.2	-10.1	252	2.8
2	44.2	Small	32.0	II	Poor	9.1	-3.3	31.2	-1.5	262	-8.0
3	37.8	Small	30.0	III	Poor	8.8	-24.1	22.9	-16.1	195	-11.8
4	43.0	Small	28.0	I	Poor	7.6	6.4	25.6	5.9	245	14.3
5	33.0	Small	22.0	III	Poor	10.1	-18.2	37.2	-8.8	213	-25.1
6	32.8	Small	15.0	I	Poor	4.1	7.1	4.3	2.4	30	20.0
7	37.2	Small	23.0	I	Poor	4.9	1.5	13.9	-3.0	140	2.9
8	32.0	Small	14.0	II	Fair	5.7	2.0	17.2	-0.9	156	-3.2
9	28.0	Small	20.0	III	Fair	7.2	-19.9	26.8	-10.9	174	-18.6
10	41.0	Small	33.0	III	Fair	13.8	-12.3	34.6	-10.4	232	-8.6
11	44.2	Small	28.0	I	Fair	10.5	10.5	32.2	1.8	310	8.7
12	42.0	Small	34.0	III	Fair	14.2	-16.3	36.8	-9.4	329	-17.0
13	48.0	Small	38.0	I	Fair	9.7	3.1	32.0	0.4	298	2.7
14	31.0	Small	26.0	III	Good	7.7	-12.2	26.5	-12.2	180	-15.0
15	62.0	Medium	45.0	III	Good	24.8	-17.8	76.2	-26.0	505	-27.9
16	33.2	Small	29.0	I	Good	3.4	2.9	12.1	4.7	120	5.0
17	40.0	Small	30.0	II	Good	8.2	0.9	27.6	-0.6	256	2.3
18	48.0	Small	26.0	I	Good	10.6	8.9	36.0	0.9	329	3.0

TABLE 6.4   Continued

Storm event number	Rainfall (mm)	Storm size	AMC	HYC	Rainfall intensity (mm.h^{-1})	Runoff		Peak rate of runoff		Sediment yield	
						Observed	Dv (%)	Observed	Dv (%)	Observed	Dv (%)
Statistical parameters											
Model efficiency (E) [13]						0.8948		0.8762		0.8269	
Coefficient of residual mass (CRM)						-0.07		-0.10		-0.06	
Coefficient of determination (R^2)						0.9729		0.9791		0.9285	
Root mean square error (RMSE), ton or, kg ha^{-1} of sediment at outlet						1.52		5.11		41.7 ton (6.03 kg ha^{-1})	
t-test for difference						-2.0790		-2.0253		-1.58757	

t-table value ($t_{0.975,17}$) for two-tailed t distribution= 2.109.8

of the events but it is more than 20% but less than 30% for two or three events. The trend obtained during calibration was also found during validation, that for small rainfall events the model has a tendency to under predict and for medium and large events it over predicts. There is over prediction in most of the AMC III conditions of watershed and there is no definite trend for different hydrologic conditions.

For Tarafeni watershed during model validation the different statistical parameters such as model efficiency was found to be 0.978, 0.965, and 0.957; R^2 as 0.985, 0.993 and 0.993; RMSE as 1.10 mm, 6.18 m^3 s^{-1}, 78.51 tons (4.97 kg ha^{-1}, respectively for runoff, peak flow and sediment yield at the outlet. The calibration parameters for model option and input parameters of Tarafeni watershed were used for validation of Bhairabbanki watershed.

For Bhairabbanki watershed model efficiency was found to be 0.895, 0.876, and 0.827; R^2 as 0.973, 0.979 and 0.928; RMSE as 1.52 mm, 5.11 m^3 s^{-1}, 41.7 ton (6.03 kg ha^{-1}) respectively for runoff, peak flow and sediment yield at outlet. The CRM value was found to be negative and close to zero for all parameters and also for both the watersheds. Hence it can be said that overall the model has a tendency for slightly over prediction. The students' t-test for significant difference showed that there is no significant difference between the observed and simulated values for all the parameters and for both the watersheds. The graphical representation of observed versus simulated values showed that data points are very close to 1:1 line for most of the simulations except for few medium and large rainfall-runoff events (Figures 6.5–6.7). So from the whole validation study it could be understood that AGNPS model can be applied to nearby hydrologically similar watersheds having similar rainfall pattern. The statistical parameters obtained for Bhairabbanki watershed show that though the model is simulating acceptable results for Bhairabbanki, its simulation capability can be further improved by calibrating the model while applying to any other watershed of similar hydrologic and climatic conditions.

6.3.3 SENSITIVITY ANALYSIS

The results of sensitivity analysis indicated that runoff is sensitive to only curve number. The peak flow is most sensitive to CN value followed

a) Tarafeni watershed   *b) Bhairabbanki watershed*

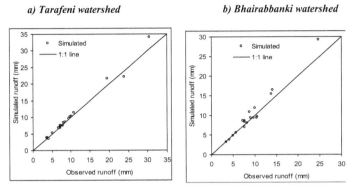

FIGURE 6.5 Observed and simulated runoff during AGNPS model validation. (a) Tarafeni watershed; (b) Bhairabbanki watershed.

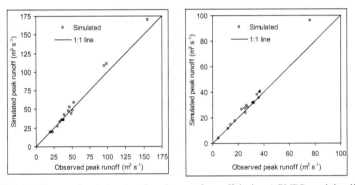

FIGURE 6.6 Observed and simulated peak rate of runoff during AGNPS model validation.

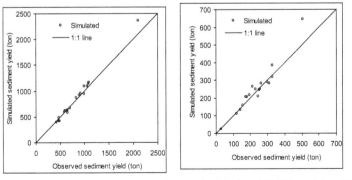

FIGURE 6.7 Observed and simulated sediment yield during AGNPS model validation.

by channel slope and field slope. The sensitivity analysis study also revealed that the variables most significantly affecting the sediment yield in descending order are CN, K, EI, P, C, LS, SL, CSS, CS-based on mean absolute deviation value. The sensitivity analysis suggests that above variables should be carefully estimated for accurate sediment yield prediction. Since the runoff is calculated by CN technique and peak flow and sediment yield is also dependent on it, this parameter should be carefully calibrated for individual watersheds before applying AGNPS model for runoff, peak flow or sediment simulations. The findings of this sensitivity analysis are similar to the results obtained by [26] (developer of the model).

6.3.4 ASSESSMENT OF CRITICAL AREA AND BEST MANAGEMENT PRACTICES (BMP)

After calibration and validation of the AGNPS model, it was used for assessment of critical areas in the study watersheds using 24 h-25 year rainfall. For finding out 24 h-25 year rainfall amount, rainfall frequency analysis was done for 22 years of data available for the study watersheds. The annual maximum one-day rainfall was fitted to different probability distributions and the data were found to follow log-normal distribution. 24 h rainfall amount was found out for different return period and critical areas were found out considering a 24 h 25-year rainfall amount. The grids were classified as slight (0–5 t ha^{-1}), moderate (5–10 t ha^{-1}), high (10–20 t ha^{-1}), very high (20–40 t ha^{-1}), severe (40–80 t ha^{-1}), and very severe (>80 t ha^{-1}) erosion depending upon the cell erosion. The same analysis was made for both the watersheds for present land use scenario and also for 1989 to find out critical areas.

Considering the socio economics of the people residing in the watershed and different programs under government and non-government organizations nine set of BMPs (mostly biological measures) were proposed to find out whether it is possible to improve the watershed condition or any engineering measures would still be required. These BMPs suggest conversion of degraded forest to open forest and open forest to dense forest; converting waste land to degraded forest with tall erosion resistant grass and fallow land to agriculture; all agricultural land are to be contour bunded if the slope is less than 6% in a grid and applying all above combinations.

The critical areas in both the watersheds were found out by simulating for 236.4 mm of rainfall (24 h-25 year) for average condition of watershed (AMC II and fair hydrologic condition) and rainfall energy intensity calculated by the model for a USDA type II rainfall (as similar rainfall distribution characteristics is found for high rainfall events in the watershed). This resulted into 166.4 mm of runoff, 809.41 m^3 s^{-1} of peak flow, 77254 ton (4.89 t ha^{-1}) of sediment and 6.37 kg ha^{-1} of nutrient yield at Tarafeni watershed outlet. The nutrient includes both nitrogen and phosphorus in both forms (adsorbed to sediment and soluble in runoff). Similarly for Bhairabbanki watershed with present land use scenario and same rainfall amount with similar assumptions resulted 166.6 mm of runoff, 519.9 m^{3}s^{-1} of peak flow, 19397 ton (2.81 t ha^{-1}) of sediment and 5.86 kg ha^{-1} nutrient yield at its outlet.

In Tarafeni watershed 19.2% of total area is under very severe erosion, 45.2% is under severe erosion, 27.4% is under very high erosion, 7.8% under high erosion and only 0.4% area is under moderate erosion. In Tarafeni watershed 9.9% area is under permissible soil erosion range, hence rest 91.1% area needs soil conservation measures to reduce water quality and land degradation. Similarly in Bhairabbanki watershed 3.2% of total area is under very severe erosion, 36.16% under severe erosion, 42.56% under very high erosion, 16.7% under high erosion, and 1.37% of total area under moderate erosion class. Based on USDA recommendation only 22.20% of total area of Bhairabbanki watershed is under permissible erosion range. Hence remaining 77.80% area needs soil conservation measures to reduce water quality and land degradation. However overall Bhairabbanki watershed is in a better condition than Tarafeni in terms of area under permissible erosion limit.

In Tarafeni watershed by changing degraded forest to open forest and open forest to dense forest, there is not much reduction in runoff, peak flow, sediment or nutrient yield. By converting fallow land to agriculture reduces runoff by 5.04%, peak flow by 5.02%, sediment yield by 24.76% and nutrient yield by 19.22%. By adopting contour bunding in agriculture field, runoff is reduced by 2.9%, peak flow by 2.96%, sediment yield by 10.54% and nutrient yield by 7.83%. Implementing all the management practices together results in highest reduction in runoff by 10.08%, peak flow by 10.02%, sediment yield by 37.14% and nutrient yield by 29.54%.

Hence it is obviously the best option as it includes all the management practices when there is no resource constraint. But applications of all the BMPs are not economical for the farmers of the watershed or by any other agencies (government or non-government organizations). Hence it can be fairly recommended that for Tarafeni watershed the most economical BMP is that, the farmers should do contour bunding and the second economical option is to convert the fallow land to agriculture, because large area in watershed is under fallow class, which produces more sediment from it. Since the area under wasteland is not much, by managing that land and converting it to degraded forest with tall erosion resistant grass does not change the soil and nutrient loss at outlet appreciably.

Similar results were also found for Bhairabbanki watershed. By changing degraded forest to open forest and open forest to dense forest, results reduction in runoff by 7.77%, peak flow by 7.54%, sediment yield by 3.9% and nutrient yield by 3.49%. By converting fallow land to agriculture reduces runoff by only 3.96%, peak flow by 3.81%, but sediment yield by 25.84% and nutrient yield by 20.93%. Implementing all BMPs together (BMP-9) results in highest reduction in all values, such as runoff by 11.74%, peak flow by 11.37%, sediment yield by 35.0% and nutrient yield by 29.07%. Hence obviously it is the best option as it includes all the management practices when there is no resource constraint. But applications of all the BMPs through BMP-9 are not economical for the farmers of the watershed or by any other agencies (government or non-government organizations). Hence it can be fairly recommended that for Bhairabbanki watershed the most economical BMP is that, the farmers should convert the fallow land to agriculture and the second economical option is that farmers should do contour bunding in their agricultural lands to reduce the runoff, peak flow, sediment and nutrient yield.

The above results indicate for planning and execution of engineering measures of erosion control besides the BMP-9. The suggested BMPs are in commensurate with the present watershed management policy which suggests the application of agronomical/biological erosion control measures is to be applied first and later only those areas of watershed should be treated with combination of biological and engineering measures which cannot be improved with the biological erosion control measures alone.

6.4 CONCLUSIONS

Hydrologic modeling with AGNPS model gives better prediction for study watersheds by considering AGNPS method of runoff calculation; default k coefficient or percentage of runoff prior to peak for hydrograph shape factor; geomorphic way of channel parameter estimation; and energy intensity calculated by break pint or time intensity rainfall method.

For successful applications of AGNPS model calibration of the curve numbers, crop management factors, surface condition constants and Manning's roughness coefficient depending upon different stages of crop growth are essential. A very close prediction with the observed values of runoff, peak flow and sediment yield is obtained after calibration.

The AGNPS model is very well validated for Bhairabbanki watershed with calibrated values of Tarafeni watershed indicates that the model can be successfully adopted for other watersheds of similar hydrologic and climatic conditions.

The sensitivity analysis of AGNPS model shows that the curve number to be the most sensitive parameter followed by soil erodibility factor, rainfall energy intensity value, conservation practice factor, crop management factor, land slope, slope length, channel side slope and channel slope for sediment yield prediction. For runoff and peak flow the curve number is most sensitive, hence caution should be taken while using curve number values and should be calibrated most precisely.

Application of AGNPS model for critical area assessment of study watersheds shows that 9.9% and 22.0% of area is under permissible erosion for Tarafeni and Bhairabbanki watershed respectively, hence best management practices (BMPs) which include biological measures along with engineering measures are desirable to change this scenario.

The most economical BMP for Tarafeni watershed is contour bunding in agricultural lands followed by converting fallow lands to agriculture, where as for Bhairabbanki it is conversion of fallow land to agriculture followed by contour bunding in agricultural lands.

AGNPS model is proved to be a successful model for hydrologic evaluation of agricultural watersheds using remote sensing and GIS techniques in the sub-humid sub-tropical region.

6.5 SUMMARY

Use of mathematical models for hydrologic evaluation of watersheds is the current trend and extraction of watershed parameters using remote sensing and geographical information system (GIS) in high-speed computers is the aiding tools and techniques for it. Agricultural Nonpoint Source Pollution (AGNPS) model is one such mathematical, hydrologic, distributed, event based model used as a watershed analysis tool for evaluating non-point source pollution from agricultural watersheds.

The present study of hydrological modeling was undertaken in two agricultural watersheds namely Tarafeni and Bhairabbanki and its sub watersheds located respectively in Midnapore and Bankura district of West Bengal state in India. Indian remote sensing satellite images for the year 1989 (IRS-1A-LISS-II-B1-20-52, date of pass (DOP) 21–02–1989 and 13–11–1989 and IRS-1D-LISS-III-107–56, DOP 22–02–2000 and 14–11–2000 were used for land use land cover classification and updating drainage network. The different image processing and GIS software used in this study are ERDAS/IMAGINE and ARC/INFO. AGNPS version 5.00 was used to simulate runoff and sediment yield from different storms for the watershed. The AGNPS model is very well validated for Bhairabbanki watershed with calibrated values of Tarafeni watershed and the study indicates that the model can be successfully adopted for other watersheds of similar hydrologic and climatic conditions. The sensitivity analysis of AGNPS model shows that the curve number is most sensitive parameter for prediction of runoff, peak flow and sediment yield prediction. Application of AGNPS model for critical area assessment of study watersheds shows that 9.9% and 22.0% of area is under permissible erosion for Tarafeni and Bhairabbanki watershed respectively.

KEYWORDS

- **AGNPS**
- **agricultural land**
- **calibration**

- chemical nutrients
- chemical transport
- contour bunding
- erodibility factor
- erosion
- GIS
- hydrologic modeling
- India
- model
- non-point source pollution
- overland flow
- remote sensing
- runoff
- sediment
- sediment yield
- sensitivity analysis
- software
- USDA – SCS
- validation
- watershed
- yield

REFERENCES

1. Bagnold, R. A., 1966. An approach to the sediment transport problem from general physics. *Proc. Paper 422-J.* U.S. Geological Survey, Reston, VA.
2. Chahor, Y., Casalí, J., Giménez, R., Bingner, R. L., Campo, M. A., and Goñi, M., 2014. Evaluation of the An AGNPS model for predicting runoff and sediment yield in a small Mediterranean agricultural watershed in Navarre (Spain). *Agricultural Water Management*, 134:24–37.
3. Chakraborty, A. K., 1993. Strategies for watershed management planning using remote sensing techniques. *J. Indian Soc. of Remote Sensing*, 21(2):87–98.
4. Emili, L. A., and Greene, R. P., 2013. Modeling agricultural nonpoint source pollution using a geographic information system approach. *Environmental Management*, 51(1):70–95.

5. Foster, G. R., Lane, L. J., Nowlin, J. D., Laflen, J. M., and Young, R. A., 1981. Estimating erosion and sediment yield on field sized areas. *Trans. of the ASAE,* 24(50):1253–1262.

6. Koelliker, J. K., and Humbert, C. E., 1989. Application of AGNPS model for water quality planning. *ASAE Paper No. 89–2042,* ASAE, St. Joseph, Michigan.

7. Koohafkan, A. P., 2000. Land resources potential and sustainable land management-An overview. Lead paper of the *International conference on Land resource management for food, employment and environmental security* during November 9–13, 2000, New Delhi, India. p. 1–22.

8. Lane, L. J., 1982. Development of procedure to estimate runoff and sediment transport in ephemeral streams. In: *Recent Developments in the Explanation and Prediction of Erosion and Sediment yield. Proc. of Symp.* IASH, 137:275–282.

9. Lee, M. T., 1987. Verification and applications of a non point source pollution model. In: *Proc. of the Natl. Eng. Hydrology Symp.,* ASAE, New York, NY.

10. Lo, K. F. A., 1995. Erosion assessment of large watersheds in Taiwan. *J. Soil and Water Cons.,* 50(2):180–183.

11. Mitchell, J. K., Engel, B. A., Srinivasan, R., and Wang, S. S. Y., 1993. Validation of AGNPS for small watersheds using an integrated AGNPS/GIS system. *Water Resources Bulletin,* 29(5):833–842.

12. Momm, H. G., Bingner, R. L., Wells, R. R., and Wilcox, D., 2012. AGNPS GIS-based tool for watershed-scale identification and mapping of cropland potential ephemeral gullies. *Applied Engineering in Agriculture,* 28(1):17–29.

13. Nash, J. E., and Sutcliffe, J. V., 1970. River flow forecasting through conceptual models, Part-1: A discussion of principles. *Journal of Hydrology,* 10(3):282–290

14. Perrone, J., and Madramootoo, C. A., 1997. Use of AGNPS for watershed modeling in Quebec. *Trans. of the ASAE,* 40(5):1349–1354.

15. Simons, D. B., and Senturk, F., 1976. The motion of bed material. In: *Sediment Transport Technology, Water Resour. Publication.,* Fort Collins, CO, p. 566.

16. Singh, G., Babu, R., Narain, P., Bhushan, L. S., Abrol, I. P., 1992. Soil erosion rates in India. *J. of Soil and Water Cons.,* 47(1):97–99.

17. Smith, R. E., and Williams, J. R., 1980. Simulation of surface water hydrology. W. G. Knisel (ed.) In: *CREAMS: A field scale model for chemicals, runoff, and erosion from agricultural management systems. Cons. Res. Report 26,* USDA, Washington, D.C., 13–35.

18. Taguas, E. V., Yuan, Y., Bingner, R. L., and Gómez, J. A., 2012. Modeling the contribution of ephemeral gully erosion under different soil managements: A case study in an olive orchard micro attachment using the AnnAGNPS model. *Catena,* 98:1–16.

19. USDA Soil Conservation Service, 1972. Section 4: Hydrology In: *National Eng. Handbook.* Washington, DC.

20. USDA Soil Conservation Service, 1986. Urban hydrology for small watersheds, *Technical Release 55,* Eng. Division, Washington, DC.

21. Velayutham, M., 2000. Status of land resources in India. Lead paper of the International conference on *Land Resource Management for Food, Employment and Environmental Security* during November 9–13, 2000, New Delhi, India, p. 67–83.

22. Wischmeier, W. H., Johnson, C. B., and Cross, B. V., 1971. A soil erodibility nomograph for farmland and construction sites. *J. Soil and Water Cons.,* 26(5):189–193.

23. Wishmeier, W. H., and Smith, D. D., 1965. Predicting rainfall erosion losses from cropland east of the Rocky Mountains – Guide for selection of practices for soil and water conservation. *USDA Handbook No. 282.*

24. Wishmeier, W. H., and Smith, D. D., 1978. Predicting rainfall erosion losses – A Guide to Conservation Planning. *USDA, Agric. Handbook No. 537*, Washington, DC.

25. Young, R. A., Onstad, C. A., and Bosch, D. D., 1986. Sediment transport capacity in rills and small channels. In: *Proc. Fourth Federal Interagency Sediment Conference, Subcomm. On Sedimentation of the Interagency Advisory Comm. on Water Data*, Vol. 23, Washington, DC. pp. 6–25 to 6–33.

26. Young, R. A., Onstad, C. A., Bosh, D. D., and Anderson, W. P., 1987. AGNPS: Agricultural non point source pollution model: A watershed analysis tool. *USDA ARS, Conservation Research Report No. 35,* Washington DC, 77 p.

27. Young, R. A., Onstad, C. A., Bosch, D. D., and Anderson, W. P., 1994. AGNPS Version 4.03, *AGNPS User's Guide.* Agril. Res. Service, USDA, Morris, MN and St. Paul, Minn.: Minnesota Pollution Control Agency Minnesota.

28. Young, R. A., Onstand, C. A., Bosch, D. D., and Anderson, W. P., 1995. *AGNPS User's Guide Version 5.00.* Agril. Research Service, USDA, Morris, MN.

PART II

RESEARCH INNOVATIONS IN SOIL AND WATER ENGINEERING

CHAPTER 7

SOCIAL-BASED EXPLORATORY ASSESSMENT OF SUSTAINABLE URBAN DRAINAGE SYSTEMS (SUDS)

HOHUU LOC,[1] K. N. IRVINE,[2] and NIRAKAR PRADHAN[3]

[1]*Water Engineering and Management Program, Asian Institute of Technology, Pathumthani, 12120, Thailand; Research Center for Environmental Quality Management, Kyoto University. 1–2 Yumihama, Otsu, 5200811, Japan, E-mail: ho.huu.45z@st.kyoto-u.ac.jp*

[2]*National Institute for Education, Nanyang Technological University, Nanyang Walk, Singapore*

[3]*Environmental Engineering and Management Program, Asian Institute of Technology, Pathumthani, 12120, Thailand*

CONTENTS

7.1   Introduction 170
    7.1.1   Background of Sustainable Urban
         Drainage Systems (SUDS) 170
7.2   Methodology 172
    7.2.1   Study Area 172
    7.2.2   Interviews 173
    7.2.3   Hypothesis Testing 173
    7.2.4   Numerical Modeling 174

　　　　　7.2.4.1　Model Calibration 174
　　　　　7.2.4.2　SUDS Evaluation 174
7.3　Results... 176
　　　7.3.1　Interviews... 176
　　　　　7.3.1.1　Testing Assumptions 177
　　　　　7.3.1.2　Results for Hypothesis Testing...................... 178
　　　7.3.2　Pcswmm Model.. 179
　　　　　7.3.2.1　Model Calibration and Validation 179
　　　　　7.3.2.2　SUDS Evaluation 179
7.4　Discussion.. 182
7.5　Conclusions... 183
7.6　Summary.. 183
Keywords 184
References... 184

7.1　INTRODUCTION

7.1.1　BACKGROUND OF SUSTAINABLE URBAN DRAINAGE SYSTEMS (SUDS)

Urban flooding problems have conventionally been addressed using hard engineering solutions such as expanding sewage networks, building sluices, etc. which are drainage-efficiency driven but not necessarily sustainable and eco-friendly. They, nevertheless, could affect the natural water cycle processes, such as infiltration, evapotranspiration, surface and sub-surface flows [5]. Some of the drawbacks of the traditional drainage systems are:

- Increased frequency, magnitude of storm discharge, resulting from the direct connection of sewerage systems to receiving water bodies without attenuation measures [22],
- Amplified volume of total runoff resulting from the depletion of transpiration that is caused by vegetation surface reduction [8],
- Increased occurrence of low-magnitude flows [9],
- Reduced lagging time of storm water [4].

All of the aforementioned suggest that there should be other methods that consider the urban flooding problem with more sustainable approaches. By changing the concept of handling storm water from a nuisance to an utilizable resource, not only can the flooding problem be solved, other benefits also can be captured. This mind-set change is where the concept of Sustainable Urban Drainage Systems (SUDS) emerged [16]. Table 7.1 summarizes fundamental differences between these two mind-sets in dealing with urban flooding.

There are several SUDS technologies available. Within the scope of this study, four of the most popular SUDS technologies (Figure 7.1) are considered:

   (i) Rainwater harvesting – which can be a supplement for water supply sources; reduce direct discharge to the drainage system and prevent urban flooding [10, 18],

   (ii) Green roofs – have numerous benefits, including: reduction of runoff peaks and volumes, resulting in lower urban flood risks, insulation of heat transfer, resulting in lower cost for air conditioning, and reduction of the heat island effect [11, 29], absorption

TABLE 7.1   Comparison Between Conventional and SUDS Approach [16]

Aspects	Conventional Measures	SUDS
Quantity	Storm water is a nuisance and should be removed from the urban substantially and efficiently	Storm water is attenuated and partially utilized using appropriate measures
Quality	Storm water will be collected with wastewater and be treated together	Storm water is collected at the sources and treated locally
Handling method	Hard engineering methods	Soft, nature-mimicking methods
Utilization	Not concerned	Rainfall could be harvested to support domestic water supply, irrigation
Recreational values	Not available	Usually constructed in harmony with local cityscapes

Source: [16]

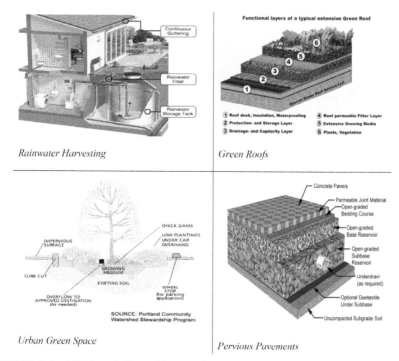

FIGURE 7.1 SUDS techniques [Source: http://www.bluegranola.com; http://landscape online.com; Portland Community].

of air pollutants [30], provision of wildlife habitat for birds and general enhancement of environment for the area [3, 12, 22],

(iii) Urban green space provides improved resiliency in runoff management and multiple other ecosystem services [26],

(iv) Pervious pavements – a technology that both enhances infiltration and improves surface runoff quality [2, 23].

7.2 METHODOLOGY

7.2.1 STUDY AREA

The NhieuLoc – ThiNghe (NL – TN) basin (Figure 7.1) is located in the central part of HCMC and occupies an area of approximately 33 km², and stretching across 7 city districts (1, 3, 10, PhuNhuan, Tan Binh, Go Vap and BinhThanh). The population of the Basin is about 1.2 million people (20% of the total Ho Chi Minh City population), representing a population density of 290 people per hectare. Land use is mixed, with 49.3% being residential and the remaining representing commercial, public and industrial uses. Elevation within the Basin is variable, with the north and northwest sections being up to 8 m above sea level, while the southern part of the Basin averages only 1.3 m above sea level.

In the Vietnamese content, the implementation of SUDS could not have been successful with the support of local communities. Therefore, social based assessment is the major part of this study. Additionally, a numerical hydrologic and hydraulic model also was developed to assess the water management efficiency from the technical sense.

7.2.2 INTERVIEWS

In order to obtain the preference of local people for SUDS solutions, face-to-face interviews were conducted using questionnaires. The locations for the interviews were chosen based on the annual inundation reports from Ho Chi Minh City Flood Control Centre (FCC). The purpose of this survey was to evaluate the preference level of the community for different kinds of SUDS alternatives in terms of esthetics. After completing the survey, hypothesis-testing techniques were used to analyze and consolidate the results. The survey was done based on systematic sampling method. Because of the difficulties in collecting response through emails or phone calls, direct face-to-face interviews were done over 6 weeks in May to June 2013. Questionnaires were categorized by high, middle and low-income groups. The classification was done by evaluating interviewees' household conditions. In order to quantify the preference level of the interviews, a Likert scale method was used with 5 level of likeness, where 1 represents the least favored and 5 represents the favorite [15].

7.2.3 HYPOTHESIS TESTING

In order to make stronger conclusions from the survey results, hypotheses testing techniques were employed. More specifically, z tests for mean and proportion are used to support or reject all the claims that could be derived from the survey primary analysis. Showing below are the formula employed for such a test:

$$z = \frac{\overline{X} - \mu}{\sigma / \sqrt{n}} \tag{1}$$

7.2.4 NUMERICAL MODELING

The drainage system of the NL-TN Basin was simulated using PCSWMM which was adapted from an earlier SWMM model that was developed by Camp, Dresser and MacKee [6], but updated to more accurately represent surface slope using DEMs in ArcGIS; represent current land use characteristics and percent imperviousness; and consider new bathymetric data for the NL-TN Canal. In total, 228 sub-catchments, 333 conduits and 228 junctions were included in the model (Figure 7.3).

7.2.4.1 Model Calibration

The locations of meteorological and hydrological stations (Tan Son Hoa and Phu An, respectively), from which the boundary conditions of the model were collected (Figure 7.2). Because sewer flow is not routinely monitored in Vietnam, an alternative approach was used to calibrate the model. A 90-mm design storm (Figure 7.4) was used as input and model results for the water level along the NL-TN Canal under a similar rainfall event were compared to observed levels [15].

7.2.4.2 SUDS Evaluation

Simulation of the SUDS technologies was done using the LID editor in PCSWMM that explicitly represents the structure and hydrology of these technologies [7, 20, 24, 28]. For each of the Rain Harvesting, Green Roof, Rain Garden, and Pervious Pavement, information required for input to the model included surface storage depth, roughness, and slope, vegetation

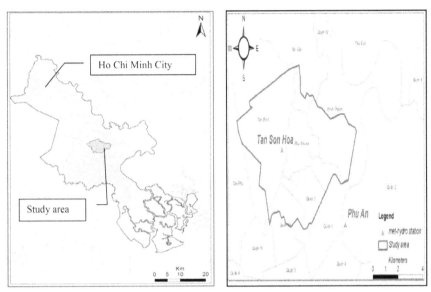

FIGURE 7.2   NhieuLoc – ThiNghe basin.

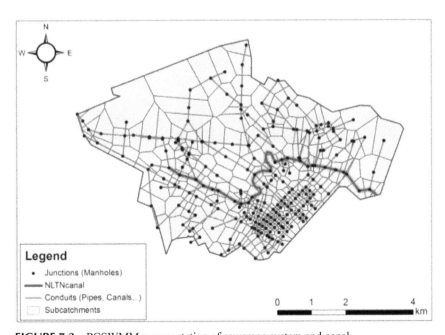

FIGURE 7.3   PCSWMM representation of sewerage system and canal.

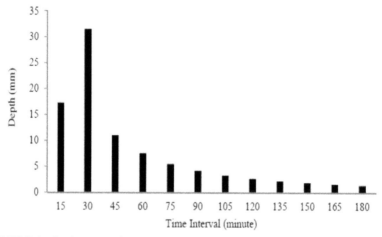

FIGURE 7.4   Design storm for PCSWMM model.

coverage, thickness and hydraulic conductivity of the underlying sub-strates, and underdrain characteristics. Details about these elements are discussed in Ref. [27]. The efficacy of SUDS was evaluated based on two main criteria; flood attenuation capacity and pollutant removal rates. The study has in total more than 200 sub-catchments, as discussed above. Within the scope of this study, three representative sub-catchments were chosen to simulate the effect of SUDS. The selected sub-catchments are representative and distinguished in land use patterns, area and were not located close to the watercourse (to avoid the confounding factor of flood-ing due to tides). The model was then simulated with 5-year return period storm [14].

7.3   RESULTS

7.3.1   INTERVIEWS

Figure 7.5 shows the Likert scale scoring for SUDS technologies.

There were no SUDS that were predominantly chosen. The highest value was 3.72 (high income preference for urban green space). The pri-mary analysis results also show that previous pavements are the favorite for almost everyone, followed by urban green space and rainwater harvesting.

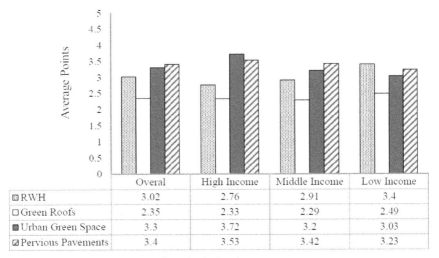

	Overal	High Income	Middle Income	Low Income
⊞ RWH	3.02	2.76	2.91	3.4
☐ Green Roofs	2.35	2.33	2.29	2.49
▣ Urban Green Space	3.3	3.72	3.2	3.03
▨ Pervious Pavements	3.4	3.53	3.42	3.23

FIGURE 7.5 Average SUDS Likert Scale Scoring.

Green roofs consistently had the lowest scores. However, the results seem to vary greatly with different groups of socio-economic groups and therefore hypothesis tests were conducted.

7.3.1.1 Testing Assumptions

The z test for mean was conducted using the values taken from Figure 7.5. Simultaneously, the proportions test to further support the confidence of any conclusions. For these tests, some additional assumptions were adopted:

(1) A transformation scheme was adopted to clearly classify the people's preference into three distinctive layers in which, smaller than 3 denote *Reject,* greater than 3 denotes *Accept* and 3 denote *Neutral* (Figure 7.6).
(2) The hypothesized mean is chosen at 33.33%.

Applying two assumptions above, the approval percentage for each kind of SUDS were calculated and are shown in Figure 7.7.

Interviewees, in general, showed substantial support for urban green space and pervious pavements. However, significant differences could be

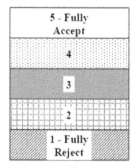

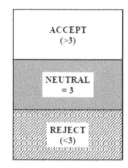

FIGURE 7.6   Proportion test assumption.

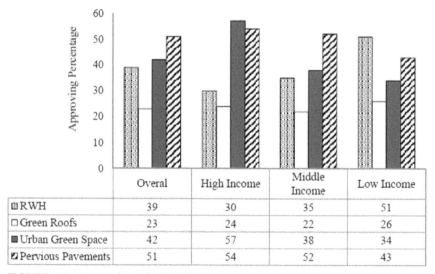

	Overal	High Income	Middle Income	Low Income
⊞ RWH	39	30	35	51
▢ Green Roofs	23	24	22	26
■ Urban Green Space	42	57	38	34
▨ Pervious Pavements	51	54	52	43

FIGURE 7.7   Approval rate for SUDS.

seen from different groups. More specifically, Low income people showed more support for Rain Water Harvesting whereas for different socio-economic groups, this type of SUDS was only preferred to Green Roofs, which is the least favored one.

7.3.1.2   Results for Hypothesis Testing

The lowest accepted confidence level was 60%. Anything smaller than that will be denoted as Neutral. And the confidence level would then be labeled as Not Applicable (N/A). After completing all the needed tests, the preference level of different group for Rainwater Harvesting, Green Roofs, Urban Green Space and Pervious Pavements are generated with the respective confidence level and the results are displayed in Table 7.2.

The results show that people prefer methods applied in communal areas, i.e., pervious pavements and urban green space to those installed inside their houses, i.e., Rainwater Harvesting and Green Roofs. Pervious Pavements was the most favored technique. The confidence level of approval for this SUDS are between 80 and 99%. Urban Green Space had the second approval rating on the list. High and middle-income people show support for green space development with a confidence level ranging from 75 to 99%. The low-income group, however, gave a Neutral response about the application of this type of SUDS. The third position was RWH. This is one of the most ubiquitous SUDS solutions in developed countries. However, it has never been so in developing mega cities such as Ho Chi Minh City or Bangkok. Middle and high-income groups did not favor this SUDS solution with the confidence level to disapprove ranging from 60 to 85, respectively. Interestingly low-income residents showed support for RWH with 80 to 95% confidence level. This type of technique appears to be too strange to be accepted by the majority of people. When asked, people have shown substantial doubts about the applicability of a vegetation coverage on their roofs, resulting in consistent disapproval response with 85–99% confidence level [19].

7.3.2 PCSWMM MODEL

7.3.2.1 Model Calibration and Validation

Details on model validation and calibration were presented in [14].

7.3.2.2 SUDS Evaluation

The model simulation results are shown in Table 7.3 below.

Green roof technology, in general, was the best performer followed by urban green space, pervious pavement and rainwater harvesting when flood attenuation capacity was considered as the assessment criteria.

TABLE 7.2 Interviews Analysis Results

Mean test results

Groups					Alternatives						
	Rain water harvesting		Green roofs		Urban green space		Pervious pavements	
	Response	Confidence	Response	Confidence	Response	Confidence	Response	Confidence
Overall	Neutral	N/A	Disapprove	95%	Approve	90%	Approve	99%
High	Disapprove	75%	Disapprove	95%	Approve	99%	Approve	99%
Middle	Disapprove	70%	Disapprove	99%	Approve	90%	Approve	99%
Low	Approve	85%	Disapprove	95%	Neutral	N/A	Approve	80%

Proportion test results

Groups					Alternatives						
	Rain water harvesting		Green roofs		Urban green space		Pervious pavements	
	Response	Confidence	Response	Confidence	Response	Confidence	Response	Confidence
Overall	Neutral	N/A	Disapprove	99%	Approve	95%	Approve	99%
High	Disapprove	85%	Disapprove	99%	Approve	99%	Approve	99%
Middle	Disapprove	60%	Disapprove	95%	Approve	75%	Approve	99%
Low	Approve	95%	Disapprove	85%	Neutral	N/A	Approve	85%

TABLE 7.3 PCSWMM Model Simulation Results

Parameters	Sub-catchment			Mean
	1	**2**	**3**	
Current situation				
Peak Runoff (m³/s)	29.11	4.17	10.45	14.58
TSS (kg)	573	71	177	273.67
Rain water harvesting system				
Peak Runoff (m³/s)	27.26	3.91	9.8	13.66
F (%)	6.36	6.21	6.23	6.27
TSS (kg)	557	68	171	265.33
TE (%)	3.06	2.81	3.38	3.06
Green roofs				
Peak Runoff (m³/s)	24.29	3.18	7.74	11.74
F* (%)	16.54	23.7	25.94	22.06
TSS (kg)	478	57	141	225
TE** (%)	16.58	19.72	20.9	19.07
Urban green space				
Peak Runoff (m³/s)	14.8	3.26	9.09	11.59
F (%)	21.32	3.7	9.76	14.86
TSS (kg)	355	60	167	192.33
TE (%)	38.91	15.49	5.65	20.02
Pervious pavement				
Peak Runoff (m³/s)	26.81	2.62	9.79	13.07
F (%)	7.89	37.15	6.57	17.13
TSS (kg)	552	38.3	167	244
TE (%)	3.66	46.06	5.65	16.20

The ranking for pollutant removal capacity was quite similar, although urban green space rated higher in trap efficiency than green roofs in this assessment category.

Despite an increasing body of research related to SUDS technologies, Marsalek et al. [19] observed that implementation at the municipal level has been limited. Most of the research that has been done for SUDS technologies focuses on temperate climates [24], although Silveira [25] and

Goldenfum et al. [13] reviewed the challenges to implementing sustainable urban drainage, particularly in developing countries having a tropical climate. There is some concern that SUDS technology will be less effective for larger storm events (or greater rainfall as experienced in tropical climates), however, the experience in Singapore [17] seems to counter this concern and similarly, an unpublished report by Drexel University for a test area in Cambria Heights, New York City, showed that SUDS performed exceedingly well in controlling runoff from Super Storm Sandy and Hurricane Irene.

7.4 DISCUSSION

The results from the social survey (aggregated over all socio-economic classes) and the numerical model, based on flood reduction criteria, are compared and shown in Table 7.4.

In general, there are considerably great differences between the results from Social Survey and PCSWMM. There is only one agreement of the two, which is the 2nd rank for Urban Green Space. Green Roofs, despite being the least favored by the community, was shown as be the best flood attenuation technology. On the contrary, Pervious Pavements was only the 3rd in the model performance, nevertheless, was the favorite SUDS. Rainwater harvesting although being the most popular SUDS in the literature review, was neither preferred by the community nor able to show competitiveness in the model results.

The different results in this chapter suggested that there might be considerable gaps between engineers, decision makers and public awareness for the same problem. Conventionally, this issue would be given very little

TABLE 7.4 Comparison of Results

SUDS ranking	Social survey	PCSWMM
Rain water harvesting	3	4
Green roofs	4	1
Urban green space	2	2
Pervious pavements	1	3

concern or even ignored because hard engineering projects are decided and executed solely by local authorities. This might not be a successful approach for SUDS implementation for which the success may be dependent on public awareness and support. SUDS require not only initial investment but routine maintenance as well, i.e., periodically cleaning for rainwater tank, or irrigation treatment for vegetation covers of green roofs, urban green space, etc. These tasks should be best performed by local communities. This target would only be achieved only if comprehensive understanding and full support of the respective communities were achieved.

7.5 CONCLUSIONS

Authors applied social survey with direct interviews and numerical modeling approaches, and each approach presented a different viewpoint for SUDS applicability, local residences' and engineers', respectively. In other words, what engineers believe to be the best solution might not be accepted by local people. This gap typically has not been given enough concern. Therefore, it is essential to bring these measurements together onto the same platform to come up with the ultimate solution. The main challenge of this task is to integrate the qualitative measurement of the survey with the quantitative results of the numerical model results.

7.6 SUMMARY

Through the a combination of social survey and numerical modeling, the objective of this study was to evaluate the applicability of SUDS in Ho Chi Minh City as adaptive measures for inundation mitigation as well as urban environment improvement. Four types of SUDS are considered: Rainwater Harvesting, Green Roofs, Urban Green Space and Porous Pavements. The survey revealed that different opinions regarding the desirability of various SUDS options. More specifically, High and Middle Income groups preferred pervious pavements to other SUDS whereas Low Income group prefer Rainwater Harvesting Systems. Additionally, a numerical model using PCSWMM was adapted to compare flood reduction and pollutant trap efficiencies of each types of SUDS from technical point of view. Interestingly, the model simulation results gave different assessment for SUDS. More specifically, Green Roofs and Urban Green Space are the best technologies followed by Pervious Pavements and Rain Water Harvesting.

KEYWORDS

- low impact development (LID)
- sustainable urban development
- sustainable urban drainage systems (SUDS)
- Thailand
- water utilization

REFERENCES

1. Berndtsson, J. C., 2010. Green roof performance towards management of runoff water quantity and quality: A review. *Ecological Engineering*, 36:351–360.
2. Brattebo, B. O. and Booth, D. B., 2003. Long-term storm water quantity and quality performance of permeable pavement systems. *Water Research*, 37:4396–4376.
3. Brenneisen, S., 2003. The benefits of biodiversity from green roofs key design consequences. Conference Proceedings. *Greening Rooftops for Sustainable Communities*, Chicago, USA.
4. Burns, D., Vivtar, T., McDonnel, J., Hasset, J., Duncan, J., and Kendall, C., 2005. Effects on suburban development on runoff generation in the Croton River basin, New York, USA. *Journal of Hydrology*, 311:266–281.
5. Burns, M. J., Fletcher, T. D., Walsh, C. J., Ladson, A. R., and Hatt, B. E., 2003. Hydrologic shortcomings of conventional urban storm water management and opportunities to reform. *Landscape and Urban Planning*, 105:230–240.
6. Camp Dresser and MacKee International (CDM), 1999. Feasibility Study and Preliminary Design Project for NhieuLoc – ThiNghe Basin. *Project Final Report*.
7. Chaosakul, Kootatep and Irvine, 2013. Low Impact Development Modeling to assess localized flood reduction in Thailand. Conference Proceedings: *2012 Storm water and Urban Water System Modeling*. Toronto, Canada. ISBN: 978-0-9808853-8-5.
8. Cuo, L., Lettenmaier, D. P., Alberti, M. and Ricey, J. E., 2009. Effects of a century of land cover and climate change on the hydrology of the Puget Sound basin. *Hydrological Processes*, 230:903–933.
9. Degasperi, C. L., Berge, H. B., Whiting, K. R., Berkey, J. L., Cassin, J. L., and Fuerstenberg, R., 2009. Linking hydrologic alteration to biological impairment in

urbanizing streams of the Puget Lowland, Washington, USA. *Journal of America Water Resources Association*, 45:512–533.

10. Eroksuz, E. and Rahman, A., 2010. Rainwater tanks in multi-unit buildings: A case study for three Australian cities. *Resources Conservation Recycling*, 54:1449–1452.

11. Fang, C. F., 2008. Evaluating the thermal reduction effect of plant layers on rooftops. *Energy Build*, 40:1048–1052.

12. Gedge, D. and Kadas, G., 2005. Green roofs and biodiversity. *Biologist*, 3:161–169.

13. Goldenfum, J. A., Tassi, R., Meller, A., Allasia, D. G., and Silveira A. L., 2007. Challenges for the sustainable urban storm water management in developing countries: From basic education to technical and institutional issues. Conference proceedings: *6th International Conference on Sustainable Techniques and Strategies in Urban Water Management*, Novatech 2007, Lyon.

14. Ho, H. L., Babel, M. S., Sutat, W., Irvine, K. N., and Duyen, P. M., 2015. Exploratory assessment of SUDS feasibility in NhieuLocThiNghe basin, Ho Chi Minh City, Vietnam. *British Journal of Environment and Climate Change*, 5(2):91–103.

15. Ho, H. L., Das Gupta, A., Duyen, P. M., and Rajbhandary, J., 2014. Social aspects of the application of SUDS for the case of Nhieu LocThi Nghe basin, Ho Chi Minh City, Viet Nam. Conference proceedings: *International Conference on Green Technology and Sustainable Development*. Ho Chi Minh City, Vietnam. ISBN: 978-604-73-2817-8.

16. Hoyer, J., Dichaut, W., Kronawitter, L., and Weber, B., 2011. Water Sensitive Urban Design, principles and inspiration for sustainable storm water management in the City of the Future. *SWITCH project*. From http://www.switchurban.eu.

17. Irvine, K. N., Chua, L. H. C., and Eikass, H. C., 2014. The Four National Taps of Singapore: A holistic approach to water resources management from drainage to drinking water. *Journal of Water Management Modeling*. DOI: 10.14796/JWMM.C375.

18. Kim, R. H., Lee, S., Kim, Y. M., Lee, J. H., Kim, S. K. and Kim, J. G., 2005. Pollutants in rainwater runoff in Korea: their impacts on rain water utilization. *Environment Technology*, 26:411–420.

19. Marsalek, J. and Schreier, H., 2009. Innovation in storm water management in Canada: The way forward. *Water Quality Research Journal of Canada*, 44(1):5–10.

20. McChutcheon and Wride, 2013. Shades of Green: Using SWMM LID Controls to Simulate Green Infrastructure. Conference Proceedings: *2012 Storm water and Urban Water System Modeling*. Toronto, Canada. ISBN: 978-0-9808853-8-5.

21. Mentens, J, Raes, D. and Hermy M., 2006. Greenroofs as a tool for solving the rainwater runoff problem in the urbanized 21st century. *Landscape Urban Plan*, 77:217–226.

22. Philip, R., Anton, B. and Van der Steen, P., 2011. SWITCH Training Kit: Integrated Urban Water Management in the City of the Future. *SWITCH project*. http://www.switchurban.eu.

23. Pratt, C. J., Newman, A. P., and Bond, P. C., 1999. Mineral oil degradation within a permeable pavement: Long term observations. *Water Science and Technology*, 39:109–130.

24. Rossman L. A., 2010. Modeling low impact development alternatives with SWMM. In *Dynamic Modeling of Urban Water Systems, Monograph* 18, by James, W., Irvine, K. N., Li J. Y., McBean, E. A., Pitt, R. E., Wright, S. J., eds., Ontario.

25. Silveira, A. L., 2001. Problems of urban drainage in developing countries. Conference proceedings: *International Conference on Innovative Technologies in Urban Storm Drainage*, Novatech, Lyon.
26. Tzoulas, K., Korpela, K., Venn, S., Yli-Pelkonen, V., Kazmierczak, A., Niemela J., and James, P., 2007. Promoting ecosystem and human health in urban areas using Green Infrastructure: A literature Review. *Landscape and Urban Planning*, 81:167–178.
27. US–EPA, 2010. Storm Water Management Model: User's Manual. US-EPA 600/R-05/040. Cincinnati, OH, USA.
28. VanWoert, N. D., Rowe, D. B., Andresen, J. A., Rugh, C. L., Fernandes, R. T., and Xiao, L., 2005. Green roofs storm water retention: Effects of roof surface, slope and media depth. *Journal of Environmental Quality*, 34:1036–1044.
29. Wong, N. H., Chen, Y., Ong, C. L., and Sia, A., 2003. Investigation of thermal benefits of roof top garden in the topical environment. *Building Environment*, 38:261–270.
30. Yang, J., Yu, Q., and Gong, P., 2008. Quantifying air pollution removal by Green roofs in Chicago. *Atmos. Environment*, 42:7266–7273.

CHAPTER 8

CLIMATE CHANGE IMPACTS ON PLANNING AND MANAGEMENT OF WATER RESOURCES

G. B. SAHOO[1] and S. G. SCHLADOW[2]

[1]Department of Civil and Environmental Engineering and Tahoe Environmental Research Center, University of California, One Shields Avenue, Davis, CA 95616, USA, E-mail: gbsahoo@ucdavis. edu, Phone: (530) 752-1755, Fax: (530) 752-7872

[2]Department of Civil and Environmental Engineering, University of California, One Shields Avenue, Davis, CA 95616, USA

CONTENTS

8.1   Introduction 188
8.2   Study Site 191
8.3   Mathematical Models 193
    8.3.1   Lake Hydrologic Budget 193
    8.3.2   Lake Heat Budget 194
8.4   Input Data to DLM-WQ 194
    8.4.1   Meteorological Data Input 194
    8.4.2   Flows and Pollutant Loadings 195
    8.4.3   Lake Data 196
8.5   Methodology 197
    8.5.1   Development of Precipitation
            Contour Lines on Lake Tahoe 197

8.5.2 The DLM-WQ Model 198

8.5.3 Calibration and Validation of the Dlm-Wq Model.......... 198

8.6 Results and Discussions... 200

8.6.1 Lake Water Temperature 200

8.6.2 Lake Surface Water Level 201

8.6.3 Water Balance.. 202

8.6.4 Effects of Continued Warming.. 205

8.7 Conclusions.. 206

8.8 Summary... 208

Acknowledgements.. 209

Keywords 210

References...211

8.1 INTRODUCTION

Lakes and reservoirs are important freshwater resources for human use, e.g., flood control, water supplies to agriculture and municipalities, fish and wildlife management, endangered species, recreation, hydropower generation, transportation, low-flow augmentation and in-stream users. These systems are also valuable as they host unique and magnificent aquatic habitats. Although water demands from stakeholders always exceed supplies, the conflicts over water use grow during drafts and summer due to increasing competition between beneficiaries. Water conflicts become acute during droughts and managers responsible for lake operations need a detailed water budget to explore potential lake release policies. Detailed estimates of hydrologic budgets are essential for the pursuit of sustainable water use, development of lake ecosystem management schemes, and future water use plans under uncertainty related to climate change [19]. For example, the Upper Klamath Basin, USA gained national attention during the 2001 crop-growing season, when irrigation water from approximately 800 farms and ranches were diverted to aid in the survival of three endangered fish species and aquatic habitat. In response to the 2001 events, less restrictive endangered species act (ESA) releases were applied during the summer of 2002. As a consequence, over 33,000 adult salmon died in the Klamath River due to low DO, high water temperature, and parasite blooms [3]. The parasite blooms and low DO were the direct result of high stream water temperature due to low flows in the Klamath River [5]. This elucidates the lack of information on accurate estimation of available water resources and sustainable planning.

Ensuring sustainable water allocation to all stake-holders requires an understanding of the hydrologic budget. Lake water budgets provide quantitative estimates of hydrologic cycles and allow prediction of the effects of natural- or human-induced changes in the hydrologic cycle and lake supplies to users. It is hard to estimate each hydrologic component of a large lake accurately because temporal and spatial meteorology (e.g., wind and precipitation) vary significantly over lake surface area. Evaporation is an important component of the hydrologic and heat budget. However, it is difficult to take evaporation measurements spatially and continuously over the water surface of a large and dynamic lake. In most cases, mathematical models are applied to quantify evaporation [1, 14–21]. Lake water temperature is a prerequisite for estimation of evaporative losses. The change in lake water temperature depends on the heat budget, lake dynamics, and water qualities that determine light extinction (i.e., retention of shortwave radiation) along vertical depth. Thus, a complex mathematical model that includes thermodynamic, hydrodynamic, and water quality sub-modules is required for simulation of lake dynamics. The UC Davis-developed Dynamic Lake Model with Water Quality (DLM-WQ) (Figure 8.1) demonstrates the ability to simulate dynamics and physical process of large lakes with adequate accuracy [17, 18, 22, 25].

Lake operating rules are prepared based on a lake's surface elevation and downstream demands (i.e., hydrologic budget). Thus, robust and accurate estimates of the heat and hydrologic budgets are essential. In addition, the heat transfer coefficients in the heat budget model reported in the literature vary greatly depending on site. Blanc [2] compared ten different bulk methods and found the different formulations for the exchange coefficients to cause variations of 40–120% in estimates of the heat, moisture and momentum fluxes. Figure 8.2 illustrates the processes influencing storage of heat and water, fluxes to and from the water surface, and the resulting issues in estimating and predicting evaporation rates. Due to the natural feedback mechanisms of heat flux across an air-water interface, an accurate

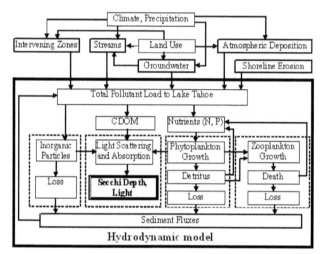

FIGURE 8.1 Schematic of lake clarity model (DLM-WQ). Note: The double lines box includes all in-lake processes and a hydrodynamic model. The four broken line boxes from left to right represent for fine inorganic particle, light, phytoplankton and zooplankton sub-model, respectively. Shown are external sources [streams, intervening zones, atmosphere, groundwater, and shoreline erosion (thick line boxes on the top of double line box)] and internal source [sediment fluxes (thick line box inside the double line box)] of the pollutant loads. CDOM represents colored dissolved organic matters. N and P represent for nitrogen and phosphorus.

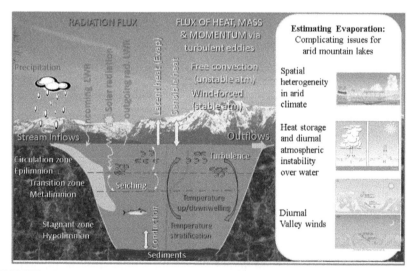

FIGURE 8.2 Processes and patterns of storage of heat and water and fluxes of heat, water and momentum.

estimate of water surface temperature is possible. On the other hand, mass fluxes in the hydrologic budget (e.g., evaporation and precipitation which vary spatially over a lake) have no such feedback and measurement errors may continue to grow in an unbounded fashion unless a water balance can be closed by adjusting fluxes against accurate measurements of lake level (water storage). Thus, the objectives of this study were to:

1. Develop a methodology to estimate the heat and hydrologic budget of the Lake Tahoe, CA – NV, USA using DLM-WQ model,
2. Analyze the estimated heat and hydrologic budget components over an 15-year period from 1994 to 2008,
3. Estimate the lake water level under climate change scenarios, and
4. Identify different lake water management scenarios.

To demonstrate the validity of the model, estimated water temperatures and lake water elevations were compared to event-based measured records and prediction performance of the model were estimated. The calibrated DLM-WQ model was applied to examine lake water level under global warming scenarios.

8.2   STUDY SITE

Lake Tahoe, CA-NV, USA has a maximum depth of 501 m, a mean depth of 305 m and the ratio of watershed (813 km^2) to lake surface (495 km^2) area is low at 1.64 (Figure 8.3). Lake water level is controlled by a dam. The 5.5 m high dam can increase the lake's capacity by approximately 0.9185 km^3. Releases are controlled based on flood protection for downstream areas, water supply for Reno/Sparks and Carson City, irrigation water for Truckee-Carson Irrigation District (TCID) and Newlands Irrigation District, hydropower, and environmental water for Pyramid Lake, Stillwater Wildlife Refuge, and instream flows [7]. When the lake level falls below the natural rim, there is no flow to the Truckee River for water supply, irrigation, or fishing. Therefore, it is important to develop a detailed methodology to critically correctly estimate hydrologic components of the lake for the purpose of water management, including preparedness for changes in climatic conditions.

The Lahontan Regional Water Quality Control Board (Lahontan) and Nevada Division of Environmental Protection (NDEP) TMDL technical

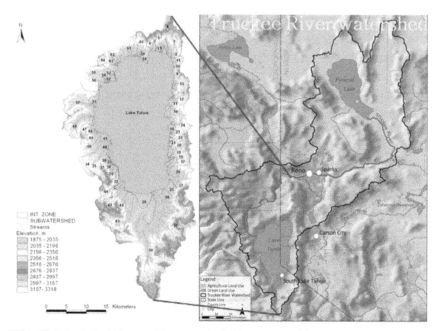

FIGURE 8.3   Lake Tahoe and its watershed. The stream without a map ID represents Truckee River, the only outflow to Lake Tahoe.

report [12] shows that precipitation on Lake Tahoe varies significantly from the shore to the middle of the lake, with a large rain-shadow between the wetter California side and the drier Nevada side (Figure 8.3). High-resolution spatial (e.g., 500 m) and temporal (e.g., hourly) detailed meteorological records over the lake are not available. Although meteorological data recorded at the shore is often considered representative for the entire lake for heat and hydrologic estimates this introduces errors that must be adjusted to develop accurate lake heat and hydrologic budgets.

Lake Tahoe never freezes although the night-time air temperature during winter remains below zero degree centigrade. Its minimum temperature during extreme winter climate remains approximately 5°C because of (1) the Lake's constant movement and (2) large volume of water (151 km^3) that has enough heat content to maintain the lake temperature at approximately 5°C. Historical, in situ lake water temperature records from 1970 to 2006 shows that the lake has warmed at 0.015°C per year and the lakes summer density stratification is becoming more stable [6, 23]. Sahoo et al. [20] reported from the one-year running average of the daily meteorological data from the downscaling exercise of the Geophysical Fluid Dynamics Laboratory (GFDL) Global Circulation Model output, over the 21st Century, along with the best fit trend lines that shortwave radiation remains largely unchanged. Air temperature and longwave radiation are projected to increase approximately 2–4.5°C, and 5–10%, respectively. The wind speed is projected a decline trend of 7–10%. Continued climate change at Lake Tahoe is expected to significantly alter the hydrologic and heat budget, which, in turn, is likely to affect water resource planning.

8.3 MATHEMATICAL MODELS

The heat and hydrologic budgets require all components be estimated with sufficient accuracy. The calculation of heat and hydrologic budgets allows us to evaluate the performances of the selected bulk formulae for estimation of evaporation since an accurate hydrologic budget must be balanced.

8.3.1 LAKE HYDROLOGIC BUDGET

Water level is directly related to the water budget and was estimated based on the following equation:

$$WL_t = WL_{t-1} + S_t + GW_t + R_t - E_t - O_t \qquad (1)$$

where, WL_t and WL_{t-1} are the water level at current and previous time step t and t–1, respectively; S_t and GW_t are the stream and groundwater inflow between time steps t–1 and t, respectively expressed as an equivalent height of water at the surface; R_t, E_t and O_t are the direct precipitation on the lake, the evaporation and the outflow from the lake, respectively, between time steps t–1 and t, expressed as an equivalent height of water at the surface.

8.3.2 LAKE HEAT BUDGET

The heat budget (kJ m^{-2} hr^{-1}) equation is given by:

$$
\left(1 - R_{SW}\right)SW_I + \left(1 - R_{LW}\right)LW_I - LW_O - H_E \pm H_S
$$
$$
- \Delta S + H_{Qinf} - H_{Qout} = 0 \qquad (2)
$$

where, SW_I, LW_I and LW_O are the incident shortwave radiation, the incident longwave radiation, and the outgoing longwave radiation from the water surface, respectively; H_E and H_S are the evaporative (i.e., latent) heat loss and the sensible heat, respectively; ΔS is the stored heat and is observed in the change in lake water temperature; R_{SW} and R_{LW} are albedo from water surface for shortwave and longwave radiation, respectively; and H_{Qinf} and H_{Qout} are heat of stream inflow [24] and lake outflow, respectively.

For deep lakes like Tahoe, we neglect the heat fluxes resulting from heat exchanges across the sediments at the bottom of the water body and from biological processes.

8.4 INPUT DATA TO DLM-WQ

The input data for DLM-WQ includes daily meteorological data, daily stream inflow and associated water quality data, daily lake outflow, atmospheric deposition, groundwater inflows, shoreline erosion pollutant loading. The DLM-WQ requires lake hypsographic data (i.e., depth to area and volume relation), lake physical data (boundary conditions, initial conditions), physical model parameters, and water quality model parameters. The DLM-WQ physical parameters and water quality parameters are described by Sahoo et al. [25].

8.4.1 METEOROLOGICAL DATA INPUT

Daily meteorological data include solar short wave radiation (kJ m^{-2}d^{-1}), incoming longwave radiation (kJ m^{-2}d^{-1}), vapor pressure (mbar) or relative humidity (%), wind speed (m s^{-1} at 10 m above the ground surface) and precipitation (mm). Data from 1994 and 2009 were collected at the meteorological station near Tahoe City SNOTEL (SNOwpack TELemetry) maintained by the United States Natural Resources Conservation Services (US – NRCS). Because of the large altitudinal difference between the SNOTEL sites and the lake, combined with the boundary layer difference between these forest based sites and the lake surface, other meteorological stations were not considered to be appropriate for use in the lake modeling. Because the high temporal resolution (hourly) detailed weather records since 1989 are available at the Tahoe City meteorological station, weather records of this station were considered to be representative of the lake. Shortwave radiation, wind speed, air temperature, dew point temperature, and precipitation are directly measured data, whereas longwave radiation and vapor pressures are estimated values using the aerodynamic formulae described in TVA [28].

DLM-WQ is a one-dimensional model, so it needs weather records of one meteorological station. DLM-WQ utilizes a daily averaged wind speed (direction data are not utilized because of the one-dimensional assumption) located on a dock approximately 100 m from the shore. Other meteorological parameters (Shortwave radiation, longwave radiation, air temperature and relative humidity) can be reasonable assumed to be uniform over the lake surface. The one parameter that is known to display significant spatial variability is precipitation, and hence we have adjusted precipitation distribution over the lake on the basis of the isohyetal lines to provide a more realistic input. The isohyetal map of Lake Tahoe used for the TMDL [12, 26] shows that precipitation over the lake varies nearly 50% from the shore to the mid-lake.

8.4.2 FLOWS AND POLLUTANT LOADINGS

Water flow into Lake Tahoe includes channelized runoff from 54 streams, runoff from 10 intervening zones (direct discharge to the lake without a channel), precipitation directly on the lake surface, and groundwater flux. Most of the urban zones drain into intervening zones. The stream without a map ID in Figure 8.1 represents the Truckee River, the only outflow from Lake Tahoe. The USGS records are available for the daily outflow from the lake and inflows of 10 Lake Tahoe Interagency Monitoring Program (LTIMP) streams [4]. These 10 LTIMP tributaries are estimated to account for up to 40% of the total stream input. Because measured data for all 64 streams are not available, we used the LSPC watershed model to estimate streamflows. The LSPC is a U.S. Environmental Protection Agency-supported watershed modeling system (http://www.epa.gov/athens/wwqtsc/html/lspc.html) that includes Hydrologic Simulation Program – FORTRAN (HSPF) algorithms for simulating watershed hydrology and general water quality processes on land as well as stream transport model.

Using weather data from nine SNOTEL meteorological stations and National Climatic Data Center (NCDC) around the lake (http://www.wcc.nrcs.usda.gov/snow/), Sahoo et al. [22] generated the streamflow and associated pollutant loads for the period 1994 to 2009 using the LSPC watershed model. Except during low precipitations years (2001, 2007 and 2008), LSPC estimated flows are within ±8% of USGS measurements. LSPC estimated daily cumulative flows and USGS recorded daily cumulative flows overlap each other. Thus, estimated low flows on a few occasions do not have significant impact on the long-term water balance.

The available LSPC estimated stream inputs and directly measured weather data are distributed over time. However, groundwater load and shoreline erosion data are the same for all the simulated years because of the lack of adequate, long-term loading data from these two sources.

8.4.3   LAKE DATA

Lake data are required to provide initial conditions for the DLM-WQ model runs. UC Davis Tahoe Environmental Research Center (TERC) collects vertical profiles of temperature and concentration of chlorophyll-α, DO, biological oxygen demand (BOD), soluble reactive phosphorous (SRP), particulate organic phosphorus (POP), dissolved organic phosphorus (DOP), nitrate (NO_3) and nitrite (NO_2), ammonium (NH_4), particulate organic nitrogen (PON), dissolved organic nitrogen (DON), and seven classes of inorganic fine particles (0.5–1.0, 1.0–2.0, 2.0–4.0, 4.0–8.0, 8.0–16.0, 16.0–32.0, and 32.0–63.0 μm) at two lake stations (index station with 160 m depth and mid-lake station with 460 m depth). Lake fine particle data are available for the period 1999–2002 and 2006–2009 [11]. To run the DLM-WQ model starting from January 1994, we prepared a fine particle data-set for that month that is very close to average particle concentration for the month of January of other years. The other state variables are measured data. Data from the mid-lake station in the deeper part of the lake (460 m depth) were used to provide the initial conditions. The lake profile data recorded on January, 1994 was used for the initial condition.

The DLM-WQ requires lake hypsographic information. We used the surface area and cumulative volume as functions of elevation of the lake data reported in Gardner et al. [9]. We estimated the elevation of each stream before it enters the lake from GIS DEM and used along with stream and lake water temperature to estimate the plunging depth of the stream discharge. The elevation of a spillway constructed at the lake outlet is approximately 1899 m National Geodetic Vertical Datum (NGVD). Lake discharges water above 1899 m the Truckee River. Bottom elevation of the lake is approximately 1400 m NGVD.

8.5 METHODOLOGY

Temporal and spatial weather records over the lake are not available. However, since Lake Tahoe is located in a west-east rain shadow, a high-resolution precipitation contour map is required for an accurate hydrologic budget.

8.5.1 DEVELOPMENT OF PRECIPITATION CONTOUR LINES ON LAKE TAHOE

High-quality 0.041667° spatial resolution (latitude and longitude) monthly precipitation data are available at PRISM climate group (http://www.prism. oregonstate.edu/) for the United States. The monthly PRISM data for the period 1990 to 2011 are averaged to annual precipitation at each grid point over Lake Tahoe. The grid points over Lake Tahoe corresponding to PRISM data (PRISM latitude and longitude data) were identified using Google Earth. The USGS lake bathymetry data (http://tahoe.usgs.gov/bath.html) available at 10 m spatial resolution are used to draw lake map to overlay the gridded annual precipitation. Precipitation contours using PRISM data over the lake are drawn using MATLAB software and gridded annual precipitation values.

The average annual precipitation directly on the lake was estimated by (1) dividing the lake into 2×2 km^2 grids, (2) assuming the precipitation in each grid can be determined by the enclosed isohyetal lines, (3) summing precipitation in all the lake grids, (4) dividing the sum total of grids' precipitation by the number of grids, and (5) comparing the estimated precipitation with the isohyetal line passing through Tahoe city. It was found that precipitation should be reduced by approximately 35% if the Tahoe city meteorological data would be used as representative of Lake Tahoe in the DLM-WQ hydrologic simulation.

8.5.2 THE DLM-WQ MODEL

The DLM-WQ [25] was modified to couple the turbulent diffusive heat transfer model while other features of the model were kept unchanged. The hydrodynamic component of the model is based on the original DYRESM (DYnamic REservoir Simulation Model) that is described by Hamilton and Schladow [10]. Fleener [8] added the river plunging algorithms in the hydrodynamic module. The primary hydrodynamic model is one-dimensional (1-D) and is based on a horizontally mixed Lagrangian layers approach [10]. However, the stream inflows and mixing due to stream turbulence are two-dimensional (2-D). Figure 8.1 shows the conceptual design of DLM-WQ. All the ecological modules are incorporated into the 1-D hydrodynamic model (double line box). The hydrodynamic model simulates stratification, mixing, the transport of all pollutant in the vertical direction, and determines the stream plunging depths. The ecological modules simulate transformation processes associated with algal photosynthesis. Flows and pollutants (nutrients and fine particles) from the atmosphere, streams and intervening zones (both urban and non-urban), groundwater and shoreline erosion into the lake are shown at the top of the double line box. Land use and weather drive flow and pollutant loading.

8.5.3 *CALIBRATION AND VALIDATION OF THE DLM-WQ MODEL*

The first twelve months of simulation was considered as the model initialization (warm-up) period. This is reasonable because deep mixing occurs during February and March, snow melting occurs during May to July, stratification builds up during June to October, and epilimnion layer deepens during November to mid-January. Therefore, DLM-WQ is able to estimate the surface water temperature and water level close to measured records in twelve months.

The predictive performances of DLM-WQ are measured using four different statistical efficiency criteria: the Nash-Sutcliffe efficiency (NSE), coefficient of determination (R^2), root mean square error (RMSE), and mean error (ME). The NSE is an index used for assessing the predictive accuracy of a predictive model [13]. The R^2 measures the strength and the direction of a linear relationship between observed and predicted values. The ME indicates the average of the total model errors and is used to measure how close model predictions are to observed values. The RMSE indicates an overall (global) discrepancy between the observed values and predicted values. The mathematical expressions for NSE, R^2, RMSE, and ME are shown below:

$$\text{NSE} = 1 - \frac{\sum_{i=1}^{n}(O_i - P_i)^2}{\sum_{i=1}^{n}(O_i - \overline{O})^2} \tag{3}$$

$$R^2 = \left[\frac{\sum_{i=1}^{n}(O_i - \overline{O})(P_i - \overline{P})}{\sqrt{\sum_{i=1}^{n}(O_i - \overline{O})^2 \sum_{i=1}^{n}(P_i - \overline{P})^2}} \right]^2 \tag{4}$$

$$\text{RMSE} = \sqrt{\frac{1}{n}\sum_{i=1}^{n}(O_i - P_i)^2} \tag{5}$$

$$\mathrm{ME} = \frac{1}{n}\sum_{i=1}^{n}(O_i - P_i) \tag{6}$$

where, O_i and P_i are observed and predicted value at time i, respectively; $\bar{O}$ and $\bar{P}$ are the mean of the observed and predicted values, respectively; and n = total number of observations.

Although ranges of NSE, R^2, ME, and RMSE vary between $-\infty$ to 1, -1 to 1, $-\infty$ to ∞, and $-\infty$ to ∞, respectively, yet the model predictions are considered to be precise if values of NSE, R^2, ME, and RMSE are close to 1, 1, 0, and 0, respectively.

8.6   RESULTS AND DISCUSSIONS

The calibration and validation of the DLM-WQ was performed to match the DLM-WQ estimated value as close as to measured lake water temperature and surface level with R^2 value as the objective function. The heat exchange coefficients were found optimum at 1.82×10^{-6} of C_{EN}, 3.00×10^{-6} of C_{HN} and 1.3×10^{-3} of C_{DN}, respectively for Lake Tahoe. DLM-WQ estimated 36% reduction of Tahoe City precipitation, which closely matches with the estimates using precipitation contour lines.

8.6.1   LAKE WATER TEMPERATURE

Lake water temperature changes dynamically due to weather and lake motion (Eq. (2)). In situ water temperature measurement records were available once a month at the surface (0 m) and at depths of 10 m, 50 m, 100 m, 150 m, 200 m, 250 m, 300 m, 350 m, 400 m, and 450 m. All heat budget components (including incoming longwave radiation, outgoing longwave radiation, evaporation heat loss, and sensible heat exchanges) directly affect only surface water. Shortwave radiation penetrates below the surface and decays exponentially thus affect the internal heat storage. Simulated and measured water temperatures below 50 m are approximately 5°C, thus, are not shown. The estimated water temperature in Figure 8.4 demonstrates a close match with measured temperature values with $R^2 = 0.968$, NSE = 0.911, RMSE = 1.487°C and

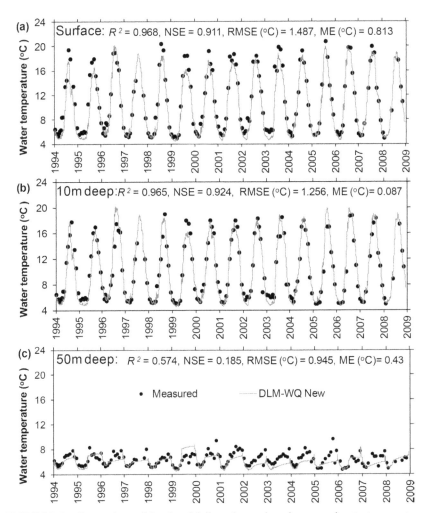

FIGURE 8.4 Comparison of simulated daily and event based measured water temperature. Top: surface, Center: 10 m depth, and bottom: 50 m depth.

ME = 0.813°C. The prediction performance efficiencies of DLM-WQ for water temperature at all measured depths (0 m, 10 m, 50 m, 100 m, 150 m, 200 m, 250 m, 300 m, 350 m, 400 m, and 450 m) were excellent with R^2 = 0.972, NSE = 0.942, RMSE = 0.697°C and ME = 0.061°C. This indicates that DLM-WQ can simulate lake vertical thermal dynamics adequately.

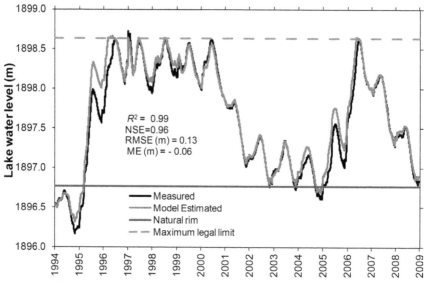

FIGURE 8.5   Comparison of simulated and measured daily lake surface water level.

8.6.2   *LAKE SURFACE WATER LEVEL*

The DLM-WQ estimated daily water level was compared with the USGS lake level measurements (Figure 8.5). The prediction efficiency of the DLM-WQ is high with coefficient of determination (R^2) and NSE were 0.99 and 0.96, respectively with RMSE and ME values close to 0 at 0.13 m and –0.06 m, respectively.

8.6.3   *WATER BALANCE*

Figure 8.6(a) shows that precipitation is lowest during summer (July to September) and highest during December to February. However, stream inflow is highest during April to June due to snow melting and lowest during August to October. Highest monthly evaporation occurs in August and lowest evaporation occurs during winter (January to May).

Figure 8.6(b) shows the percentage contribution to lake hydrologic budget of each source for the period 1994–2008. The average contribution

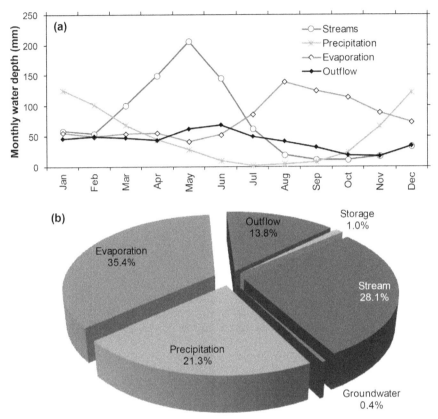

FIGURE 8.6 (a): Monthly average based on daily estimates during period 1994–2008 and (b): Relative contribution based on daily estimates of the period 1994–2008.

from stream, groundwater, precipitation directly onto the lake, evaporation and outflow during 1994–2008 are 28.1, 0.4, 21.3, 35.4 and 13.8%, respectively. Evaporation is the largest contributor in the hydrologic budget and its accurate estimation is therefore critical for water supply to downstream users and communities around the lake, as well as ecosystem management. Stream flow and precipitation components depend on climate forcing. Since the lake area does not change significantly when lake water elevation fluctuates between 1896.77 m and 1898.63 m, annual evaporation is taken to be approximately 937 ± 81 mm (Table 8.1). The inter-annual lake outflow and storage (= inflow + precipitation – outflow – evaporation) variation is significant (Table 8.2). Lake storage depends on outflow

TABLE 8.1 Estimated annual equivalent water depth (mm) of the lake hydrologic budget components

Year	Stream inflows	Precipitation over lake	Evaporation from the lake	Ground water	Outflow of the lake	Storage change	WS change
1994	300.12	446.93	837.78	10.44	0.16	−80.46	−90.40
1995	1679.40	807.67	703.85	10.52	91.83	1701.91	1661.80
1996	1463.34	1045.17	908.42	10.55	1362.73	247.91	244.50
1997	1396.08	515.28	845.41	10.52	1336.39	−259.92	−444.50
1998	1268.71	772.08	776.17	10.52	1081.97	193.18	191.10
1999	1085.65	487.65	898.35	10.52	908.35	−222.87	−228.10
2000	669.75	508.71	993.12	10.55	375.80	−179.91	−188.10
2001	302.57	453.37	1015.43	10.52	467.53	−716.49	−710.10
2002	554.86	479.43	1053.68	10.52	207.03	−215.89	−223.90
2003	642.99	490.76	923.75	10.52	200.84	19.69	9.60
2004	532.58	516.77	1056.95	10.55	120.45	−117.50	−149.80
2005	1033.34	924.47	927.56	10.52	189.42	851.35	820.20
2006	1396.12	682.49	1044.20	10.52	656.07	388.87	346.10
2007	357.85	409.50	1092.28	10.52	383.13	−697.54	−695.20
2008	395.22	558.78	990.22	10.55	281.85	−307.51	−399.40
Ave	871.91	606.60	937.81	10.53	510.90	40.32	9.59
SD	480.48	193.98	112.36	0.03	453.71	607.00	607.21

WS, Ave, and SD represent water surface level, average, standard deviation, respectively.

from Lake Tahoe, which is based on the operating policy of TCID to meet downstream users' requirement. Table 8.1 shows those consecutive wet years (1996–1999) allow the managers to release more water for the downstream users whereas consecutive dry years (2001–2004) restrict the higher withdrawal of water from the lake. This is important for future planning, especially as it relates to uncertainties due to climate change.

For outflow to occur, the lake water level must be within the lake natural rim (elevation 1896.77 m) and maximum legal limit (elevation 1898.63 m). Although the lake water level fluctuated within 1.86 m, yet the volume of water due to evaporation and precipitation were large because of the lake's large surface area (495 km²). A large volume of water within the lake operating limits (i.e., 1.86 m) can be stored during winter and released as per the stake holders' requirement during spring and summer given that estimates of the hydrologic components are available to the lake manager(s).

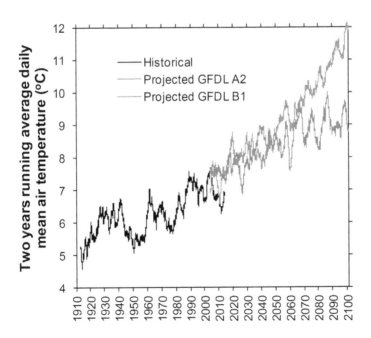

FIGURE 8.7   Historical and projected daily average air temperature.

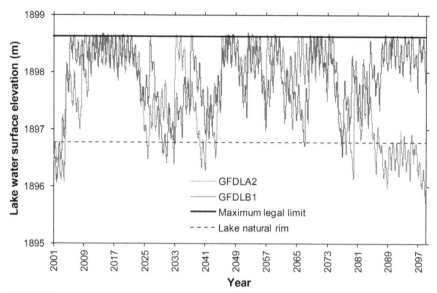

FIGURE 8.8 Simulated daily lake water level for GFDLA2 and GFDLB1 scenarios. Shown are the lake maximum legal limit and natural rim level. X-axis values represent the beginning of the year.

8.6.4 EFFECTS OF CONTINUED WARMING

Air temperature is widely expected to increase under all future climate scenarios. Figure 8.7 illustrates that annual average air temperature at Lake Tahoe is projected to increase by additional approximately 2–4.5°C by the year 2100 depending on the radiative forcing scenario. Figure 8.7 shows that scenario A2 is the direct extension of the historical trend. Figure 8.8 presents water level of the lake for both the GFDL A2 and GFDL B1 scenarios. The lake model suggests that climate change will drive the lake surface level down below the natural rim after 2086 for the GFDL A2 but not the GFDL B1 scenario. Outflow is zero when the lake level falls below the natural rim. The lake level dips down below the natural rim when evaporation rate is higher than sum total of stream inflows, groundwater contributions and on-lake precipitation over the lake (Figure 8.9). As long as the lake level is below the rim, the effects of annual evaporation and inflow are cumulative, and cannot be influenced by gate operation. The

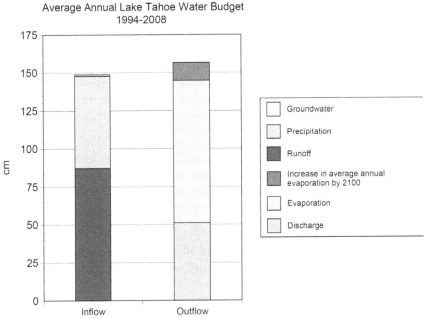

FIGURE 8.9 Annual average water budget during 1994 to 2008 and additional evaporation due to global warming.

results indicate that continued climate changes could pose serious threats to the water supplies from the lake that are most highly valued.

8.7   CONCLUSIONS

The calibrated and validated watershed model, LSPC, was used to generate daily the streamflows and associated pollutant loads of the entire watershed. LSPC estimated streamflow for the 10 LTIMP tributaries agreed well with the measured values during 1994–2008. Isohyetal lines over the lake were developed using the PRISM gridded monthly data during 1990–2011. The annual precipitation contour lines indicate a reduction of approximately 35% of Tahoe City precipitation. DLM-WQ estimated 36% reduction of Tahoe City precipitation.

The site-specific and stability-dependent bulk transfer coefficients (C_{EN}, C_{HN}, and C_{DN}) and precipitation inputs were calibrated using predictive performance values R^2 as the objective functions. Sensitivity analysis on bulk transfer coefficients, hydrologic inputs and model assumptions were conducted to ascertain that calibrated values were optimal. C_{EN} variation results in a greater change in lake water level compared to C_{HN} and C_{DN}. The DLM-WQ estimated water temperatures and lake water level were in excellent agreement with those of measured records for the period 1994–2008 with R^2 equal to 0.97 and 0.99, respectively. Evaporation was found to be the largest contributor in the hydrologic budget. The average stream inflow, groundwater fluxes, precipitation over the lake, evaporation from the lake and outflow from the lake during 1994–2008 are 28.1, 0.4, 21.3, 35.4, and 13.8 percent, respectively. The DLM-WQ estimated average annual evaporation rate is approximately 937 ± 81 mm with highest monthly evaporation rate in August and lowest evaporation during winter (January–May).

Because of the lake's large surface area, the evaporation and precipitation contribute approximately 63.5% on the hydrologic budget and spatial time series direct measurements are not available, accurate estimation of these components are important for the sustainable water management. A large volume of water within the lake operating limits (i.e., 1.86 m) can be stored during winter and released as per the stake holders' requirement during summer given that accurate estimates of the hydrologic components are available to the lake manager(s).

Projected global warming indicate that annual average air temperature at Lake Tahoe would increase by additional approximately 4.5°C by the year 2100 for A2 scenario. As a consequence, the lake model estimates that climate change will drive the lake surface level down below the natural rim after 2086 for the GFDL A2. The results indicate that continued climate changes could pose serious threats to the Lake water resources and ecosystem. Future water resources planning should take these results into account. The methodology presented has obvious relevance for sustainable management of lake water resources with respect to both future climate change and burgeoning human needs due to population growth in the region. The modified DLM-WQ can be applied to any small to medium size lakes, reservoirs and wetlands around the world for estimation of hydrologic budget, which is important for sustainable water management.

8.8 SUMMARY

We employed the modified UC Davis Dynamic Lake Model with Water Quality (DLM-WQ) model with the goal of correctly estimating the heat and hydrologic budget for Lake Tahoe (California-Nevada). We developed isohyetal lines using PRISM gridded precipitation data during 1990–2011 to estimate the precipitation distribution directly on the lake. The annual precipitation contour lines indicate a reduction of approximately 35% of Tahoe City precipitation. We utilized the watershed model, Loading Simulation Program in C++ (LSPC), to generate the stream flows and associated pollutants loadings to the lake. The modified DLM-WQ estimated lake surface water level and temperatures were in excellent agreement with those of measured records for the period 1994 to 2008 with R^2 equal to 0.99 and 0.97, respectively. Although latent heat loss (17%) is small in the heat budget, the effect on water balance due to inaccurate estimates of evaporative loss is very large because evaporation (35.2%) is the largest contributor in the hydrologic budget. Meteorology is the driving force for lake internal heating, cooling, and mixing. Using the modified DLM-WQ together with the downscaled climatic data of the two emissions scenarios (B1 and A2) of the Geophysical Fluid Dynamics Laboratory (GFDL) Global Circulation Model, we found that that higher rate of evaporation due to climate change would drive the lake surface level down below the natural rim after 2085 for the GFDL A2. The results indicate that continued climate changes could pose serious threats to the Lake water resources. Future water resources planning should consider these results.

ACKNOWLEDGEMENTS

This work was partially supported by an EPA/NSF – sponsored Water and Watersheds grant through the National Center of Environmental Research and Quality Assurance (R826282), the USEPA sponsored

(R819658 & R825433) Center for Ecological Health Research at University of California Davis, and grants 01-174-160-0 and 01-175-160-0 from the Lahontan Regional Water Quality Control Board. The research described has not been subjected to any EPA review and therefore does not necessarily reflect the views of the Agency, and no official endorsement should be inferred. The authors also acknowledge the contributions of TERC and USGS staff and students, who have collected and maintained the long-term water quality and streamflow databases that make modeling studies such as this possible.

KEYWORDS

- California
- climate change
- Dynamic Lake Model with Water Quality model, DLM-WQ
- evaporation
- flood protection
- Geophysical Fluid Dynamics Laboratory, GFDL
- Global Circulation Model, GCM
- hydro – meteorological inputs
- hydrodynamics
- isohyetal lines
- lake dynamics
- lake ecosystem
- lake stability
- lake Tahoe
- lake/reservoir heat budget
- lake/reservoir water budget
- land use
- Load Simulation Program in C++, LSPC
- Nevada
- PRISM data
- reservoir management
- reservoir operation
- SNOTEL (SNOwpack TELemetry)
- sustainable water management
- Tahoe Environmental Research Center, TERC
- thermodynamics
- Turbulent diffusive heat transfer model, TDHFM
- United State Geological Survey, USGS
- United States Natural Resources Conservation Services, US – NRCS

- **University of California, UC**
- **water quality**
- **watershed hydrology**

REFERENCES

1. Assouline, S., Tyler, S. W., Tanny, J., Cohen, S., Bou-Zeid, E., Parlange, M. B., and Katul, G. G., 2008. Evaporation from three water bodies of different sizes and climates: measurement and scaling analysis. *Advances in Water Resources*, 31:160–172.

2. Blanc, T. V., 1985. Variation of bulk-derived surface flux, stability, and roughness results due to the use of different transfer coefficients schemes. *Journal of Physical Oceanography*, 15:650–669.

3. Boehlert, B. B., and Jaeger, W. K., 2010. Past and future water conflicts in the Upper Klamath Basin: An economic appraisal. *Water Resources Research*, 46: W10518.

4. Boughton, C., Rowe, T., Allander, K., and Robledo, A., 1997. Stream and groundwater monitoring program, Lake Tahoe Basin, Nevada and California. *U. S. Geological Fact Sheet FS-100–97*, 6 pp.

5. California Department of Fish and Game (CADFG), 2003. Klamath river fish kill: Preliminary analysis of contributing factors. Yreka, California. September 2002

6. Coats, R., Perez-Losada, J., Schladow, G., Richards, R., and Goldman, C., 2006. The warming of Lake Tahoe. *Climatic Change*, 76:121–148.

7. Cobourn, J., 1999. Integrated watershed management on the Truckee River in Nevada. *Journal of the American Water Resources Association*, 35(3):623–632.

8. Fleener, W. E., 2001. *Effects and Control of Plunging Inflows on Reservoir Hydrodynamics and Downstream Releases*. Dissertation Thesis, University of California, Davis – CA.

9. Gardner, J. V., Larry, A. M., and Clarke, J. H., 1998. *The bathymetry of Lake Tahoe, California-Nevada*. US Geological Survey, Open-File Report 98–509. http://tahoe.usgs.gov/bath.html.

10. Hamilton, D. P., and Schladow, S. G., 1997. Prediction of water quality in lakes and reservoirs, Part I – model description. *Ecological Modeling*, 96:91–110.

11. Heyvaert, A. C., Nover, D., Caldwell, T., Towbridge, W., Schladow, G., and Reuter, J. E., 2010. *Assessment of Particle Size Analysis in the Lake Tahoe Basin*. Desert Research Institute, Reno, NV and UC Davis Tahoe Environmental Research Center, Incline Village, NV. Prepared for USFS – Pacific Southwest Research Station, Berkeley, CA. 75 p.

12. Lahontan Regional Water Quality Control Board (Lahontan) and Nevada Division of Environmental Protection (NDEP), 2010. *Lake Tahoe Total Maximum Daily Load Technical Report*. (http://ndep.nv.gov/BWQP/tahoe3.htm). Lahontan Water Board,

South Lake Tahoe, California, and Nevada Division of Environmental Protection, Carson City, NV. 340 p.

13. Legates, D. R. and McCabe, G. J., 1999. Evaluating the use of goodness-of-fit measures in hydrologic and hydroclimatic model validation. *Water Resources Research*, 35(1):233–241.

14. Lüers, J. and Bareiss, J., 2010. The effect of misleading surface temperature estimations on the sensible heat fluxes at a high Arctic site – the Arctic Turbulence Experiment 2006 on Svalbard (ARCTEX-2006). *Atmospheric Chemistry and Physics*, 10:157–168.

15. Momii, K., and Ito, Y., 2008. Heat budget estimates for Lake Ikeda, Japan. *Journal of Hydrology*, 36:362–370.

16. Porporato, A., 2009. Atmospheric boundary-layer dynamics with constant Bowen ratio. *Boundary-layer Meteorology*, 132:227–240.

17. Sahoo, G. B., and Schladow, S. G., 2014. Heat and Hydrologic Budget Estimates for Upper Klamath Lake Oregano, USA. *Water Resources Management*, 28:1395–1414.

18. Sahoo, G. B., Nover, D., Schladow, S. G., Reuter, J. E., and Jassby, D., 2013c. Development of updated algorithms to define particle dynamics in Lake Tahoe (CA-NV) USA for total maximum daily load. *Water Resources Research*, 49: doi: 10.1002/2013WR014140.

19. Sahoo, G. B., Schladow, S. G., and Reuter, J. E., 2013b. Hydrologic Budget and Dynamics of a Large Oligotrophic Lake Related to Hydro-meteorological Inputs. *Journal of Hydrology*, 500:127–143.

20. Sahoo, G. B., Schladow, S. G., Reuter, J. E., Coats, R., Dettinger, M., Riverson, J., Wolfe, B., and Costa-Cabral, M., 2013a. The Response of Lake Tahoe to Climate Change. *Climatic Change*, 116:71–95.

21. Sahoo, G. B., Nover, D., Reuter, J. E., Heyvaert, A. C., Riverson, J., and Schladow, S. G., 2013a. Nutrient and particle load estimates to Lake Tahoe (CA-NV, USA) for total maximum daily load establishment. *Science of the Total Environment*, 444:579–590.

22. Sahoo, G. B., Riverson, J., Wolfe, B., Reuter, J. E., and Schladow, S. G., 2012. *Development of a water quality modeling toolbox to inform pollutant reduction planning, implementation planning and adaptive management.* Technical Report, page 301. Prepared for USDA Forest Service-Pacific Southwest Research Station, TCES Suite 320, 291 Country Club Drive, Incline Village, NV 89451.

23. Sahoo, G. B., and Schladow, S. G., 2008. Impacts of Climate Change on Lakes and Reservoirs Dynamics and Restoration Policies. *Sustainability Science*, 3(2):189–199.

24. Sahoo, G. B., Schladow, S. G., and Reuter, J. E., 2009. Forecasting stream water temperature using regression analysis, artificial neural network, and chaotic non-linear dynamic models. *Journal of Hydrology*, 378(3–4):325–342.

25. Sahoo, G. B., Schladow, S. G., and Reuter, J. E., 2010. Effect of Sediment and Nutrient Loading on Lake Tahoe (CA-NV) Optical Conditions and Restoration Opportunities Using a Newly Developed Lake Clarity Model. *Water Resources Research*, 46. doi: 10.1029/2009WR008447.

26. Simon, A., 2008. Fine-Sediment Loadings to Lake Tahoe. *Journal of the American Water Resources Association (JAWRA)*, 44(3):618–639.

27. Tahoe Environmental Research Center (2011. *Tahoe: State of the Lake Report 2011.* http://terc.ucdavis.edu/stateofthelake/StateOfTheLake2011.pdf.
28. Tennessee Valley Authority (TVA), 1972. Heat and mass transfer between a water surface and the atmosphere. Water Resources Research Laboratory Report 14, Report No. 0–6803. Norris – Tennessee.

CHAPTER 9

REMOTE SENSING AND GIS APPLICATIONS FOR WATER RESOURCES PLANNING IN MICRO-WATERSHED

P. K. ROUT, J. C. PAUL, and B. PANIGRAHI

Department of Soil and Water Conservation Engineering, College of Agricultural Engineering and Technology, Orissa University of Agriculture and Technology, Bhubaneswar, Odisha, India. Phone: +91 9437762584, E-mail: jcpaul66@gmail.com; kajal_bp@yahoo.co.in

CONTENTS

9.1   Introduction 218
    9.1.1   Watershed as a Geographical Unit of Study 219
    9.1.2   Watershed Management 220
    9.1.3   Objective of Watershed Management 221
    9.1.4   Activities of Watershed Management.......................... 221
    9.1.5   Technology in-Use for Watershed Development/
               Management Remote Sensing (RS) 222
          9.1.5.1   Geographical Information System (GIS) 223
    9.1.6   Research Objectives 224
9.2   Review of Literature 224
    9.2.1   Watershed Characterization and Prioritization................. 225

9.2.2   Remote Sensing and GIS for Land Use/Land Cover Planning 226

9.2.3   Remote Sensing and GIS for Watershed Hydro-Geomorphology, Groundwater Prospects and Morphometric Analysis 229

9.2.4   Remote Sensing and GIS for Watershed Management, Land/Soil and Water Resource Action Plans 231

9.3   Materials and Methods 234

9.3.1   Study Area 234

9.3.2   Data for Location 235

     9.3.2.1   Spatial Datasets 235

         9.3.2.1.1   Satellite Data 235

         9.3.2.1.2   Geo-Referencing Satellite Data Products 236

         9.3.2.1.3   Topo Sheets 236

         9.3.2.1.4   Village Boundary Map 236

     9.3.2.2   Attribute Data 237

     9.3.2.3   Agriculture Use (Crop Classification) 237

     9.3.2.4   Ground Truth 237

9.3.3   System and Software 237

     9.3.3.1   System and Peripheral Used in GIS: 237

     9.3.3.2   Software Used in GIS 239

9.3.4   Methodology for Thematic Database Creation 239

     9.3.4.1   Delineation of Watershed 239

     9.3.4.2   Preparation of Village and Road Network Map 240

     9.3.4.3   Preparation of Land Use/Land Cover Map 240

     9.3.4.4   Preparation of Hydrogeomorphological Map 241

     9.3.4.5   Preparation of Drainage and Surface Water Body Map 242

9.3.4.6   Preparation of Slope Map.................................... 242

9.3.4.7   Preparation of Soil Resources Map................... 243

9.3.5   Watershed Morphometry Analysis 243

9.3.6   Methodology Adopted to Generate the Action
Plan in GIS 249

9.3.6.1   Use Of Thematic Maps for Integration 249

9.3.6.2   Digital Gis Database of Thematic Maps 249

9.3.6.3   Working Procedure in Arc/Info GIS................... 250

9.3.6.4   Working Procedure in Arc/Info
Package Environment... 250

9.3.6.4.1   Overlaying Analysis 250

9.3.6.4.2   Proximity Analysis (Buffer
Operations)..................................... 251

9.3.6.4.3   Tabular and Statistical Analysis
and Database Query......................... 251

9.3.7   Generation of Action Plan 251

9.3.7.1   Water Resource Development Plan.................... 251

9.3.7.2   Land Resources Development Plan 254

9.4   Results and Discussion 255

9.4.1   Index Map of Puincha Micro-Watershed 255

9.4.2   Land Use/Land Cover Map of Puincha
Micro-Watershed.. 257

9.4.3   Hydrogeomorphological Characteristics of
Puincha Micro Watershed 259

9.4.4   Morphometric Analysis of Micro Watershed 261

9.4.5   Slope Attributes of Micro-Watershed................................ 263

9.4.6   Soil Characteristics of Micro-Watershed 263

9.4.7   Water Resources Development Plan 265

9.4.8   Land Resources Development Plan 265

9.5   Recommendations... 268

9.5.1   Optimally Used Land 268

9.5.2   Afforestation/Forest Densification 268

9.5.3   Agro-Forestry/Agro-Horticulture With
Appropriate Soil Conservation Measures.......................... 268

9.5.4   Forest Plantation.. 269

9.5.5   Dry Land Cropping 269

9.5.6   Intensive Agriculture 270

9.5.7   Agricultural Plantation 270

        9.5.7.1   Tree Plantation 270

        9.5.7.2   Slope Plantation 270

9.6   Conclusions... 270

9.7   Summary.. 272

Acknowledgements.. 272

Keywords 273

References.. 274

9.1   INTRODUCTION

During the past decades, more and more of the complex environmental challenges have been addressed by using a watershed approach. According to the U.S. Environmental Protection Agency (EPA), environmental management using a watershed approach constitutes "a coordinating framework for environmental management that focuses public and private sector efforts to address the highest priority problems within hydrologically defined geographic areas."

The National Research Council report also noted that a watershed approach "uses sound, scientifically based information from an array of disciplines to understand the factors influencing the aquatic and terrestrial ecosystems, human health, and economic conditions of a watershed." The watershed is considered to be the integrating focus, the most appropriate spatial arrangement and functional unit for managing complex environmental problems. For example, managing issues of bio-complexity in the environment on a watershed basis offers the potential benefit of balancing the competing demands placed on natural and human systems.

Because of the highly complex nature of human and natural systems, the ability to understand them and project future conditions using a watershed approach has increasingly taken a geographic dimension. The advent of geoinformatics technology in past two decades has provided an opportunity of obtaining reliable information of the natural resources of the watershed and therefore has become a tool of either original mapping or updating of the existing maps.

Geoinformatics has been described as "the science and technology dealing with the structure and character of spatial information, its capture, its classification and qualification, its storage, processing, portrayal and dissemination, including the infrastructure necessary to secure optimal use of this information" or "the art, science or technology dealing with the acquisition, storage, processing production, presentation and dissemination of geoinformation." Branches of geoinformatics include cartography, remote sensing, web mapping, spatial analysis and geographic information systems.

Geographic Information Systems (GIS) technology has played critical roles in all aspects of watershed management, from assessing watershed conditions through modeling impacts of human activities on water quality and to visualizing impacts of alternative management scenarios. The field and science of GIS have been transformed over the last two decades. Advancements in computer hardware and software, availability of large volumes of digital data, the standardization of GIS formats and languages, the increasing interoperability of software environments, the sophistication of geo-processing functions, and the increasing use of real-time analysis and mapping on the internet have increased the utility and demands for the GIS technology. In turn, GIS application in watershed management has changed from operational support to prescriptive modeling and tactical or strategic decision support system.

9.1.1   WATERSHED AS A GEOGRAPHICAL UNIT OF STUDY

A watershed represents a topographically defined area that is drained by a stream system, representing a smaller upstream catchment, which is a constituent of a larger river basin. This landscape encompasses both surface and groundwater supplies, in addition to related terrestrial and community resources. The watershed is being viewed as a place based and ecological entity, as well as a socioeconomic and political unit to be utilized for management planning, conservation strategies and implementation purposes.

Watershed is thus a development unit in which all the natural resources like soil, water, geomorphology and land use are in harmony there by facilitating adoption of holistic approach to problem solving. Watershed is considered to be the ideal unit for analysis and management of natural resources for planning. The soil, vegetation and water are the basic resources, which interact and establish in a watershed. Hence, all these three resources have to be managed collectively and in an integrated way. The physiography of the land, slope and nature of soil cover, land use/cover, hydro geomorphology, climate, socio-economic and legal aspects, etc. and the hydrological features of the land area determine the productive interaction between these natural resources. Watersheds can also be repositories of global environmental benefits, such as biodiversity and carbon sequestration. Moreover, upper watersheds are linked, through water flows to downstream land and coastal areas far from the steep terrains where water flows are generated.

9.1.2 WATERSHED MANAGEMENT

Watershed management may be defined as an integrated approach of greenery for a better environment. The scientist of the world believe that an environment catastrophe is occurring by way of global warming, change in climate and hazard to health due to green house effect. Ozone depletion is another feature of concern in response to increase chlorofluorocarbons. These ominous effects are caused by several phenomena like carbon dioxide accumulation and their interplay. Some optimistic veterans question these effects on the premise of inconclusive nature of the available data. Everyone agrees that greenery consumes the superfluous carbon dioxide, stores the deleterious gas, releases oxygen much in demand, provide the basic needs, plays a role in restoring climate and thus revives a better environment. Watershed management applied locally for developing green foliage, enriches environment globally in due course of time.

Watershed degradation in the third world countries threatens the livelihood of millions of people and constrains the ability of countries to develop a healthy agricultural and natural resource base. Increasing populations of people and livestock are rapidly depleting the existing natural recourses the soil and vegetation system cannot support the present level of use. In a sense the carrying capacity of these lands is being exceeded. As the population continues to rise, the pressure on forests, community lands and marginal agricultural lands leads to inappropriate cultivation practices, forest removal and grazing intensities that leave a barren environment yielding unwanted sediment and damaging stream-flow to downstream communities.

9.1.3   OBJECTIVE OF WATERSHED MANAGEMENT

- Proper utilization of different land use patterns such as up gradation of forest cover by afforestation, increasing the productivity of Agricultural lands, converting the waste land into arable/grass land within watershed boundary.
- Reduction of the groundwater table fluctuation, water logging and salinity problem.
- Proper availability of soil moisture in pre-monsoon and post-monsoon seasons.
- Provision of a low cost schedule for maintenance, reclamation and conservation programs.
- Minimization of damaging runoff & sediment yield along with flood protection.
- Pasture development either by itself or in conjunction with plantation.
- Agro-forestry and horticultural development.
- Adoption of multi-disciplinary approaches to derive optimum benefits from cultivated crops, grasses, forestry, conservation engineering and animal husbandry, etc.

9.1.4   ACTIVITIES OF WATERSHED MANAGEMENT

- To reduce damages by floods and sediment deposit.
- To protect and control the output of water from the Watershed.
- To check the erosion of reclaimed soil and to deteriorate the by controlling quality of runoff water.
- To stabilize areas contributing to soil erosion and sediment produce there by reducing the sediment concentration in the runoff water.

- To improve the quality of grass plot, woodlands and wild life of forests by protecting soil from erosion.
- To improve the quality of human life living on the Watershed by improving the infrastructure facilities and reaped the output from land.

9.1.5 TECHNOLOGY IN-USE FOR WATERSHED DEVELOPMENT/MANAGEMENT REMOTE SENSING (RS)

Literally remote sensing means obtaining information about an object, area or phenomenon without coming in direct contact with it. However, modern remote sensing means acquiring information about earth's land and water surfaces by using reflected or emitted electromagnetic energy. The remote sensing is basically a multi-disciplinary science, which includes a combination of various disciplines such as optics, spectroscopy, photography, computer, electronics and telecommunication, satellite launching, etc. All these technologies are integrated to act as one complete system in itself, known as Remote Sensing System. There are a number of stages in a remote sensing process, and each of them is important for successful operation. These stages are:

- Emission of electromagnetic radiation, or EMR (sun/self-emission).
- Transmission of energy from the source to the surface of the earth, as well as absorption and scattering.
- Interaction of EMR with the earth's surface: reflection and emission.
- Transmission of energy from the surface to the remote sensor.
- Sensor data output.
- Data transmission, processing and analysis.

Remote sensing which is now universally recognized as a highly effective and versatile technology for mapping, estimating, and monitoring and integrated planning is adopted widely for watershed monitoring/management. It has been used for various operational mapping and monitoring projects by various organizations under state and Central Government. The advantages are:

- It provides reliable, near real time base line information.
- Topographic and thematic map preparation instead of ground survey.

- Characteristic of the object or land features (Landuse, Rock type, vegetation condition, soil and environmental quality, etc.).
- Accurate discrimination of spatial units and its study on the spatial distribution covering large area because of advantages in synoptic coverage.
- Photomaps generation (ortho-photo) for topographic/resource/environmental data generation (as a map alternative).
- Capable of operating in portions of electro-magnetic spectrum which are beyond the photographic emulsion sensitivity.
- Detectors having wider dynamic range than photographic emulsions for identifying subtle change in scene radiance (feature discrimination).
- Relatively fast and economical for map and statistics generation in comparison to other survey or data collection methods.
- Terrain data is available in both analog and digital form (Digital terrain model).
- Amenable to computer processing and compatible to GIS packages makes the data easy to handle along with socio-economic and environmental data.
- Temporal data receiving facilities makes it useful for monitoring purposes.
- Introduction of sub-meter resolution satellite data provides opportunity to map plot level watershed information.

9.1.5.1   Geographical Information System (GIS)

GIS is defined as the computerized information system or integrated database management system in which large volume of geo-referenced spatial data (location of a layer or coverage defined by the co-ordinate referencing system) derived from a variety of source (like RS, GPS, aerial photography), etc. is efficiently stored and organized, manipulated and analyzed, retrieved and displayed/presented according to the user defined specifications. Besides this a standard GIS package establishes link between spatial and non-spatial data. By adopting GIS technique, various spatial data like paper map, charts, working drawings, physical survey maps, satellite imageries, aerial photographs, etc. can converted to digital format for operations like data linking, map joining, map over laying, clipping, generation of new maps, etc. easily and quickly for any analysis and assessment. This digital process helps the agricultural scientists, engineers, planners, decision makers and administrators to prepare action plans/micro plans for watershed development planning and to prepare schemes conveniently and accurately.

Taking the present day importance of watershed development and the advantages of Remote Sensing and GIS techniques in watershed management, a study was carried out on action plan preparation of Puincha micro-watershed in Banki block of Cuttack district of Odisha, India using Remote Sensing and GIS Technology with the following objectives:

9.1.6   RESEARCH OBJECTIVES

1. To prepare land use/land cover, hydro geomorphology, soil resource and slope map of micro-watershed.
2. To study and evaluate the potential of existing natural resources, groundwater prospect, soil type and slope aspects of the micro-watershed.
3. To study the morphometry of the micro-watershed and
4. To develop a land and water management action plan of micro-watershed using geospatial data through geoinformatics technology.

9.2   REVIEW OF LITERATURE

Satellite remote sensing techniques provide an efficient and economical means of acquiring timely data for the development and management of natural resources. Considering this, Remote sensing techniques have been gainfully used for deriving the various natural resources information such as land use/land cover, soils, hydro-geomorphology, drainage, etc., all over the world during the last three decades. Remote sensing technology has proved to be rapid and provides timely information with repetitive coverage of a given area in 5–22 days. The potential of satellite based remote sensing data for watershed characterization, prioritization, action plan preparation, monitoring and impact assessment have been highly explored in the Indian context. Several remote sensing application projects of the national, regional and local levels have been taken up for research, development and program implementation.

9.2.1 WATERSHED CHARACTERIZATION AND PRIORITIZATION

A study was carried out employing land sat data and aerial photography to map soil type, land use, geology and land forms of various watersheds in India by Gunjal [8]. The National Remote Sensing Agency (NRSA) and Space Application Center (SAC) used Landsat data and aerial photos for land use information at a reconnaissance level for a number of catchments. All India Soil and Land Use Survey (AIS and LUS) conducted watershed priority delineation surveys using 1: 50,000 scale aerial photos and topo – maps on a watershed basis for the major states. Through these surveys, very high and high sediment yielding priority watersheds were identified. Land degradation studies had also been reported for various parts of the country.

Chakraborthy [3] studied the strategies for watershed classification using remote sensing techniques and opined that decision support system for management planning requires scientific knowledge of resources information, expected runoff and sediment yield. It also requires priority classification of watersheds for conservation planning, monitoring of watershed for environmental impact assessment and technologies of GIS for data base creation, scenario development and appropriate decision making.

Shah [29] determined priority classes of sub watersheds in a part of Song river watershed, based on spatial erosional soil loss estimates using IRS-1A LISS –II digitally classified physiographic, Soil, Land use/land cover map, terrain slope information and rainfall data following USLE. The results indicated that out of fifteen sub-watersheds, nine sub-watersheds belong to high to very high priority classes covering 36.2% area of the watershed. The remaining six sub- watersheds covering 63.8% area of the watershed were classified as low to moderate priority categories.

Sidhu et al. [31] were used remote Sensing and GIS for prioritization of the upper Machkund watershed in Andhra Pradesh, which covers an area of 16,111 ha. The division of the watershed and sub-watershed areas was carried by visual interpretation of the geocoded IRS-IB (bands 2, 3 and 4),

FCC prints (scale 1:50,000) and the drainage pattern of watershed area, following the Atlas of AIS and LUS. The watershed area was classified as belonging to water region 4, Godavari Basin. Based on secondary and tertiary drainage pattern, the watershed areas were further sub-divided into 8 sub watersheds. Union modules (GIS), hydro-geomorphology, land use, land cover and slope maps were combined to generate erosion intensity and composite maps. Watersheds were prioritized following the sediment yield index (SYI) approach.

Biswas et al. [1] prioritized nine sub watersheds of Nayagram block in Midnapore district of West Bengal based on morphometric analysis of the drainage basin. It was observed that sub watershed 8 has got the highest priority because of high erosion intensity, which was also confirmed through SYI model. Thus, morphometric analysis could be used for prioritization of sub watersheds even without the availability of reliable soil maps of the area.

Ramesh et al. [24] generated information on natural resources using IRS 1C LISS III data in Dakshina Kannada district. They prioritized the micro-watersheds using an elimination technique and considering factors like current vegetation cover, waste lands, soil type, erosion status, slope, scope for development, etc., Each of these factors are given weightages in spatial domain using ARC-Info GIS to arrive at very low, low, medium, high and very high priority micro-watersheds.

Srivastava et al. [34] prioritized mini watersheds in Badrigad watershed using GIS technique. The present study demonstrates the usefulness of the GIS for morphometric analysis and prioritization of the mini watersheds. This technique is suitable mostly for un-gauged watersheds of the hilly areas for the identification of critical areas and implementation of watershed management programs.

9.2.2 REMOTE SENSING AND GIS FOR LAND USE/LAND COVER PLANNING

Jayarama et al. [11] revealed that more accurate soil maps in terms of boundary delineation and composition of soil mapping units could be prepared by visual interpretation. They also stated that the method could thus be used in revising and improving the existing reconnaissance soil maps prepared by conventional methods.

Srivastava and Rao [33] prepared land use/land cover map of Jharia coalfield using IRS 1A LISS II data consisting of nine land use categories such as settlements, mining area, agriculture, open scrub, etc., They concluded that using remote sensing techniques, basic data on land use can be obtained easily to assess the changes in an area over a period of time.

Goyal et al. [7] utilizing remote sensing data, aerial photographs and Survey of India topo sheets, delineated various geomorphic units like Aeo fluvial plains, Recent Sahibi flood plain and Aravally hills, Rock-out crops and pediments. Representative profiles were studied and soils were classified as Typic-Ustipsamments, TypicUstochrepts and TypicUstorthents in Rewari district, Haryana.

Chakraborthy et al. [2] utilized IRS-IA Satellite data and derived land use/land cover information together with conventional field data which were used as inputs into USLE model to predict soil loss on a sub-watershed-wise basis and then prioritized into 18 sub-watersheds according to the order of magnitude of their soil loss potentials.

Krishna et al. [17] applied the skill and knowledge of remote sensing and GIS for canopy mapping in deltaic region of West Godavari in Andhra Pradesh. They found that established relationships of cropping pattern with field data were helpful in resource inventory at micro-level.

Chakraborthy et al. [4] characterized and evaluated the vegetative dynamics and land use/land cover types of Birantiya Kalan watershed in western Rajasthan. They found that conventional methods of detecting land use/land cover is costly, low in accuracy, time consuming and particularly difficult in large areas. Remote Sensing, due to its capability of synoptic viewing and repetitive coverage, provides useful information on land use/land cover as well as vegetative vigor dynamics.

Sharma et al. [30] used remote sensing to detect the land use/land cover change and its climatic implications in the Godavari deltaic region. Interpretation has revealed changes in land use/land cover in the study area during the past 26 years. The study suggested that human activities in terms of land use/land cover alterations might be responsible for the local level climatic changes.

Joshi and Gairola [13] studied the land cover dynamics pattern in Balkhila sub-watershed situated in Garhwal, Himalayas. They used IRS-1D LISS III data of 01 February 2003 for land use/land cover classification. The results revealed that the land cover dynamics is dependent on the sun's illumination. The altitude and slope are no more a barrier for resource extraction and the human activity zone is shifting towards higher altitudes and slopes. The changes are also defined along the roads and settlements.

Mohammed et al. [19] obtained the soil data from soil resource inventory; land and climate were derived from the remote sensing satellite data (Landsat TM, bands 1 to 7) and were integrated in GIS environment to obtain the soil erosion loss using USLE model for the watershed area. The priorities of different sub-watershed areas for soil conservation measures were identified. Land productivity index was also used as a measure for land evaluation. Different soil and land attribute maps were generated in GIS, and R, K, LS, C and P factor maps were derived. By integrating these soil erosion map was generated. The mapping units, found not suitable for agriculture production, were delineated and mapped as non-arable land. The area suitable for agricultural production was carved out for imparting the productivity analysis; the land suitable for raising agricultural crops was delineated into different mapping units as productivity ratings good, fair, moderate and poor. The analysis performed using remote sensing and GIS helped to generate the attribute maps with more accuracy and the ability of integrating these in GIS environment provided the ease to get the required kind of analysis.

Farooq et al. [6] introduced a proper land use planning in the district Rajouri of state Jammu and Kashmir through Geo-spatial Techniques. The study suggested that good quality grass should be planted in the grasslands to improve the output and quality of the grass so that further expansion of grasslands and removal of forest can be stopped. It has been noticed that in this area meadows have decreased by more than 75% followed by forest cover 43%, barren land 16% and mixed vegetation 2.5%.

Kannan et al. [15] prepared land use map of Cauvery delta zone using satellite data and it revealed that distribution of crops were classified into *kharif +rabi* crops (124927 ha), *kharif* crop (1068 ha), *rabi* crop (17461 ha) and aquaculture (721 ha). It was suggested to raise groundnut on 13.2% area not suitability for paddy. Sugarcane, gingerly, brinjal, chilly crops were recommended in pre rainy season period in fine textural soils depending on the water availability.

Singh et al. [32] introduced a proper land use planning in the Semi-arid Region of Madhya Pradesh in India through Geo-spatial Techniques. The land Resource Development map was prepared which suggests the intensive agriculture, double crop, horticulture, agro-horticulture and silvi-pasture.

Srivastava et al. [34] studied about the temporal change in land use of Himalayan watershed using remote sensing and GIS in Badrigad watershed. It was found that agricultural area over a period of 43 years had increased slightly but area under barren land had reduced by 60 percent, which has converted mainly into forest land.

Jayaraju and Khan [12] studied that remote sensing (RS) integrated with geographical information system (GIS) provides an effective tool for analysis of land use and land cover changes at a regional level. The geospatial technology of RS and GIS holds the potential for timely and cost-effective assessment of natural resources. These techniques have been used extensively in the tropics for generating valuable information on forest cover, vegetation type and land use changes. In the present study, RS and GIS have been used to assess land cover patterns in Pulivendula–Sanivaripalli area of south India. With this in view, an assessment has been made on some of the natural resources and environmental potential of Pulivendula–Sanivaripalli area of south India. To achieve these, three thematic maps (land use and land cover, drainage and slope) were prepared through image interpretation and limited checks. The land use-land cover pattern falls under the broad categories of agricultural land, forest land and wasteland.

9.2.3 REMOTE SENSING AND GIS FOR WATERSHED HYDRO-GEOMORPHOLOGY, GROUNDWATER PROSPECTS AND MORPHOMETRIC ANALYSIS

Tikekar et al. [37] prepared hydro-geomorphological map by integrating various thematic maps derived using information related to lithology, structure, geomorphology and hydrology of the terrain in Bhandra district, Maharashtra. In order to demarcate the ground water potential zones of Marudaiyar basin in Tamil Nadu, different thematic maps such as lithology, land forms, lineaments, surface water bodies, drainage density, slope and soil maps were integrated and a ground water potential zone map was prepared which was in agreement with the bore well yield data collected in the field.

Rao [25] conducted Hydro-geomorphological studies in Niva river basin, Chittoor district of AP using Landsat FCC. The basin was classified into different zones covered by denudational hills, residual hills, inselbergs, pediments, pediplains with moderate and shallow weathered zones and valley fills. The ground water prospects ranged from poor in hills to good in fractured zones and valley fills.

Nag [21] was of the opinion that remote sensing technique is an indispensable tool in morphometric analysis and ground water studies. The results indicate that moderately weathered pediplains and valley fills are good prospective zones of ground water exploration. Sarkar et al. [27] used GIS to evaluate the ground water potentiality of Shamri micro watershed in Shimla taluk, Himachal Pradesh. The study established that GIS is a potential tool for facilitating the generation and use of thematic maps to identify the ground water potentialities of an area.

Jaishankar et al. [10] applied Remote Sensing technique for ground water exploration in Agnigundala mineralized belt in Andhra Pradesh, by using IRS-IB data. Based on erosion and depositional characters, various geomorphic units like hills, pediments, buried pediments, plains and valley fills have been identified. The ground water potentials of individual geomorphic units were evaluated to obtain a complete picture. Magesh et al. [18] studied about the morphometric characteristics in Bharathapuzha river basin of Kerala The present study has proved that the geoprocessing technique used in GIS is an effective tool for computation and analysis of various morphometric parameters of the basin and helps to understand various terrain parameters such as nature of the bedrock, infiltration capacity, surface runoff, etc. The quantitative analysis of linear, relief and aerial parameters using GIS is found to be of immense utility in river basin evaluation, basin prioritization for soil and water conservation and natural resource management. The geo-processing techniques employed in this study will assist the planner and decision makers in basin development and management studies.

Kinthada et al. [16] generated various geospatial information themes for multi-criteria modeling and analysis for informed decision-making and preparing an integrated action plan for sustainable development of land and water resource in the Domaleru watershed in Andhra Pradesh. Integrated modeling and analysis of various geospatial themes, like, topography, geomorphometry, geomorphology and hydrology in the GIS environment helped in reconstructing the hydrogeomorphological scenario of the Domaleru watershed. The methodology helped in defining a strategy used for considerable rise in the levels of groundwater as well as storage capacity of surface water bodies and growth in vegetation cover over reclaimed wasteland areas.

9.2.4 REMOTE SENSING AND GIS FOR WATERSHED MANAGEMENT, LAND/SOIL AND WATER RESOURCE ACTION PLANS

Murthy et al. [20] integrated thematic maps on geomorphology, geology, soil, land use/land cover, forest or vegetation, drainage and slope by to suggest suitable land and water resources development plans. The water resources development plan depicted the zones of exploitation through tube well, dug wells, development and conservation through rainwater harvesting structures. Land resources development plan depicted the alternative land use practices through double cropping, horticulture, etc. Decisions regarding the use of land and water resources of a backward region depend mostly on their productive potential and local priorities.

Uday et al. [38] were integrated spatially by Land use/land cover, soil, hydro-geomorphology, drainage, slope and transportation network maps to arrive at composite mapping units which are unique combination of various resources. This action plan comprising of alternate land use practices and a comprehensive plan for soil conservation and water harvesting structures like check dams, nala bunds, etc., was suggested to improve the productivity of Hirehalla watershed in Bijapur district.

Dwivedi et al. [5] gave an action plan for sustainable development of land and water resources was generated by upon integration of landforms, soils, land use/land cover, slope with socio-economic and meteorological data, peoples' aspirations, etc., in a GIS environment. Using high resolution IRS 1C PAN and LISS III satellite data, natural resources inventory was carried out on 1:25,000 scale. By integrating the thematic information, site-specific recommendations were given in the form of an action plan, which aimed at conservation, up-gradation and optimal utilization of natural resources to attain sustainable agricultural production.

Reddy et al. [26] integrated remote sensing and GIS for identification of sites for artificial recharge. They used geocoded IRS-IC LISS III FCC data for analysis and generation of thematic maps. These thematic maps were assigned based on their hydrological properties and were integrated in ILWIS 2.1 GIS. Based on cumulative weightage index, the composite map classes were divided into seven ground water potential zones. These identified sites were used for construction of water harvesting structures like check dams, percolation tanks and rock fill dams.

Tikekar et al. [37] followed integrated approach using geology, geomorphology, land use/land cover, soil characteristics and surface water system for optimum use of water resources. They concluded that upgrading of socio-economic situation of the region (Maharashtra) is possible if the developmental programs are planned, considering the available geomorphic units, surface water resources and exploitation of locked up groundwater potential zones particularly in riverine sediments. Integrated approach of remote sensing and GIS provides further insight into the hydrogeological regime of the area, which can be utilized for site selection for artificial recharge and facilitates in decision making for efficient planning of ground water management.

Paul et al. [23] applied Remote Sensing and GIS techniques for the integrated watershed planning of Bajpur watershed in Khurdasadar block of Khurda district (Orissa). Thematic maps like land use/land cover map, hydrogeomorphology map, soil map, slope map, drainage and surface water body maps were prepared by using both satellite imageries and survey of India toposheet. By integrating all these maps in GIS, water and land resource management plans were developed. The water resources development plan indicates the sites for groundwater exploitation and sites for surface water development including sites for different soil and water conservation structures. Eleven action items were suggested under land resource action plan with specific sites, aerial locations and maps. Thus remote sensing technology demonstrates the usefulness for providing up-to-date, reliable and accurate information on different natural resources of the watershed which is a pre-requisite for an integrated approach to identify the suitable sites for water harvesting structures, check dams, farm ponds, percolation tanks, nala bunds, etc. The GIS technique is helpful to integrate the information into a composite land unit development map to generate alternate land use system for sustainable development of the watershed.

Kalgapurkar et al. [14] developed a soil and water conservation plan using RS and GIS. The soil and water conservation plan is found to be logical and effective. It also scientifically optimizes the intensity of the conservation measures as the type of conservation measures based on depth of soil, allowable erosion rate and land slope. GIS proved be a very effective, effective and useful tool for integration of layers and incorporating the decisions rules for developing the plan.

Patel et al. [22] found that Geographical information system and remote sensing are proven to be an efficient tool for locating water harvesting structures by prioritization of mini-watersheds through morphometric analysis. In this study, the morphometric analysis and prioritization of ten mini-watersheds of Malesari watershed, situated in Bhavnagar district of Saurashtra region of Gujarat state, India, are studied. For prioritization of mini-watersheds, morphometric analysis is utilized by using the linear parameters such as bifurcation ratio, drainage density, stream frequency, texture ratio, and length of overland flow and shape parameters such as form factor, shape factor, elongation ratio, compactness constant, and circularity ratio. The different prioritization ranks are assigned after evaluation of the compound factor. Digital elevation model from Shuttle Radar Topography Mission, digitized contour, and other thematic layers like drainage order, drainage density, and geology are created and analyzed over ArcGIS 9.1 platform. Combining all thematic layers with soil and slope map, the best feasibility of positioning check dams in mini-watershed has been proposed, after validating the sites through the field surveys.

Wakode et al. [39] studied that Geographical Information System (GIS) have proved to be an efficient tool in the delineation of drainage pattern for water resources management and its planning. In this study, GIS and image processing techniques have been adopted for the identification of morphological features and analyzing the properties of the upper catchment of Kosi River. The basin area includes the high-altitude Himalayan Mountains, including Mount Everest and Kanchenjunga peaks. This basin is the main contributing area for devastating floods in 2008 in the Bihar state of India. The catchment can be divided into three sub-catchments, namely, Arun, Sunkosi, and Tamur. A morphometric analysis shows the nature of drainage in the upper catchment of Kosi River and some causes behind the high-intensity floods by comparing the properties of these three sub-catchments.

9.3   MATERIALS AND METHODS

This chapter mainly deals with the introduction to study area (location, physiography and socio-economic status of the study area), the different types of data used, the methods as well as materials used for carrying out the project work.

9.3.1   STUDY AREA

Puincha micro-watershed lies geographically between 20°16'0"N to 20°18'0"N latitude and 85°23'E to 85°25'E longitude. It is situated in Banki block of Cuttack district of Odisha, India, which is about 60 km distance from district head quarter. The location map is given in Figure 9.1. The micro watershed has an area of approximately 646.67 ha and is identified in survey of India topo-sheet number 73H/7 in 1:50,000 scale. The unique code assigned to this micro watershed is 0407010201110101.

Puincha micro-watershed comes under the Western undulating Zone near to river Mahanadi. Rock of this area is mostly of granitic gneiss and alluvium type. The micro watershed is elongated shape with flat topography. The climatic of the study area is sub tropical monsoon type, major rainfall being received by southwest monsoon. The mean annual rainfall is 1710 mm and the mean annual temperature is 27.5°C.

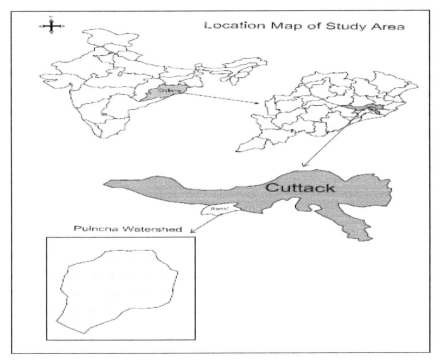

FIGURE 9.1   Location map of study area.

9.3.2   DATA FOR LOCATION

9.3.2.1   Spatial Datasets

The spatial datasets include satellite data, topographic data and village boundary maps. Thematic resource maps from secondary sources are also used.

9.3.2.1.1   Satellite Data

IRS-1C sensor data in LISS-III-FCC (False Color Composite refers to the composite image generated by remote sensing observation in green, red and infrared spectral bands by assigning complementary color, i.e., blue, green; red respectively to above said observation bands of Indian Remote Sensing Satellite), IRS-1D-LISS-III-FCC and IRS-1D-PAN Panchromatic data) were used for generation of thematic maps like land use/land cover, hydro geomorphology map, soil resources map and drainage map of the micro watershed. The satellite data in 1:50, 00/10,000 scales of 2007–10 (Multi-season data) are used.

9.3.2.1.2 Geo-Referencing Satellite Data Products

It involves three activities, i.e., GCP data collection, Image geo-referencing and rectification, and Image fusion. Geo-referencing and image rectification involve the removal of random and systematic errors in the image and later transforming image to UTM projection system and WGS84 datum. The GCP are collected using GPS receivers in the field. About 10 to 15 well-distributed GCP's for the watershed is used for georeferencing. The Geo-eye image is rectified using GCPs in second order polynomial transformation with RMS accuracy better than one pixel at each checkpoint. Higher order polynomial transformation allows keeping the registration error to the minimum. ERDAS image processing software package was used for this operation.

9.3.2.1.3 Topo Sheets

The survey of India topographical numbering 73H/7 on a scale of 1:50,000 with contour interval of 20 m was used to prepare the base map and to finalize the transport network, land use/land cover classes, hydro-geo-morphological units, slope classes and drainage pattern of the watershed.

9.3.2.1.4 Village Boundary Map

The village boundary map (1:4000) of the watershed was collected from district collector, Cuttack, local Tehsil office and from Odisha Sampad. The map has been prepared with the use of database collected from Census department and Revenue department, Govt. of Odisha. The maps were combined to generate the complete village maps. These maps served as important guides for preparation of different thematic maps.

9.3.2.2   Attribute Data

The attribute data includes datasets relating to rainfall, land features and agriculture use data. The monthly rainfall data of Bankisadar block for last 20 years (1993–2012) was collected from the Cuttack Sadar block office, Cuttack for present study. The net geographical area of Puincha micro-watershed is about 646.67 ha out of which 82.77 ha is forest area. The forest area is mainly divided into two parts; one part is hilly while other lies in the bank of river, which is devoid of any sort of vegetation. About 387.46 ha of land are under agriculture, which is completely rainfed where mainly a *kharif* crop is taken. The land features are given in Table 9.1.

9.3.2.3   Agriculture Use (Crop Classification)

The villages come under semiarid zone. So farmers grow crop according to the rainfall. As water shortage is a big issue, no crop can be taken in summer. Overall only one or two crops are taken in the year. It can say that monsoon decides cropping patterns of Jokiladal, Jokiladesh and Puincha village. Total 387.25 ha area is cultivated in *kharif* whereas in *rabi* because of scanty rainfall and since there is no irrigation facility, only 173.10 ha are cultivated.

9.3.2.4   Ground Truth

Ground truth has been carried out in the areas, which are found doubtful during the interpretation of satellite imageries. The thematic maps related to natural resources of the watershed were checked on the ground. Soil samples have been collected from the sites to finalize the soil maps.

TABLE 9.1 Land Features of Puincha Micro Watershed (area in ha)

Village	Geographical area	Forest area	Community land	Non agriculture area	Permanent pasture	Miscellaneous tree	Uncultivable waste land	Cultivated area
Jokiladal	257.47	–	3.80	12.90	2.34	16.39	10.20	193.00
Jokiladesh	0.85	–	0.10	–	–	–	–	0.02
Puincha	388.35	82.77	3.54	12.49	3.84	31.54	22.20	194.44
Total	646.67	82.77	7.44	25.39	6.18	47.93	32.40	387.46

9.3.3   SYSTEM AND SOFTWARE

9.3.3.1   System and Peripheral used in GIS

1. Black and white CCAL COMP scanner
2. Personal Computer with Pentium –IV dual core micro Processor
3. HP-810 INKJET Color Printer

9.3.3.2   Software used in GIS

1. R2V-2.1 version (Digitizing software)
2. ARC/INFO-VERSION (Editing and labeling)
3. ARC/VIEW VERSION (Designing and overlaying software)
4. MS-OFFICE (Text and Table addition software)

9.3.4   METHODOLOGY FOR THEMATIC DATABASE CREATION

The steps followed for image interpretation are:

1. Using the interpretation key prepared, the thematic maps are prepared by using onscreen interpretation procedure.
2. Satellite image(s) are displayed on the computer screen after proper rectification, geo-referencing and enhancement. In order to enhance the image appearance, proper LUT was applied to highlight the boundaries.
3. Onscreen interpretation is carried out in a separate layer in shape file format using the geo-referenced digital cadastral base.
4. Shape files are developed for all types of thematic maps.
5. Thematic map class codes in numeric format are used for labeling using the specific codes provided in this document.
6. Textual errors while manual label entering required additional but avoidable effort to rectify them. Preferably, customized GUI having drop down list of the labels, which appear along with text was used.

7. Integration of other layers such as watershed boundary/road/ settlement/water body, etc. was carried out subsequently. As per the need of the research study, the census/attribute data/field photo/ attributes, etc. are linked to the database.

9.3.4.1 Delineation of Watershed

The boundary up to watershed level has been delineated from the atlas of 1:4,00,000 scale prepared by AIS & LUS and maps and also from Odisha Samapd prepared by ORSAC. The region, catchment, sub catchment, watershed of region 4 was then enlarged by optical instruments to 1:250,000 survey of India maps. The transferred boundaries were corrected by the help of survey of India contour and drainage network as depicted in 1:50,000 scale. The catchment, sub catchment and watershed boundaries were then transferred to 1:50,000 scale Survey of India topo-maps. The boundaries up to watershed levels were then modified/corrected by overlaying the contour and drainage information of survey of India topo-sheets and geo-coded satellite image in 1:50,000 scale. The boundary of Puincha micro watershed was delineated from SOI topo-sheet 73H/7 in 1:50,000 and fixed for use in this study.

9.3.4.2 Preparation of Village and Road Network Map

The village and road network map of the Puincha micro watershed was prepared from SOI top sheet 73H/7 of 1:50000 scale and the village map. The transparent tracing film was placed over the topo sheet and the base items like National Highways (NH) metal roads, railway line, prominent river, etc. were drawn. Village and forest boundaries generated and incorporated in the base map.

9.3.4.3 Preparation of Land Use/Land Cover Map

Land use refers to men's activities and various uses, which are carried on land. Land cover refers to natural vegetation, water bodies, artificial cover and other resulting due to land transformations. The following steps were adopted to prepare the land use/land cover map of Puincha micro watershed:

1. The micro watershed boundary was delineated using SOI topo sheet in 1:50,000 scale.

2. Initial base map of the study area was prepared from SOI topo sheet indicating watershed boundary and few control points like state highway, metallic and nonmetallic roads, prominent rivers and water bodies, etc.

3. The base map was superimposed on multi-season satellite FCC data. Boundaries of various land use/land cover classes were delineated by means of visual interpretation technique whose fundamental is based on size, shape, shadow, tone, texture, and pattern and association characteristics of images.

4. After preliminary image interpretation, the findings were compared with ground truth.

5. Corrections and modifications were made wherever found necessary and final land use/land cover map of Puincha micro watershed was prepared.

6. The area/spatial distribution of different units under land use/land cover map were calculated.

9.3.4.4 Preparation of Hydrogeomorphological Map

Hydrogeomorphological map depicts different aspects like land form characteristics, geological information, etc. Information about different land form characteristics is a vital input for land management, soil mapping, etc. whereas information about different geology like lithology/rock types is an indispensable source to identify the occurrence of ground water potential zones. The following steps were adopted to prepare the land use/ land cover map of Puincha micro water-shed.

1. The micro watershed boundary was delineated using SOI topo sheet in 1:50,000 scale.

2. Initial base map of the study area was prepared from SOI topo sheet indicating watershed boundary and few control points like state highway, metallic and nonmetallic roads, prominent rivers, etc.

3. The base map was superimposed on multi-season satellite FCC data. Boundaries of various hydrogeomorphological units were demarcated by means of visual interpretation technique whose fundamental is based on size, shape, shadow, tone, texture, pattern and association characteristics of images. The structural information like folds, fractures, lineaments, etc. were incorporated.

4. After preliminary image interpretation, the findings were compared with ground truth.

5. Corrections and modifications were made wherever found necessary and final hydrogeomorphological map of Puincha micro watershed was prepared.

6. The area/spatial distribution of different units under hydrogeomorphological map were calculated.

9.3.4.5   Preparation of Drainage and Surface Water Body Map

Drainage and surface water body map refers to the different drainage lines likes major rivers, streams, streamlets, *nalas, etc.* and the presence of different water bodies likes tanks, reservoirs, ponds, etc. The following steps were adopted to prepare the land use/land cover map of Puincha micro water-shed.

1. The micro watershed boundary was delineated using SOI topo sheet in 1:50,000 scale.

2. Initial base map of the study area was prepared from SOI topo sheet indicating watershed boundary and few control points like state highway, metallic and nonmetallic roads, prominent rivers, etc.

3. The base map was superimposed on multi-season satellite FCC data. Boundaries of various hydrogeomorphological units were demarcated by means of visual interpretation technique whose fundamental is based on size, shape, shadow, tone, texture, pattern and association characteristics of images. The structural information like folds, fractures, lineaments, etc. were incorporated.

4. After preliminary image interpretation, the findings were compared with ground truth.

5. Corrections and modifications were made wherever found nec-
    essary and final drainage map of Puincha micro watershed was
    prepared.

9.3.4.6 Preparation of Slope Map

Information on slope is an important parameter in preparing action plans
for management of land and water resources of micro watershed. Survey of
India topo maps on 1:50,000 scales were used for preparing the slope map.
The vertical drop is estimated/measured from the contour intervals and the
horizontal distance between the contours were measured from maps by
multiplying the map distance with the scale factor. Close space contours
on the map indicating higher percentage slope and was compared to sparse
contours in the same space. The density of contours on the map was used
for preparing the slope map that gives various groups/categories of slopes.
The categories of slopes based on contour spacing are given in Table 9.2.

9.3.4.7 Preparation of Soil Resources Map

Soil is a three-dimensional body and cannot be directly interpreted from
satellite image. For extracting information on soils, study of pedons in
ground condition followed by physio-chemical analysis of pedons in the
chemical laboratory is a must in conjunction with the satellite image analysis. Genesis of soils is directly influenced by climate, relief, organize matter, parent material and time. Physiographic analysis carried out through

TABLE 9.2 Land Slope Classes

Sl. No.	Slope category	Slope, %	Lower and upper limit of contour spacing, cm
1	Nearly level	0–1	More than 4 cm
2	Very gentle sloping	1–3	More than 1.33 cm and upto 4 cm
3	Gently sloping	3–5	More than 0.8 cm and upto 1.33 cm
4	Moderately sloping	5–8	More than 0.4 cm and upto 0.8 cm
5	Strongly sloping	8–15	More than 0.26 cm and upto 0.4 cm

the use of satellite image indicates the spatial distribution of variety of soils occurring in a micro-region.

For preparing the soil resource map of the Puincha micro water-shed, the satellite image has been visually interpreted using standard interpretation elements. Pedons have been studied in selective physio-graphic location. Parameters like soil depth, soil permeability, erosion condition, etc. were studied for individual soils. Sample pedons were collected from variety of physiographic locations for further analysis in the chemical laboratory. Incorporating the laboratory analysis data, the field findings of soil and terrain parameters with the physiographic details from satellite image, the final soil resource map for the water-shed has been prepared.

9.3.5 WATERSHED MORPHOMETRY ANALYSIS

It includes the analysis of description of watershed geometric and its channel system to measure linear aspects of drainage network and relief aspect of channel network. After preparing drainage map of watershed, the computation of morphometric parameters included: Basin area (A), basin perimeter (P), basin length (L_b), basin width (B), stream length (Lu), mean stream length (L_{sm}), bifurcation ratio (R_b), drainage density (D_d), texture ratio (R_t), stream frequency (F_s), elongation ratio (Re), circularity ratio (R_c), form factor (F_f), and relief ratio (R_R).

Basin length (L_b)
Basin length is the longest dimension of a basin to its principal drainage channel.

Average basin width (B)
It is calculated as below:

$$B = \frac{A}{L_b} \qquad (1)$$

where, A= basin area, km², and L_b= basin length, km.

Bifurcation ratio (R_b)

The bifurcation ratio is the ratio of the number of stream segments of given order to the number of segments of next higher order.

$$R_b = \frac{N_u}{N_{u+1}} \tag{2}$$

where, N_u = total number of stream segments of order u, and N_{u+1} = number of stream of segment of next higher order.

Horton [9] considered the bifurcation ratio as index of relief and dissertation. Strahler [35] demonstrated that bifurcation ratio shows a small range of variation for different regions or for different environment except where the powerful geological control dominates. It is observed that the R_b is not same from one order to its next order, and these irregularities are dependent upon the geological and lithological developments of the drainage basin [36]. The bifurcation ratio is dimensionless property and generally ranges from 3.0 to 5.0. The lower values of R_b are characteristics of the watersheds, which have suffered less structural disturbances [36] and the drainage pattern has not been distorted because of the structural disturbances [21].

Stream Length (L_u)

The stream length (L_u) has been computed based on the law proposed by Horton [9]. Stream length is one of the most significant hydrological features of the basin as it reveals surface runoff characteristics. The stream of relatively smaller length is characteristics of areas with larger slopes and finer textures. Longer lengths of streams are generally indicative of flatter gradient. Generally, the total length of stream segments is maximum in first order stream and decreases as stream order increases. The numbers of streams are of various orders in a watershed are counted and their lengths from mouth to drainage divide are measured with the help of GIS software.

Mean stream length (L_{sm})

The mean stream length is a characteristic property related to the drainage network and its associated surfaces [36]. The mean stream length (L_{sm}) has been calculated by dividing the total stream length of order by the number of stream.

$$L_{sm} = \frac{L_u}{N_u} \qquad (3)$$

where, L_u = mean stream length of a given order, km, and N_u = number of stream segment.

Form factor (F_f)

Form factor (F_f) is defined as the ratio of the basin area to the square of the basin length.

$$F_f = \frac{B}{L} = \frac{A_u}{L_b^2} \qquad (4)$$

where, Au = Area of the basin, km², and L_b = Maximum basin length, km.

This factor indicates the flow intensity of a basin of a defined area [10]. The form factor value should be always less than 0.7854 (the value corresponding to a perfectly circular basin). The smaller the value of the form factor, the more elongated will be the basin. Basins with high form factors experience larger peak flows of shorter duration, whereas elongated watersheds with low form factors experience lower peak flows of longer duration.

Basin shape factor (S_b)

$$S_b = \frac{L_b^2}{A} \qquad (5)$$

where, L_b = maximum length of the basin along the main stream from the outlet to the most distant ridge of the basin, km, and A = area of the basin, km².

Circulatory ratio (R_c)

Circularity ratio is the ratio of the area of a basin to the area of circle having the same circumference as the perimeter of the basin.

$$R_c = \frac{4\pi A}{P^2} \qquad (6)$$

where, A = area of basin, km², and P = perimeter of basin, km.

It is influenced by the length and frequency of streams, geological structures, land use/ land cover, climate and slope of the basin.

Compactness coefficient (C_c)
Compactness coefficient of a watershed is the ratio of perimeter of watershed to circumference of circular area, which equals the area of the watershed. The C_c is independent of size of watershed and dependent only on the slope.

$$C_c = \frac{P^2}{4\pi A} \qquad (7)$$

where, A = area of basin, km², and P = perimeter of basin, km.

Elongation ratio (R_e)
Schumm [28] defined elongation ratio as the ratio of diameter of a circle of the same area as the drainage basin to the maximum length of the basin.

$$R_e = \frac{D_c}{L_b} = \frac{2}{L_b} \times \sqrt{\frac{A}{\pi}} \qquad (8)$$

where, D_c = diameter of the circle having same area as that of the basin, km, L_b = maximum basin length, km, and A = basin area, km².
Values of R_e generally vary from 0.6 to 1.0 over a wide variety of climatic and geologic types. Re values close to unity correspond typically to regions of low relief, whereas values in the range 0.6–0.8 are usually associated with high relief and steep ground slope [36]. These values can be grouped into three categories namely: (a) circular (>0.9), (b) oval (0.9–0.8), (c) less elongated (<0.7).

Texture ratio (R_t)
Drainage texture ratio (R_t) is the total number of stream segments of all orders per perimeter of that area [10].

$$R_t = \frac{N_1}{P} \qquad (9)$$

where, N_1 = total number of first order streams, and P = perimeter of basin, Km.

It depends upon a number of natural factors such as climate, rainfall, vegetation, rock and soil type, infiltration capacity, relief and stage of development.

Relief ratio (R_h)
The relief ratio (R_h) is ratio of maximum relief to horizontal distance along the longest dimension of the basin parallel to the principal drainage line [28].

$$R_h = \frac{B_h}{L_b} \tag{10}$$

where, B_h=Basin relief, km, and L_b=Basin length, km.
The R_h normally increases with decreasing the drainage area and size of watersheds of a given drainage basin. Relief ratio measures the overall steepness of a drainage basin and is an indicator of the intensity of erosion process operating on slope of the basin [28].

Relative relief (R_R)
The maximum basin relief was obtained from the highest point on the watershed perimeter to the mouth of the stream. Using the basin relief (174 m), a relief ratio was 0.006 that was computed by a method by Schumm [28]. Relative relief was also calculated using the formula:

$$R_R = \frac{H_m}{P} \tag{11}$$

where, H_m = maximum watershed relief, and P = perimeter, mm.

Drainage density (D_d)
Horton [9] introduced the drainage density (D_d) that is an important indicator of the linear scale of landform elements in stream eroded topography. It is the ratio of total channel segment length cumulated for all order within a basin to the basin area, which is expressed in terms of km/km^2.

$$D_d = \frac{L}{A} \tag{12}$$

where, L = total length of stream, km, and A = area of basin, km^2.

TABLE 9.3   Drainage Density Classes

Value of D_d	Class
Below 1km/sq.km	Extremely low density
1–2 km/sq.km	Low density
More than 2 km/sq.km	Medium density

The drainage density indicates the closeness of spacing of channels, thus providing a quantitative measure of the average length of stream channel for the whole basin. It has been observed from drainage density measurement made over a wide range of geologic and climatic type that a low drainage density is more likely to occur in region and highly resistant of highly permeable subsoil material under dense vegetative cover and where relief is low. High drainage density is the resultant of weak or impermeable subsurface material, sparse vegetation and mountainous relief. Low drainage density leads to coarse drainage texture while high drainage density leads to fine drainage texture [36]. Drainage density classes are shown in Table 9.3.

Ruggedness number (R_n)
It is the product of maximum basin relief (H_m) and drainage density (D_d), where both parameters are in the same units. An extreme high value of ruggedness number occurs, when both variables are large and slope is steep [35].

$$R_n = H_m \times D_d \tag{13}$$

where, R_n = Ruggedness number, dimensionless, H_m = maximum relief of watershed, km, and D_d =drainage density, km^{-1}.

Stream frequency (F_s)
Stream frequency (F_s) is the total number of stream segments of all orders per unit area.

$$F_s = \frac{N}{A} \tag{14}$$

where, Fs = stream frequency, km^{-2}, N = total number of stream, and A = area of basin, km^2.

9.3.6 METHODOLOGY ADOPTED TO GENERATE THE ACTION PLAN IN GIS

9.3.6.1 Use of Thematic Maps for Integration

The thematic maps prepared from the remote sensing satellite FCC data through visual interpretation technique (i.e., onscreen digitization) were converted into the digital raster image by ARC-GIS.

9.3.6.2 Digital GIS Database of Thematic Maps

Then the raster images of these thematic maps were stored in TIFF (Tagged Image File Format) fields in form of point, line and polygon, etc. The image quality of the raster images are improved by providing contrast, threshold options. Then vector co-ordinates of the point, line and polygon files were finalized and the registration through control point (latitude and longitude) of the TIFF files were done and stored in CPP files. These TIFF and CPP files were converted to Data interchangeable File or ARC format using import/extract utility.

9.3.6.3 Working Procedure in ARC/INFO GIS

The Vector co-ordinate unit from pixel value in files was transformed into the geo-coded ARC/IMG unit (in Grid Files) by using import/export utility to generate a coverage having single co-ordinate system. Then the editing of different map coverage, were done by different options like dangle node, overshoot, undershoot, etc. The topology buildings, clean command along with projection setting, etc. is different sets of provisions are provided. Then the labeling of features stored in Polygon attribute table for points polygon (PAT), Arc Attribute Table for Line (AAT) were done by providing different polygon id, line id, etc.

9.3.6.4   Working Procedure in ARC/INFO Package Environment

The different thematic ARC/INFO coverage along with attribute exported to the ARC/VIEW environment and different themes were created for different view. Different symbols, colors were given to the different attributes to classify different features of a map. Then the geographical analysis of these maps was done by different procedures, such as:

- Overlaying analysis
- Proximity analysis
- Tabular and statistical analysis
- Database query

9.3.6.4.1   Overlaying Analysis

Overlaying is done by two ways. It may be vector over laying or raster overlaying. In this analysis map features and associated attributes are integrated to produce a composite map. Logical operators (AND, OR, NOT) and conditional operators ($>$, $<$, $=$, $<>$) etc. are used for this purpose.

Firstly, hydrogeomorphological map, lithology map, slope map, drainage map were overlaid to generate the water resources utilization plan. The generated water action plan was overlaid over the land use/land cover map to prepare the integrated map. Then the integrated map was again overlaid on soil and slope map to generate the indicative action plan for land resource management.

9.3.6.4.2   Proximity Analysis (Buffer Operations)

It is mainly done to locate extent of area and know the characteristics of the area surrounding a specified location or a particular structure.

By means of buffer operation under proximity analysis, different extents of area were located surrounding a structure in the water action plan. The allocation of 50 m width along the length of lineaments, 40 ha catchment area around a nala bund or percolation tank, 25 ha upstream side catchment area about check dam and 50 ha downstream side command area about water harvesting structure, etc. were done by buffer operation.

9.3.6.4.3   Tabular and Statistical Analysis and Database Query

The attribute and geographic data of the integrated land use/land cover map and water action plan were stored in form of table with all available statistics. The database query used the individual LU/LC and buffered area id to regenerate a new id for indicative action plan for land resource management action plan. All the above said analysis was done by means of ARC/VIEW-3.2 software package.

9.3.7   GENERATION OF ACTION PLAN

9.3.7.1   Water Resource Development Plan

The water resources development plan is generated by recognizing precipitation and its hydrological cycle along with nature of terrain to enhance productivity and mitigation of drought. The terrain parameters like slope, drainage pattern, soil cover and thickness and hydrological conditions like rock types, thickness of weathered strata, fracture, depth to bed rock, etc. are analyzed and integrated for the preparation of water resources development plan of watershed. The surface water resources within the micro watershed such as drainage network, tank, pond, water harvesting structure, percolation tank, etc. are identified. The areas suitable for groundwater development are also proposed. Detailed description of water resources management action through surface and ground water development is given below.

Surface Water Development

1. **Renovation of water bodies:** This includes all the existing water bodies, which need renovation because of siltation, development of weeds and structural failure. Sites for renovation is suggested by analyzing multi-season satellite data along with the siltation, sand casting, weed growth and structure failure.

2. **Nala bunds:** *Nala* bunds are embankments constructed across nala for checking velocity of runoff, increasing water percolation and improving soil moisture regime.

3. **Check dams:** Check dams are constructed across small streams having gentle slope and are feasible both in hard rock as well as

alluvial formations. The site selected for check dam should have sufficient thickness of permeable bed or weathered formation to facilitate recharge of stored water within short span of time. The water stored in these structures is mostly confined to stream course and the height is normally less than 2 m. These are designed based on stream width and excess water is allowed to flow over the wall. In order to avoid scouring from excess run off, water cushions are provided at downstream side. To harness the maximum run off in the stream, series of such check dams can be constructed to have recharge on regional scale. For selecting a site for check dams/nala bunds, following conditions may be observed:

a. The width of nala bed should be at least 5 m and not exceed 15 m and the depth should not be less than 1 m.
b. The lands downstream of check dam/bund should have irrigable land under well irrigation (This is desirable but not an essential requirement).
c. The rock strata exposed in the pounded area should be adequately permeable to cause ground water recharge through pounded water.
d. Dams should be built at sites that can produce a relatively high depth to surface area so as to minimize evaporation losses.
e. Convenient location for user groups.
f. No soil erosion in the catchment area.

TABLE 9.4 Common Logic of Providing Different Soil and Water Conservation Structures for Surface Water Development

S. No.	Water action plan units	Logic to allocate the site
1	Nalabund	In lower order stream line (1st, 2nd order) and nearly level to gently sloping land (0–3% slope)
2	Percolation tank	Along or at the intersection of fracture/lineaments with nearly level to gently sloping land (0–3% slope)
3	Check dam	Lower order streams (1st orders) and gently to moderately sloping land (3–10%)
4	Water harvesting structure	Comparatively higher order (up to 3rd order), command area upto 50 ha and nearly level to gently sloping land (0–5%)

4. **Water harvesting structure:** Water harvesting structures are the earthen structures constructed across the streams, which harvest surface runoff during the monsoon rains. They are used to collect and impound surface runoff during monsoon rains and provide protective irrigation to the *kharif* crops and make them drought proof by overcoming the accumulated soil moisture deficits within the rainy season.

The water action plan was prepared by overlaying hydrogeomorphological map (HGM map), drainage map and slope map using ARC-GIS software packages in GIS with common logic shown below in the Table 9.4.

Following considerations were taken to suggest suitable actions for drainage line treatment:

- where the local bed slope is above 20% and where thinning operation yields adequate raw materials, brushwood checks are preferred.
- where the local bed slopes are between 5–20%.
  - if boulders are freely available go for dry boulder checks.
  - if boulders are not freely available go for boulder cum earth checks.
- where local bed slopes are less than 5%.
  - Nala bunds which serve as percolation reservoirs in the upper catchment.
  - Sand field bag structures in order to check the velocity of stream flow where sand is locally available.
  - Gabion structures where velocity and volume of peak runoff is too high for loose boulder checks.

9.3.7.2 Land Resources Development Plan

From the site conditions of the different engineering structures proposed in water action plan, it is known that surface water development can be done by constructing these structures, both in upstream catchment and downstream command area. The structures like nala bund, percolation tank and check dam enrich the surface water potential upto 40 ha and 25 ha, respectively in upstream catchment area. The surface water availability will be more in downstream command area up to 50 ha by constructing water-harvesting structures. The ground water potential in the region of lineaments, valley filled and moderately weathered buried pediment can be enhanced by constructing deep bore well, shallow dug well.

An integrated plan of land use/land cover and the stipulated area of above said water action plan, (where the water availability will be more after construction of structures and digging of wells) is prepared to know the present land use conditions in those areas. Integrated plan of Puincha micro watershed was obtained by overlaying land use/land cover map with water action map to suggest the indicative action plan for land management in the buffered or integrated area, the integrated map, soil resource map, slope map, etc. were undergone a overlaying analysis in GIS. While generating the action plan a short deliberation is made on the optimality of the present land use especially keeping in view the sustainable production and quality of ecosystem. If the present land use is considered sub-optimal, then a few possible options for such a site are discussed with an aim to achieve optimality within the overall framework of sustainability of production. Unless and otherwise the present land use is beyond the threshold limit of some land parameters, a drastically different option is not recommended since such a change will not meet with high level of acceptability. For example a land unit, which is ideally suitable for horticulture or fodder or fuel wood plantation, if presently under agricultural practice, then a modest change such as agro horticulture or agro-forestry is recommended. However,

TABLE 9.5   Criteria for Decision-Rules of Action Plan

S.No.	Land use/ Land cover	Geomorphology	Ground water condition	Slope (%)	Suggested action plan
1.	Settlements, Water bodies	On all landform	-	-	Optimally used
2.	Kharif crop	Pedi plain, Alluvial plain	Mod-good	0–5	Intensive agriculture with vegetables
3.	Kharif crop	Buried pediplan	Mod-good	0–5	Agro horticulture
4.	Open forest	Denudational hills, Residual hills, Pediments	Mod-poor	>5	Gap Plantation/ forest Plantation

TABLE 9.5 Continued

S.No.	Land use/ Land cover	Geomorphology	Ground water condition	Slope (%)	Suggested action plan
5.	Scrub forest	Denudational hills, Residual hills, Pediments	Mod-poor	>5	Afforestation
6.	Degraded Plantation	Pediments, Scrub land	Mod-good	0–5	Xerophyte-Fodder Plantation
7.	Land with/ without scrubs	Denudational hills, Residual hills, Undulating upland	Mod	>15	Mixed Plantation
8.	Land with/ without scrubs	Pediments, Alluvial plain	Mod-good	0–15	Horticulture, Hortipasture
9.	Barren/Rocky/ Stony waste	Denudational hills, Residual hills, Undulating upland	Mod	0–15	Fodder and fuel wood plantation
10.	Water bodies	Alluvial plain, nearly level land	Good	0–5	Pisciculture
11.	Scrub land	Pediplain, Pediments	Mod	0–5	Gully plugging, afforestation

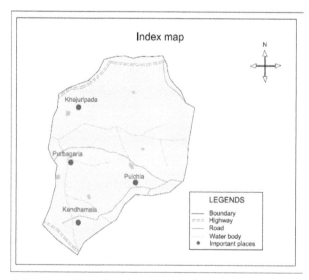

FIGURE 9.2   Index map of Puincha micro-watershed.

for a similar site if the slope is very steep then it becomes the limiting factor. Hence agricultural practice is ruled out and an altogether new land use practice like silvopasture or fodder and fuel wood plantation is recommended.

Further while making alternate recommendations for land use practice, futuristic considerations such as exploitations of ground water, if presently not exploited and possibility of adopting more efficient system of irrigation and water management and other site improvement through soil and water conservation are also kept in view. Availability of improved varieties of crops, trees, shrubs and grasses and advantages of interdependency of agriculture, livestock and other practices in case of integrated farming system that have been made available through

TABLE 9.6 Land Use/Land Cover Classification of Puincha Micro-Watershed

Map unit	Land use/land cover class	Area, ha	% of the total area
I.	Build up land settlement	12.80	1.98
II.	Agricultural land	446.27	69.01
III.	Forest	131.28	20.30
IV.	Waste land (culturable) Land with scrub	49.73	7.69
V.	Water bodies	6.59	1.02

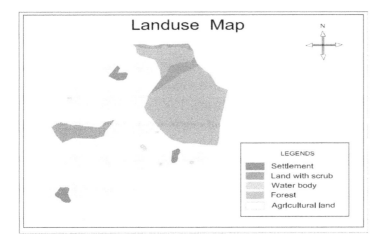

FIGURE 9.3 Land Use/Land Cover Map of Puincha Micro Watershed.

contemporary research are also considered. Thus with these conditions, finally an alternate land use practice is recommended for the site suitable for its recorded parameters. The criteria for decision rules of action plan are presented in Table 9.5.

9.4 RESULTS AND DISCUSSION

9.4.1 INDEX MAP OF PUINCHA MICRO-WATERSHED

The index map of the Puincha micro-watershed was prepared to a scale of 1:50,000 and is presented in Figure 9.2. The Index map shows different categories like watershed boundary, highways, roads, and water bodies, etc.

9.4.2 LAND USE/LAND COVER MAP OF PUINCHA MICRO-WATERSHED

Land use/Land cover map of Puincha micro watershed was prepared in a scale of 1:50,000 through visual interpretation of IRS-1C-LISS-III FCC Data and IRS-1C-LISS-III+PAN FCC data. Then the spatial data inform of land use/land cover map with its different units were classified by RS and GIS and presented in Table 9.6. The information of land use/land cover status of the micro watershed is shown in Figure 9.3.

1. Built-up land
It refers to the area of human habitation developed due to non-agricultural use and which has a cover of settlements. The built up lands in Puincha micro watershed is constituted of mainly rural settlements. The total area occupied by this land category is 12.80 ha, which is 1.98% of the total of the micro watershed.

2. Agricultural land
It refers to the land primarily used for farming and for production of food, fiber and other commercial and horticultural crops. It includes cropped land under irrigated and un-irrigated conditions, fallow land and

plantations. The agricultural land of Puincha micro watershed occupies 446.27 ha, which is 69.01% of the total area 646.67 ha.

3. Forest

It is an area bearing an association predominantly of trees and other vegetation type capable to produce timber and other products. Deciduous forest mainly shows the characteristic of shedding their leaves once in a year. Forest can be categorized into four types such as, dense forest, open forest, degraded forest and forest plantation. The area under forest is 131.28 ha, which is 20.30% area of total micro watershed.

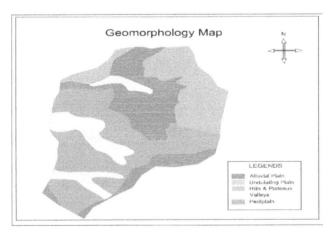

FIGURE 9.4   Geomorphology map of Puincha micro-watershed.

TABLE 9.7   Hydrogeomorphological Characteristics of Puincha Micro-Watershed

Map Unit	Geomorphic unit	Groundwater prospect	Area (ha)	% of the total area
1	Alluvial Plain	Very good	177.12	27.39
2	Hills & Plateaus	Good to moderate	154.23	23.85
3	Valleys	Very good to good	55.36	8.56
4	Pediplain	Excellent	259.96	40.20
5	Undulating Plain	Poor	–	–

4. Waste lands

The wastelands are broadly categorized into two types: (i) culturable waste land; and (ii) non-culturable waste land. The culturable wasteland is described as degraded, underutilized, deteriorated land due to lack of soil and water management or natural causes; and such land can be brought under vegetation cover with reasonable effort. This culturable waste lands mainly include the land with scrub and land without scrub which occupy relatively higher topology like upland, highland, etc. The culturable waste-land covers 49.73 ha which is 7.69% of total area.

5. Water body

This unit covers the reservoirs, ponds and tanks, etc. and covers an area of 6.59 ha which is 1.02% of the total geographical area. The result shows that Puincha micro watershed is one of the potential regions for agricultural development. Agriculture alone occupies about 69.01% of total micro watershed. Forest area constitutes of 20.30% of total area. The culturable wasteland covers 7.69% of the total area.

9.4.3   HYDROGEOMORPHOLOGICAL CHARACTERISTICS OF PUINCHA MICRO WATERSHED

The hydrogeomorphological units were identified and mapped through visual interpretation and GIS techniques in 1:50,000 scale. Figure 9.4 shows the hydrogeomorphological map of the micro watershed and Table 9.7 represents the hydrogeomorphological units of the same. Different hydrogeomorphological units represent their ground water prospects, and are discussed below.

1. Alluvial plain

Alluvial Plain is a largely flat landform created by the deposition of sediment over a long period of time by one or more rivers coming from highland regions, from which alluvial soil forms. A floodplain is part of the process, being the smaller area over which the rivers flood at a particular period of time, whereas the alluvial plain is the larger area representing the region over which the floodplains have shifted over geological time. The alluvial plain spreads an area about 177.12 ha which is about 27.39% of total area of the micro watershed. Groundwater potential is very good on these geographic units.

2. Hills and plateaus

These are relief features on the surface of the earth. It means that the earth is not a flat piece at all corners or places but is undulating, in the sense, that at places, it is raised in the form of mountains, not so steeply in the form of hills also is elevated like a table above a piece of flat land when we have a plateau. Hills and plateau are different relief features found on the surface of the earth. Though both are elevated landforms, hills are higher and steeper than plateaus. Plateaus are suddenly elevated but flat land pieces in themselves. Hills are gentler than mountains and have rounder peaks than mountains also. The hills and plateaus spreads an area about 154.23 ha which is about 23.85% of total area of the micro watershed. Groundwater potential is good to moderate on these geographic units.

3. Valleys

A valley is a depression that is longer than it is wide. The terms U-shaped and V-shaped are descriptive terms of geography to characterize the form of valleys. Most valleys belong to one of these two main types or a mixture of them, (at least) with respect of the cross section of the slopes or hillsides. The valley spreads an area about 55.36 ha which is about 8.56%

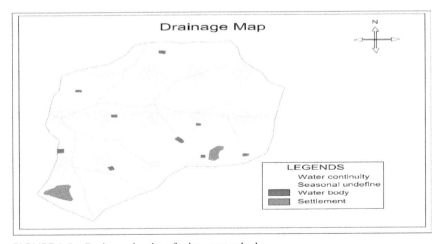

FIGURE 9.5   Drainage density of micro-watershed.

of total area of the micro watershed. Groundwater potential is very good to good on these geographic units.

4. Pediplain

The extensive slightly inclined denudation plain, which is formed under the conditions of arid and semiarid climate on the spot is earlier than the existed mountain or hilly relief by the parallel retreat of slopes from the axis of valleys. Relative to the mechanism of the formation of pediplain there is no unanimous opinion. It is considered that the main and necessary condition of forming pediplain is the long absence of the motions, which create inclines, and the fixed attitude of the basis of denudation, which determines

TABLE 9.8 Morphometric Parameters of Puincha Micro-Watershed

S.No.	Parameters	Formulae	Value
1.	Area of the basin (A) in Km^2	A	$6.46\ km^2$
2.	Perimeter of the basin (P) in Km	P	4.486 km
3.	Basin length (L_b) in Km	L_b	3.889 km
4.	Avg. Width of the basin (B) in km	$B=A/L_b$	1.661 km
5.	a) No of 1st order stream segments	N_1	13
	b) No of 2nd order stream segments	N_2	7
6.	a) Cumulative length of 1st order stream	$Lc_1=\pounds L_1$	3.431 km
	b) Cumulative length of 2nd order stream	$Lc_2=\pounds L_2$	1.234
7.	a) Mean length of 1st order stream	$Lm_1=\pounds L_1/N_1$	0.263 km
	b) Mean length of 2nd order stream	$Lm_2=\pounds L_2/N_2$	0.176 km
8.	Bifurcation Ratio (R_b)	$R_b = \dfrac{N_u}{N_{u+1}}$	13/7=1.85
9.	Drainage density (D_d)	$D_d = L/A$	$0.067\ km^{-1}$
10.	Texture Ratio (R_t)	$R_t=N_1/P$	2.897
11.	Elongation ratio (R_e)	$R_e = D_c/L_b$	0.635
12.	Circularity ratio (R_c)	$R_c = 4\pi A/P^2$	0.771
13.	Form factor (R_f)	$R_f=B/L_b$	0.427
14.	Relief Ratio	$R_r = R/L_b$	0.0033
15.	Ruggedness number (R_n)	$R_n=H_m \times D_d$	0.0067
16.	Relative Relief (R_R)	$R_R = H_m/P,$	0.022
17.	Stream Frequency (F_s)	$F_s=N/A$	3.095
18.	Compactness Coefficient (C_c)	$C_c= P^2/4\pi A$	0.248

the descending development of relief and leveling off under any climatic conditions. It is formed by the erosional works of wind. Mountains are made up of both hard and soft rocks. During erosion by wind, soft rocks are washed out, but harder rocks remained in their place. They formed some upland and are called "inselberg." The Pediplain spreads an area about 259.96 ha which is about 40.20% of total area of the micro watershed. Groundwater potential is Excellent on these geographic units.

9.4.4 MORPHOMETRIC ANALYSIS OF MICRO WATERSHED

The drainage map of the watershed was prepared by using Survey of India topographical sheets and recent false color composites derived from satellite remote sensing to a scale of 1:50,000. Figure 9.5 represents the drainage and surface water body map of the watershed. Here the seasonal undefined means seasonal flow of water comes in those 1st order streams when it rains only. Water continuity shows the 2nd order streams in the micro watershed. The morphometric parameters like basin area (A), basin perimeter (P), basin length (L_b), Basin width (B), stream length (Lu), mean stream length (Lsm), bifurcation ratio (R_b), drainage density (D_d), texture ratio (R_t), stream frequency (F_s), elongation ratio (Re), circularity ratio (Rc), form factor (Ff) and relief ratio (R_R) were computed and presented in Table 9.8.

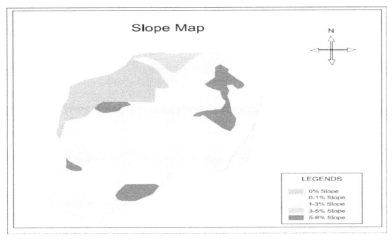

FIGURE 9.6   Slope map of Puincha Micro-watershed.

TABLE 9.9   Slope Attributes of Puincha Micro-Watershed

Map unit	Slope category	Slope (%)	Area (ha)	% of the total area
1.	Level	0	133.99	20.72
2.	Nearly level	0–1	279.69	43.25
3.	Very gently sloping	1–3	186.78	28.88
4.	Gently sloping	3–5	19.23	2.97
5.	Moderately sloping	5–8	26.98	4.17

The results indicate that the value of bifurcation ratio R_b is 1.85, which denotes the micro watershed has suffered less structural disturbances and the drainage pattern has been distorted. The value of F_f is 0.427 (which is much less than 0.7854), shows the micro watershed to be elongated nature. The value of R_c and R_e were found to be 0.771 and 0.635 (which are less than 1), respectively showing the nature of micro watershed as elongated. The values of R_t is 2.897 implies to be a moderately drained micro watershed. The value of D_d is 0.067 km^{-1}, i.e., below 1 km^{-1}, which shows the extremely low drainage density nature of micro watershed (Figure 9.5 shows drainage density of watershed). The value of F_s is found to be 3.095, i.e., below 10 showing the poor drainage frequency characteristics of the micro watershed.

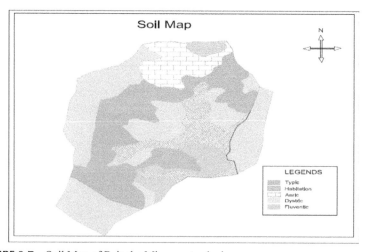

FIGURE 9.7   Soil Map of Puincha Micro-watershed.

TABLE 9.10   Soil Resources Characteristics of Puincha Micro-Watershed

Map unit	Soil unit	Soil characteristics	Area (ha)	% of the total area
1	Typic	Sandy loam to Clay loam, clay contains 17.3–56%.	218.89	33.85
2	Aeric	Fine silty to coarse silty, mixed.	73.27	11.33
3	Dystric	Coarse silty, mixed, pH less than 5.5.	270.69	41.86
4	Fluventic	Fine silty, coarse loamy, mixed.	16.95	2.62
5	Habitation	Sandy, clay, loamy, mixed.	66.87	10.34

9.4.5   SLOPE ATTRIBUTES OF MICRO-WATERSHED

The slope map of the micro watershed was to be prepared using SOI topographical sheet. Figure 9.6 represents the slope map of the sub watershed and different categories of slope and mentioned in the Table 9.9. The results indicate that the major area is nearly level and with (0–1% slope), which is 43.25% of the total area. The other land classes were: completely level (0% slope), Very gently sloping (1–3%), Gently sloping (3–5%) and Moderately sloping (5–8%) cover 20.72%, 28.88%, 2.97% and 4.17%, respectively.

9.4.6   SOIL CHARACTERISTICS OF MICRO-WATERSHED

Information generated on soil resources of Puincha micro watershed is based on the multistage approach of remote sensing based spatial information, field based observation of soil, terrain parameters and analysis of pedons in the chemical laboratory. In total different categories of soils have been recognized in the scale of 1:50,000. Figure 9.7 represents the soil resource map of the watershed. The details of individuals are discussed in Table 9.10.

The Results indicate that most of the soil texture is under Dystric unit, which covers an area about 270.69 ha (41.86% of the total area). The typic soil unit has a sandy loam to clay loam contains 17.3–56%, which covers an area about 218.89 ha which is about 33.85% of the total area. The Aeric unit, Fluventic unit and Habitation unit cover about 73.27 ha, 16.95 ha

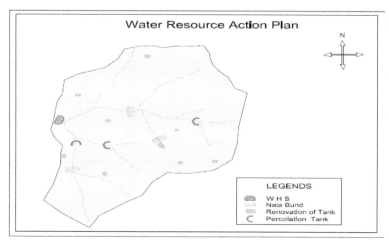

FIGURE 9.8   Water resource action and development plan map.

TABLE 9.11   Number of Proposed Structures for Water Resources Development in Puincha Micro Watershed

S. No.	Engineering Structures	Numbers
1	Renovation of water bodies	9
2	Nala bund	3
3	Percolation tank	3
4	Check dam	–
5	Water harvesting structure	1

and 66.87 ha, which is about 11.33%, 2.62% and 10.34% of the total area, respectively.

9.4.7   WATER RESOURCES DEVELOPMENT PLAN

The water resource development plan of Puincha micro watershed was obtained by overlaying hydrogeomorphological map (HGM map), Slope map (S map) and drainage map (D map) using R2V, ARC/INFO, ARC/VIEW software packages in GIS, respectively. The water resources

TABLE 9.12   Recommended Action Items for Land Resource Management

Map unit of integrated plan	Recommended land action items under catchments, command and ground water exploitation area	Total area, ha	% of total area
1.	Built up land (Settlement)	12.67	1.96
2.	Afforestation	187.86	29.05
3.	Intensive Agriculture	25.43	3.93
4.	Forest Plantation	3.16	0.49
5.	Agroforestry	37.18	5.75
6.	Agricultural Plantation	27.63	4.27
7.	Dryland cropping	330.57	51.12
8.	Slope Plantation	9.50	1.47
9.	Community Tree Plantation	12.67	1.96
10.	Slope Plantation	–	–

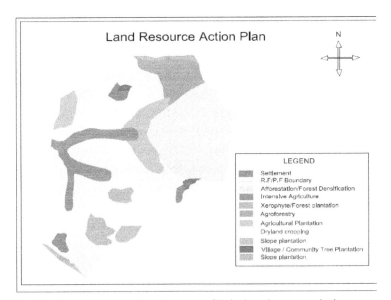

FIGURE 9.9   Land resource action plan map of Puincha micro-watershed.

management plan is generated to make the judicious and effective use of water resources of the watershed to enhance the productivity and mitigate drought. The plan indicates the sites for surface water development and sites for groundwater exploitation. Different engineering structures proposed for the water resource development is shown in Figure 9.8. The water bodies to be renovated are also indicated in this map. The number and type of proposed engineering measures are given in Table 9.11.

9.4.8   LAND RESOURCES DEVELOPMENT PLAN

In the present study IRS-1C LIIS-III and panchromatic merged data were used to delineate various spatial patterns like land use/land cover, geomorphology, soil, slope, drainage pattern, ground water potential and infrastructure detail like road/railway, water bodies and settlement, etc. of the watershed. Watershed characteristic and runoff/ morphometry were also derived and for selection in order of choice for land development program. In the land use/land cover studies various spatial details like town/cities, settlements locations, agricultural land, forest, plantation, scrub land, wastelands, barren areas, etc. were interpreted and presented in thematic maps. Statuses of ground water potential were also studied. Finally the themes are integrated using GIS software (ARC/INFO) and action plan maps were generated for Puincha watershed. The action plan maps (Figure 9.9) show recommendation for land development in Puincha watershed. The recommendation includes afforestation sites, intensive agriculture, forest plantation, agroforestry, agricultural plantation, dryland cropping, slope plantation, tree plantation, etc. Table 9.12 gives recommended action items for land resource management of Puincha micro watershed.

The study areas are viewed with the current land use/land cover status, cropping pattern, agricultural plantations, geomorphological conditions, etc. If the present land use is considered sub-optimal, then a few possible options for such a site are suggested with an aim to achieve optimality within the overall framework of sustainability of production. Decision rules for suggested land amelioration prescriptions based on land use/land cover, geomorphological features and slope are prepared and presented. Accordingly action plan maps were prepared for Puincha watershed using map-overlaying facilities of ARC/INFO package.

9.5 RECOMMENDATIONS

9.5.1 OPTIMALLY USED LAND

The productive potential of some lands under certain land use/land cover practices/activities seem to be optimum in the present environmental condition. The dense forests (with crown cover more than 40%), double cropped areas, settlements, water bodies have been included in this category. These lands need no modification at present. The area of optimally used land (settlement) is found to be 12.67 ha, which is 1.96% of the total geographical area.

9.5.2 AFFORESTATION/FOREST DENSIFICATION

Lands under scrubs and scrub forests inside forest land were selected for afforestation as those lands are degraded remnants of forests of past. The action item has been suggested in an area of 187.86 ha which is 29.05% of the total area.

9.5.3 AGRO-FORESTRY/AGRO-HORTICULTURE WITH APPROPRIATE SOIL CONSERVATION MEASURES

Shallow weathered buried pediments and pediments under single crop within 0–5% slope can be used for agro-horticulture or agro-forestry. However, the choice between adoptions of agro-horticulture and agro-forestry depends upon the land use capability class of the lands. Lands with lower topography and better soil condition can be used for agro horticulture in comparison with agro-forestry. Pediments within 0–5% slope and under scrubs can also be used either for agro-horticulture or agro-forestry.

   Agro horticulture will play an important role in these regions, as production of annual crops is insufficient. Fruit trees, if suitably integrated, would add significantly to overall agricultural production including food, fuel and fodder; conservation of soil and water; and stability to production and income. For impounding stability and providing sustainability to the farming system in marginal areas, a tree-cum-crop farming system could prove more useful. Trees and crops combination should be made considering local condition and requirement. Prior to adoption of system of agro-horticulture or agro-forestry, soil conservation measures like land leveling or grading of the lands must be carried out.

However, it must be clarified here that, both the systems of agro-horticulture and agro-forestry are the innovative ways to improve the existing land use rather than transformations of land use. Both agro-horticulture and agro-forestry will open up new opportunities for raising income levels of small farmers without putting the present agricultural practices in jeopardy. The action item has been suggested in an area of 37.18 ha which is 5.75% of the total area.

9.5.4   FOREST PLANTATION

Degraded plantation areas, scrublands and open forests are selected for forest plantation. The area of forest plantation covers 3.16 ha which is about 0.49% of the total treatable area.

9.5.5   DRY LAND CROPPING

The land with scrub under upstream (u/s) catchment and ground water exploitation area with pediplain can be recommended for dry land cropping. Soils of these dry land are sandy loam to clay loam in nature with well-drained system. The ground water condition on this area is very good to good. Lands with lower topography and better soil condition can be used for dry land cropping which increases the interaction between livestock raising/animal husbandry and the common people of the micro watershed. The action item has been suggested in an area of 330.57 ha which is 51.12% of the total integrated area.

9.5.6   INTENSIVE AGRICULTURE

The agricultural land under demonstration command and groundwater exploitation area with pediplain can be recommended for intensive agriculture practice. Soil is coarse sandy loam to fine sandy clay loam in texture and groundwater condition is good to moderate in nature. Intensive agriculture is planned on these lands by suitable crop rotation with use of inputs like high yielding variety seed, recommended fertilizer and pesticide dose. By adopting intensive agriculture minimum two crops (kharif crops and rabi crops) are to be taken for increasing cropping intensity and net return from the same unit of land. The action item has recommended in an area 25.43 ha, which is 3.93% of the total integrated area.

9.5.7   AGRICULTURAL PLANTATION

The agricultural land and ground water exploitation area with alluvial plain can be recommended for agricultural plantation. The ground water condition in this area is very good. The action item has been suggested in an area of 27.63 ha which is 4.27% of the total area.

9.5.7.1   Tree Plantation

The land with scrub under catchment is suggested for tree plantation. The tree plantation is recommended in 12.67 ha which is 1.96% of the total area.

9.5.7.2   Slope Plantation

The land with scrub and land without scrub under catchment are suggested for slope plantation. The action item has been suggested in an area of 9.50 ha which is 1.47% of the total area.

9.6   CONCLUSIONS

1. Puincha micro watershed comprised rural settlements and developed transport network.

2. Puincha micro watershed is one of the potential region for agricultural development. Agriculture alone occupies about 69.01% of total micro watershed where as forest area constitutes of 20.30% of total area.

3. False color composites (FCCs) can be effectively used for determining the ground water prospect.

4. The micro watershed has an excellent ground water prospect in an area of 259.96 ha, which is 40.20% of the total area.

5. The soil characteristics of the study area range from sandy loam to clay loam.

6. The slope category of the study area ranges from nearly level to moderately slope. But most of the land, i.e., nearly level and with (0–1% slope), which is 43.25% of the total area.

7. The result indicates the value of bifurcation ratio R_b is 1.85, which denotes the micro watershed has suffered less structural disturbances and the drainage pattern has been distorted.

8. The value of form factor (F_f) is 0.427 (which is much less than 0.7854), shows the micro watershed to be elongated nature. The value of circulatory ratio (R_c) and elongation ratio (R_e) were found to be 0.771 and 0.635 (which are less than 1) respectively showing the nature of micro watershed as elongated.

9. The value of drainage density (D_d) is 0.067 km^{-1}, i.e., below 1 km^{-1}, which shows the extremely low drainage density nature of micro watershed. The value of stream frequency (F_s) is found to be 3.095, i.e., below 10 showing the poor drainage frequency characteristics of the micro watershed.

10. There are nine types of water bodies are proposed for renovation in the water action plan of the micro watershed.

11. Different soil and water conservation structures like three nala bunds, three percolation tanks and one water harvesting structure is proposed in the water action plan for erosion control and ground water recharge in the micro watershed.

12. In total 9 action items are recommended to use physical environmental condition in a judicious manner for sustainable development of the micro watershed.

13. Dry land cropping is the action item has been suggested in most of the area of 330.57 ha which is 51.12% of the total area.

14. Intensive agriculture is the action item has recommended in an area 25.43 ha, which is 3.93% of the total area.

9.7   SUMMARY

The watershed boundary of Puincha micro watershed was delineated by using SOI topographical sheet and IRS-1C-LISS-III sensor and IRS-ID-(LISS-III+PAN) sensor FCC data in 1:50,000 scale. The present analog land use/land cover map of the watershed was prepared through visual interpretation technique from FCC data to study the land use pattern. The analog hydrogeomorphological map was prepared by demarcating the geomorphic units along with lineaments to study the ground water prospects. The analog drainage map was prepared and morphometric parameters like bifurcation ratio, stream length, form factor, basin shape factor, circulatory ratio, circulatory index, compactness coefficient, elongation ratio, texture ratio, relief ratio, relative relief, drainage density, ruggedness number and stream frequency, etc. were computed to know the shape and characteristics of the micro watershed. The analog slope map and soil map was prepared by using toposheets. Different slope categories were delineated. All those above said thematic maps were scanned, digitized, edited and labeled and analyzed in GIS (by the use of different software packages like R2V-2.1, ARC/INFO and ARC/VIEW packages, etc.) to prepare the final maps. By integrating hydrogeomorphological map, slope map and drainage map in GIS the water action plan was prepared. From the site conditions and buffered area allocated for soil conservation engineering structures, an integrated map was prepared by overlaying land use/land cover map and water action plan in GIS. Eventually, the integrated plan was overlaid over soil resource, slope map, etc. in GIS to generate indicative action plan for different alternative suggestive measures for land management.

ACKNOWLEDGEMENTS

The authors are thankful to Dr. Md. K. Khan, Dean, CAET, OUAT, Bhubaneswar, India and Dr. R. Behera, Head, I-SPACE, Bhubaneswar, India for providing facilities for this study. The help provided by

Dr. D. Mishra, Scientist, ORSAC, Bhubaneswar, India is gratefully acknowledged. This chapter is an edited version of MTech (Agri. Eng.) SWCE thesis of Orissa University of Agriculture and Technology, Bhubaneswar, India entitled "Geoinformatics Based Land and Water Management Plan Preparation of Watershed – a case study" of first author.

KEYWORDS

- agriculture
- agro forestry
- arc GIS software
- basin length
- basin perimeter
- basin width
- bifurcation ratio
- check dam
- circulatory ratio
- drainage density
- drainage map
- elongation ratio
- false color composites
- geographical information center
- geographical information system
- geoinformatics
- ground truth
- hydrogeomorphological map
- India
- intensive agriculture
- kharif
- land slope
- land use/land cover map
- morphometric analysis

- **nala bund**
- **overlaying analysis**
- **pediplain**
- **pedons**
- **percolation tank**
- **proximity analysis**
- **rabi**
- **remote sensing**
- **satellite image**
- **silvi-pasture**
- **slope map**
- **soil and water conservation measure**
- **soil resource map**
- **stream frequency**
- **stream length**
- **stream order**
- **survey of India**
- **tabular and statistical analysis**
- **thematic map**
- **topo-sheet**
- **waste land**
- **water harvesting structure**
- **water resource development plan**
- **watershed**

REFERENCES

1. Biswas, S., Sudhakar, S., and Desai, V. R., 1999. Prioritization of sub watersheds based on morphometric analysis of drainage basin – A Remote Sensing and GIS approach. *Journal of Indian Society of Remote* Sensing, 27(3):155–156.
2. Chakraborthy, B., 1993. Strategies for watershed planning using Remote Sensing Technique. *Journal of the Indian Society of Remote Sensing*, 21(2):87–96.
3. Chakraborthy, B., Kumar, S., and Pandey, N. G., 2003. Water Resources Management through Water Harvesting – a case study. *Proceedings of IWIWM– 2001*, June 21–23, 2001, Bangalore, p. 15.
4. Chakraborthy, D., Dutta, D., and Chandra, S. H., 2001. Land use indication of a watershed in arid region, Western Rajasthan using remote sensing and GIS. *Journal of Indian Society of Remote Sensing*. 29 (3): 115–128.
5. Dwivedi, R. S., Reddy, P. R., Sreenivas, K., and Ravishankar, G. 1999. A Remote Sensing and GIS based integrated approach for sustainable development of land and water resources. *Proceedings of ISRS National Symposium*. Jan. 19–21, 1999, Bangalore, pp. 352–360.
6. Farooq, M., Rashid, G., and Arora, S., 2008. Spatio-temporal change in land use and land cover of district Rajouri-A geo-spatial approach. *Journal of Soil and Water Conservation*. 7(4):20–24.
7. Goyal, V. P., Kuhad, M. S., and Sangwan, P. S., 1999. Utilization of remote sensing data for soil characterization and preparation of geomorphic-soil maps of Rewari district of Haryana, *Proceedings of ISRS National Symposium*, January 19–21, 1999, Bangalore, pp. 140–144.
8. Gunjal, S. P., 1990. An overview of remote sensing application in soil and land resource management in India. *Proceedings of National Symposium on Remote Sensing for Agricultural Applications*, December 6–8, New Delhi, p. 10.
9. Horton, R. E., 1945. Erosional Development Of Streams And Their Drainage Basins; Hydro physical Approach To Quantitative Morphology, Bulletin Of The Geological Society Of America, Vol. 56, 275–370.
10. Jaishankar, G., Jagannadha, R. M., Prakasa, R. B. S., and Jugran, D. K., 2003. Hydro-morphology and Remote Sensing application for ground water exploration in Agnigundala mineralized belt, Andhra Pradesh, India. *Journal of Indian Society of Remote Sensing*. 29(3):165–174.
11. Jayaraju, N., and Khan, A. J., 2013. Remote Sensing and Geographical Information as an Aid for Land Use Planning and Implications to Natural Resources Assessment in South India. *Developments in Soil Classification, Land Use Planning and Policy Implications*. pp. 577–590.
12. Jayaram, A., Karale, R. L., and Sinha, A. K., 1990. Remote Sensing application for geohydrological investigation in Nagpur district, *Proceedings of National Symposium on Remote Sensing for Agriculture Application*, December 6–8, 1990, New Delhi, pp. 384–388.
13. Joshi, P. K., Gairola, S., 2004. Land cover dynamics in Garhwal Himalayas – A case study of Balkhila sub-watershed. *Journal of Indian Society of Remote Sensing*, 32(2): 199–208.

14. Kalgapurkar, A. P., Mishra, K., and Tripathi, K. P., 2012. Applicability of RS and GIS in soil and water conservation measures. *Indian Journal of Soil Conservation*, 40(3):190–196.

15. Kannan, P., Sankar, M, and Prabhavath, M., 2010. Remote sensing and GIS for land use mapping and crop planning in Cauvery delta region of Tamilnadu in India. *Indian Journal of Soil Conservation*, 38(3):199–204.

16. Kinthada, N., Krosurul, S. P, Gurram, M. K., 2013. GIS and remote sensing in hydrogeomorphological mapping and integrated agro action plan development for sustainable land-water resource management in domaleru watershed, prakasam district in India. *International Journal of Advanced Technology & Engineering Research* (IJATER), 3(1):11–18.

17. Krishna, N. P. R., Maji, A. K., and Murthy, Y. V. N., and Rao, B. S. P., 2002. Remote Sensing and GIS for canopy cover mapping – deltaic region of west Godavari. *Journal of Indian Society of Remote Sensing*, 29(3):107–114.

18. Magesh, N. S., Jitheshlal, K. V., Chandrasekar, N., and Jini, K. V., 2012. GIS based morphometric evaluation of Chimmini and Mupily watersheds parts of Western Ghats in Thrissur District of Kerala. *Earth Science Informatics*, 5(2):111–121.

19. Mohammed, G. S., Kalyani, A., and Faridsalikhan, Y., 2001, Application of Geo-processing techniques to integrate the land use and Micro-watershed features of Kangeyam, Kundadam, Mulanur and Chennimalai Panchayat union of Erode district in Tamil Nadu. *Proceedings of ICORG-2000*, 2–5th, February 2001, Hyderabad, pp. 211–221.

20. Murthy, K. S. R., Amminedu, E., and Venkateswara, R. V., 2003. Integration of thematic maps through GIS for identification of groundwater potential zones. *Journal of Indian Society of Remote Sensing*, 31(3):197–210.

21. Nag, S. K., 1998. Morphometric analysis using remote sensing techniques in the chaka subbasin, Purulia district, West Bengal. *Journal of Indian Society of Remote Sensing*, 26((1&2):6–16

22. Patel, D. P., Gajjar, C.A., and Srivastava, P. K., 2013. Prioritization of Malesari mini watersheds through morphometric analysis: a remote sensing and GIS perspective. *Environmental Earth Sciences*, 69(8):2643–2656.

23. Paul, J. C., Mishra, J. N., Pradhan, P. L, and Sharma, S. D., 2008. Remote sensing and is aided land and water management plan preparation of watershed – a case study. *Journal of Agricultural Engineering*, 45(3):27–33.

24. Ramesh, K. S., Elango, S., and Adiga, S., 2001, Prioritization of watersheds of Dhakshina Kannada district, Karnataka, using Remotely Sensed data. *Proceeding of IWIWM-2001*, June 21–23, Bangalore, p. 10.

25. Rao, U. R., 1991. Remote Sensing for sustainable development. *Journal of Indian Society of Remote Sensing*, 19(4):217- 235.

26. Reddy, P. R., Vinod, K., and Sheshadri, K., 2002. Use of IRS-IC data in ground water studies. *Current Science,* 23:600–605.

27. Sarkar, B. C., Deota, B. S., Raju, P. L. N., and Jugran, D. K., 2001. A GIS approach to evaluation of ground water potential of Shamri micro-watershed in Shimla taluk, Himachal Pradesh. *Journal of Indian Society of Remote Sensing,* 29(3):151–164.

28. Schumn, S. A., 1956. Evolution of drainage systems and slopes in badlands at Perth Amboy, *New Jersey. Geol. Soc. Am. Bull.,* 67:597–646.

29. Shah, P. N., 2001, Hydro-geomorphological mapping for evaluation of ground water prospect zones in Mirzapur district in India using IRS 1A, LISS II Geocoded data. *Proceedings of ICORG-2000, Vol. II*, 2–5th Feb. Hyderabad, pp. 446–453.

30. Sharma, V. V. L. N., Muralikrishna, G., Hemamalini, B., and Nageswararao, K., 2001. Land use/Land cover detection through Remote Sensing and its climatic implications in Godavari delta region. *Journal of Indian Society of Remote Sensing*, 29(1&2):85–92.

31. Sidhu, G. S., Das, T. H., Singh, R. S., Sharma, R. K., and Ravishankar, T., 1998. Remote Sensing and GIS Techniques for prioritization of watersheds: a case study in upper Machkund watershed. *Indian Journal of Soil Conservation*, 26(2):71–75.

32. Singh, P., Thakur, J. K., and Kumar, S., 2012. Assessment of land use/land cover using geospatial techniques in a Semi-arid Region of Madhya Pradesh, India. *Geospatial Techniques for Managing Environmental Resources*, (3):152–163.

33. Srivastava, V. K., and Rao, V. M., 1992. Evaluation of IRS-IA, LISS II data for land use study of the eastern part of the Jharia coal field. *Natural Resources Management New Perspective*, NNRMS, DOS, Bangalore, pp. 356–359.

34. Srivastava, R. K., Sharma, H. C., Kumar, S., and Tiwari, A., 2013. Temporal change in land use of Himalayan watershed using remote sensing and GIS. *Journal of Soil and Water.*

35. Strahler, A. N., 1957. Quantitative analysis of watershed geomorphology. *Trans. Am. Geophys. Union*, 38:913–920.

36. Strahler, A. N., 1964) Quantitative geomorphology of drainage basins and channel networks. *In:* V. T. Chow (ed.), *Handbook of applied hydrology*. McGraw Hill Book Company, New York, pp. 4–11.

37. Tikekar, S. S., Ayyangar, R. S., and Tandale, T. D., 2005. Water resources assessment and management in metamorphic provinces of eastern Maharashtra. *Proceedings of ISRS – Symposium,* Nov. 22–24, 1995, Ludhiana, pp. 90–97.

38. Uday, R., Ramesh, K. S., Ravishankar, H. M., and Adiga, S., 1995. Integrated approach for sustainable development of natural resources on watershed basis with emphasis on water resources development. *Proceedings of workshop on Comprehensive Management of Drinking Water in Rural Areas using Remote Sensing Data and GIS,* December 21–22, Bangalore, pp. 97–105.

39. Wakode, H. B., Dutta, D., Desai, V. R., Baier, K., and Azzam, R., 2013. Morphometric analysis of the upper catchment of Kosi River using GIS techniques. *Arabian Journal of Geosciences*, 6(2):395–408.

PART III

IRRIGATION MANAGEMENT OF CROPS

CHAPTER 10

WATER PRODUCTIVITY OF RICE UNDER DEFICIT IRRIGATION

KAJAL PANIGRAHI

Department of Civil Engineering, National Institute of Technology, Rourkela, Odisha, India, Phone +91 8895997206, E-mail: kajalpanigrahi@yahoo.in

CONTENTS

10.1 Introduction.. 282
10.2 Water Balance Model of Rice 284
10.3 Materials and Methods.. 286
      10.3.1 Simulation Model of Rice Yield................................... 287
      10.3.2 Input Data... 288
10.4 Results and Discussion 291
      10.4.1 Model Validation 291
      10.4.2 Simulation of Water Balance Parameter 292
            10.4.2.1 Seepage and Percolation 292
            10.4.2.2 Actual Evapotranspiration 293
            10.4.2.3 Supplemental Irrigation (SI) 293
            10.4.2.4 Surface Runoff 295
            10.4.2.5 Effective Rainfall 295
10.5 Conclusions... 296
10.6 Summary 296
Acknowledgements... 297

Keywords 297

References.. 298

10.1 INTRODUCTION

Rice, the primary staple for more than half the world's population, is produced worldwide. Most of the rice consumers live in the developing countries. The crop occupies 33% of the world's total area planted to cereals. More than 90% of the world's rice is produced and consumed in Asia only [8]. During 2011–2012 in India, rice was cultivated in 43.97 M ha area with a production of 100 M tons and productivity of 2.372 M tons/ha. Being a semi-aquatic plant, water requirement of rice is highest among all the crops. About 50% of total irrigation water is used for rice production in Asia [3, 4]. A survey reports that to produce a kilogram of rice, 3000 liters of water is required, which is five times higher than that required to produce a kilogram of pulse crop and three times higher to produce a kilogram of oilseed crop. Because of the intensification of agriculture, per capita availability of water resources is declining day by day in many Asian countries. It is estimated that the share of water for irrigation in agriculture will dwindle to 70% from the present share of 80% now, in India by the year 2050.

The declining water resources and the reduced share of its availability for agriculture have affected all rainfed rice farmers. It is high time now to save the costly irrigated water and economize its use in agriculture. Since rice is a major water-consuming crop, it is important to save the irrigation water in rice field with new and innovative techniques of irrigation and water management. The concept of maximum yield is now changing to optimum yield for developing an efficient and economic irrigation scheduling. Accordingly, the traditional concept of continuous submergence is not necessary and optimum yield with minimum water requirement but high water use efficiency (WUE) can be obtained with intermittent irrigation [7, 22]. Since irrigation water is a scarce commodity in the rainfed farming system, water saving irrigation (WSI) is now drawing gradual attention.

Experiments conducted with different WSI techniques in various regions of the world reveal that irrigation in rice at near SAT gives comparable yield with continuous submergence and saves a lot of water [16–18]. In another approach, rice is irrigated after standing water vanishes from the field that considerably reduces seepage and percolation and hence, water requirement of rice and increases its water use efficiency [1, 12, 20].

Most sensitive period of water deficit in rice is its critical growth stage covering crop development and reproductive stage. When soil moisture content in the effective root zone of rice is depleted below 70–80% of SAT moisture content during the critical growth stage, rice yield tends to decrease [5]. Mitra [13] has reported that under upland condition, optimum yield of dry seeded rainfed rice can be obtained by maintaining soil moisture content at field capacity throughout the active growth stage.

The amount and timing of irrigation are the main concepts of irrigation scheduling that can be better studied by simulation approach. Simulation modeling techniques are now increasingly being used as an alternative to extrapolate the results of field trials that can be transferred to farmers [10]. Simulation studies by daily water balance model helps to understand the water use as well as irrigation and drainage requirement of the crops. This will help to develop appropriate strategies for efficient management of water resources for sustainable production. Simulation of water balance parameters has been studied by many researchers in low land rice [14, 15], but hardly any work is done in rainfed rice grown under up and medium land with provision of supplemental irrigation through WSI approach.

The objectives of the present study were to:

1. Formulate a water balance simulation model for rainfed rice with and without the provision of supplemental irrigation and to simulate the various model parameters,
2. Validate the developed simulated model with the observed field data and
3. Find out the best water saving irrigation technique to increase the water productivity at the farm level.

10.2 WATER BALANCE MODEL OF RICE

The various water balance parameters considered in the model are shown in Figure 10.1. The inflow to the field consists of total water supplied from rainfall and supplemental irrigation and out flow from the field consists of actual evapotranspiration, seepage and percolation, and surface runoff. The generalized water balance model is given below after considering the effective root zone of rice as a single layer and neglecting the capillary rise of groundwater in upland topo sequence where groundwater lies more than 1.5 m below crop effective root zone,:

$$SMC_i = SMC_{i-1} + P_i + SI_i - AET_i - SP_i - SR_i \tag{1}$$

where, SMC = soil moisture content, mm; P = rainfall, mm; SI = supplemental irrigation, mm; SP = seepage and percolation loss, mm; AET = actual evapotranspiration, mm; SR = surface runoff from the field, mm; and i = time index taken as 1 day in the study.

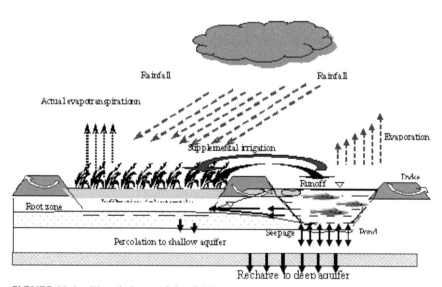

FIGURE 10.1   Water balance of rice field.

If SMC in the effective root zone of rice is more than SAT, then ponding will occur in field. Under the ponding phase, water balance in rice field is given as:

$$D_i = D_{i-1} + P_i + SI_i - AET_i - SP_i - SR_i \qquad (2)$$

where, D is the ponding depth, mm and other terms are defined as above and is given in mm.

Irrigation water is a scarce commodity in the rainfed farming system. So, WSI technique must be adopted for applying SI to rice. In the study, five different WSI techniques are considered for providing SI to rice:

1. T1–SI applied at 20% MAD of SAT during the RS;
2. T2–SI applied at 20% MAD of SAT during CD and RS;
3. T3–SI applied at 20% MAD of SAT during CD, RS and MS;
4. T4–SI applied at 20% MAD of SAT during all four stages; and
5. T5–SI applied at SAT in all the stages.

At each irrigation, 50 mm of water is applied. In all the treatments SI is provided during the prescribed growth stages as mentioned for each treatment and other stages are kept as rainfed without any SI. For all the WSI treatments, no ponding water is allowed in rice field from sowing to first 10 days after germination of rice and last 10 days before harvest. Any ponding water in the field is drained out as SR. During rest of the periods, 50 mm ponding is taken as the maximum limit and excess ponding above 50 mm is taken out as SR.

The actual evapotranspiration (AET_i) in any day 'i' is given as:

$$AET_i = Kc_i \, Ks_i \, ETo_i \qquad (3)$$

where, Kc = dimensionless crop coefficient that depends on growth stage of the crop; Ks = dimensionless crop stress coefficient that is a function of the relative available SMC in the field; and ET_o = reference crop evapotranspiration.

Daily ET_o was estimated by Penman-Monteith method for the simulation period. Values of K_c for rice are 1.05 during CE, 1.10 during both CD and RS, and 0.95 for MS [6]. The value of K_s of Eq. (3) is 1.0 under no water stress condition. But as the ponding water vanishes from the rice lands, soil moisture stress occurs that is usually provided by K_s, which consequently decreases AET. Jensen et al. [9] considered a 10% decrease in value of K_s for dry periods in rice field. In the present study, under unsaturation case, K_s is assumed to vary linearly with the ratio of SMC to SAT that is termed as relative available SMC as [1, 17]:

$$K_{si} = SMC_i/SAT \tag{4}$$

Water loss due to SP in rice fields is often inseparable and so both the terms are considered as single component [21]. The value of SP in the rice field is an extremely variable factor depending on soil and drainage condition. Under different cultural and water management practices, the values of SP are reported to vary from one to 25 mm/d [9]. For rainfed upland rice when most of the time soil in the effective root zone depth remains under unsaturated, SP is estimated by the model developed by [17] as:

$$SP_i = -16.45 + 0.145 \, (SMC_{i-1} + P_i + SI_i - AET_i - SRi) \tag{5}$$

where, all the terms have been defined earlier and are given in mm.

Under ponding stage, SP is estimated as suggested by [17] as:

$$SP_i = -16.45 + 0.145 \, (D_{i-1} + SAT + P_i + SI_i - AET_i - SRi) \tag{6}$$

where, all the terms are expressed in mm.

The value of SP in the model is computed at the end of each day where as P, SI and SR if any are assumed to occur at the beginning of the day. Water balance model for rainfed rice (treatment T0) is run by Eqs. (1)–(6) mentioned as above but the components SI and SR are taken as zero in these equations.

10.3 MATERIALS AND METHODS

The site selected for the simulation study is Hariharjore Command in Sonepore district of Odisha. It is located at a latitude of 22°19'N, longitude of 87°19'E at an altitude of 48 m above mean sea level. The study area receives about 1,570 mm mean annual rainfall of which 80% are received during the rainy season from June to September. The mean maximum and minimum temperatures are 40°C and 12°C, occurring in the month of May and January, respectively. The mean relative humidity ranges from 15.5 to 90.5%. The dominant soil group in the study area is sandy loam, acid lateritic with pH ranging from 4.8 to 5.6, and poor in organic matter. The soil has very low water holding capacity and dries up quickly after cessation of rainfall. Hence, cultivation of crop on residual moisture is difficult. Values of field capacity, wilting point and saturation moisture content of rice field in the study area within 45 cm root zone depth are 120, 42, and 170 mm, respectively.

In the present study, the basis of the modeling is DSR rice grown in upland topo sequence in wet season (rainy) without any provision of SI (rainfed) and with the provision of SI during different growth stages (WSI techniques). The dike heights of the field are adequate to check any inflow to and outflow from the field. The rice lands are considered as leveled fields. The model uses daily rainfall and other climateorological data, soil, and crop data of the study area. Effective root zone depth of rice is taken as 45 cm.

10.3.1 SIMULATION MODEL OF RICE YIELD

The yields of rice are affected by soil water deficits that might occur at any point in the growing season due to differential irrigation. Unlikely many field crops, no literature is available for prediction of rice yields that is affected by varying soil water status in the fields at different stages of its growth. The author has collected AET data at different stages of growth and rice yield from different research centers of India. Assuming all other variables of crop production remaining constant except the irrigation amount, a multilinear regression type model was developed:

$$Y_{ar} = -13.06 + 0.05\ ET_{a1} + 0.07\ ET_{a2} + 0.75\ ET_{a3} + 0.45\ ET_{a4}, R2 = 0.76$$

(7)

where, Y_{ar} = actual yield of rice (1000 kg/ha); and ET_{a1}, ET_{a2}, ET_{a3}, and ET_{a4} = values of AET during CE, CD, RS and MS, respectively (cm).

The simulation is carried out using data of 37 years (1977–2013) from the starting day of rainy season of each year when the rice is dry-seeded in the field till its harvest. The author has taken a short duration rice of 100 days as base crop for simulation of water balance model. The duration of growth stages of CE, CD, RS, and MS of the rice from the day of germination till harvest are assumed as 25, 20, 30, and 25 days, respectively allowing 3 days for germination of seed to occur after its sowing. Thus, for each year, simulation continues for 103. For first 3 days (germination period), the soil is under bare conditions. For computation of different parameters of water balance models, AET is replaced by bare soil evaporation during these 3 days, and computed as suggested by Jensen et al. [9].

A computer program was written in C language to estimate the water balance model parameters for various treatments. The model parameters were obtained on daily basis from which total values in each stage as well as that of the whole season were calculated. The seasonal total values of the parameter along-with the simulated yield of rice were fitted to ND and values were predicted at different PE levels. Values of different water balance parameters at 30, 50, and 70% PE level are shown in Table 10.1. Values of IUE (yield/total seasonal SI) and WUE (yield/total seasonal WR) were also computed for various treatments based on the values of yield, SI and WR at 50% PE, and are shown in Table 10.2.

10.3.2 INPUT DATA

For testing the water balance model parameters, field experimental data of experimental farm of the Department of Agricultural and Food Engineering, Indian Institute of Technology, Kharagpur, India was used. The experimental data of rainy season of the year 2000 carried in rice crop was used in the study. Because of limitation of scope, experiment with only two treatments (T0 and T1) were undertaken in field. Rice variety MW-10 was dry seeded in the field at the onset day of rainy season, i.e., June 10th and was harvested on September 24th for both treatments. Excess water above SAT was drained out as SR during first 10 days of seed germination and last 10 days before harvest of rice. During rest of the periods, excess of 50 mm ponding was drained out as SR. Irrigation of 50 mm was supplied when MAD value

TABLE 10.1 Simulation Results of Water Balance Parameters (mm) of Rice at Different Probability of Exceedance

PE, %	Treatment	SP	AET	WR	SI	ER	SR
30	T0	758.1	423.3	1181.4	0	1279.4	0
30	T1	747.6	430.9	1178.5	144.3	1122.4	157.0
30	T2	778.1	439.2	1217.3	212.2	1084.3	195.1
30	T3	805.8	446.5	1252.3	277.9	1045.0	234.4
30	T4	858.3	457.6	1315.9	384.7	1000.7	278.7
30	T5	1049.6	488.3	1537.9	755.8	853.7	425.7
50	T0	648.0	410.2	1058.2	0	1157.7	0
50	T1	638.2	418.2	1056.4	102.5	1040.1	117.6
50	T2	670.2	426.6	1096.8	166.4	1010.8	146.9
50	T3	698.2	434.0	1132.2	222.7	980.6	177.1
50	T4	758.5	445.0	1203.5	330.9	947.3	210.4
50	T5	955.9	472.9	1428.9	682.6	816.3	341.4
70	T0	543.3	394.2	937.5	0	1034.3	0
70	T1	527.5	405.4	932.9	86.7	945.2	89.1
70	T2	571.8	414.0	985.8	150.6	925.6	108.7
70	T3	601.2	421.6	1022.8	191.5	909.3	125.0
70	T4	645.5	434.5	1080.0	267.2	892.2	142.1
70	T5	877.3	460.8	1338.1	645.3	770.2	264.1

Note: PE = probability of exceedance, SP = seepage and percolation, AET = actual evapotranspiration, WR = water requirement, SI = supplemental irrigation, ER = effective rainfall, SR = surface runoff.

of SMC in the effective root zone depth of rice was below 20% SAT level during RS. For treatment T0, there was no provision of SI and SR.

Various instruments like tensiometer, piezometer and aquapro aceess tubes were installed in the rice field to measure different water balance model parameters in the field (Figure 10.2). Field dimensions were 40 x 20 m². In addition, drum lysimeters were also installed in the rice field to measure seepage and percolation and AET (Figure 10.3). Values of water balance parameters (SP, SR and AET) were measured daily in the field. A comparison of seasonal total water balance parameters and yield of both observed and simulated results for treatments, T0 and T1 are presented in Figure 10.4.

TABLE 10.2 Effect of Water Saving Irrigation Techniques on Yield, Water Requirement, Irrigation Use Efficiency and Water Use Efficiency of Rice

Treatment	Yield, kg/ha	SI, mm	IUE, kg/ha/ mm	Saving of SI compared to T5, %	WR, mm	WUE, kg/ha/ mm	Saving of WR compared to T5, %
T0*	2280	—	—	—	1058.2	2.15	25.9
T1	3060	102.5	29.85	85.0	1056.4	2.90	26.1
T2	3125	166.4	18.78	75.6	1096.8	2.85	23.2
T3	3250	222.7	14.59	67.4	1132.2	2.87	20.8
T4	3400	330.9	10.28	51.5	1203.5	2.83	15.8
T5	4001	682.6	5.86		1428.9	2.80	

Note: * T0 is rainfed treatment that has no provision of supplemental irrigation and surface runoff and so values of IUE and WUE are not mentioned in the table. SI = supplemental irrigation, IUE = irrigation use efficiency, and WUE = water use efficiency.

FIGURE 10.2 Tensiometer, piezometer and aquapro aceess tubes installed in the rice field.

FIGURE 10.3 Drum lysimeter installed in rice field.

10.4 RESULTS AND DISCUSSION

10.4.1 MODEL VALIDATION

The values of different water balance parameters simulated by the water balance model on daily basis were found to be close to the observed parameters. Coefficient of determination (R^2) between the observed and simulated values of various water balance parameters ranged from 0.82 to 0.91.

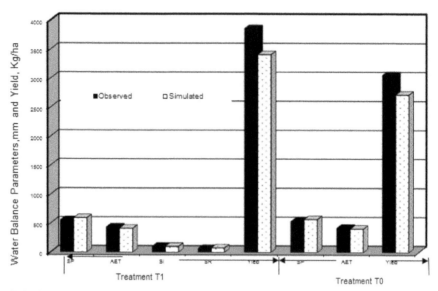

FIGURE 10.4   Comparison of performance of rainfed and WSI techniques.

Comparisons of seasonal and observed total values of different water balance parameters for treatments T0 and T1 are presented in Figure 10.4.

For treatment T1, observed and simulated seasonal values were 551.8, 589.7 for SP, 428.8, 410.5 for AET, 100.0, 100.0 for SI, 65.9, and 74.8 mm for SR, respectively. Thus, the observed and simulated seasonal values of SP and AET for treatment T0 are found to be 528.5, 565.6, 418.3, and 402.5 mm, respectively. The observed and simulated yields for treatment T1 and T0 are found to be 3851, 3408, 3050, and 2715 kg/ha, respectively. The percentage of deviation between the observed and simulated values of various water balance parameters and yield for both the treatments ranged from 0 to 13.8. The low values of deviation indicate that the models can be reasonably used for simulation of water balance parameters and yield of rice.

10.4.2   SIMULATION OF WATER BALANCE PARAMETER

10.4.2.1   Seepage and Percolation

Seasonal values of SP varied from 959.5 to 324.5, 1115.5 to 348.0, 1133.8 to 387.7, 1126.2 to 412.3, 1121.5 to 445.1, and 1329.3 to 725.8 mm for treatments T0, T1, T2, T3, T4, and T5, respectively for different PE levels ranging from 5 to 95%. Values of SP for various treatments at 30, 50, and 70% PE are shown in Table 10.1. It is observed that among all WSI treatments, T1 has lowest SP values and T5 has highest SP values at any PE level (Table 10.1). At an average chance of 50% PE, SP values for treatment T1 is 638.2 mm whereas that for T5 is 955.9 mm. Thus, there will be 33.2% less SP in rice field if it is irrigated at a SMC of 20% MAD from SAT during RS instead of maintaining SAT always. This will also save 85% of SI and 26.1% of net WR of rice in comparison to treatment T5.

If the field is frequently irrigated (that is the requirement for treatments T2 to T5), then the total profile soil moisture in the effective root zone increases as a result, the hydraulic gradient as well as unsaturated hydraulic conductivity also increases. This causes the down ward flux of water in the crop root zone to increase. It is also seen that for all the treatments, SP is the major loss in rice field that ranges from 60.4 to 66.9% of total net WR (net WR = SP + AET).

10.4.2.2   Actual Evapotranspiration

Seasonal values of AET for 30% PE level ranged from 423.3 to 488.3 mm for various treatments. As the PE level increases, values of AET for all the treatments decrease. At 50% PE, AET for all treatments varied from 410.2 to 472.9 mm and at 70% PE, it ranged from 394.2 to 460.8 mm (Table 10.1). The average values of seasonal total AET ranged from 33.1 to 39.6% of total net WR for various treatments. The percentage contribution of AET to total net WR is lowest (33.1) for treatment T5 and highest (39.6) for T1. The reason of getting lowest percentage utilization of water in meeting the AET for T5 is because of major losses of water due to SP.

10.4.2.3   Supplemental Irrigation (SI)

SI requirement of rice for all PE levels was lowest for treatment T1 than other WSI treatments of T2 to T5. At an average of 50% PE, the SI requirement for treatment T1 is 102.5 mm. In order to maintain SMC at SAT in field for all the time, at least an average of 682.6 mm of SI is needed that is 580.1 mm more than the average requirement for T1 and 565.95% more in comparison to T1 (Table 10.1).

The IUE ranged from 5.86 to 29.85 kg/ha/mm for various WSI treatments. Value of IUE is highest for treatment T1 (29.85 kg/ha/mm) and lowest for treatment T5 (5.86 kg/ha/mm). IUE for treatment T1 is 23.99 kg/ha/mm and 409.39% more than that of T5 (Table 10.2). The study reveals that there will be considerable saving of SI if the rice is irrigated by WSI technique of T1 than T5 with 23.52% reduction in yield in comparison to yield obtained by T5. Treatment T1 also requires less water (average of 1,056.4 mm) in comparison to other treatments.

In order to maintain SAT all the time in field, an average of 1,428.9 mm of water is needed that is 35.3% more as compared to T1 (Table 10.2). The range of yield simulated by various treatments is found to be from 2280 to 4001 kg/ha. The predicted yield for rainfed rice (2,280 kg/ha) is very less as compared to the various WSI treatments. WUE of rainfed treatment is also lowest (2.15 kg/ha/mm) amongst all the treatments (Table 10.2). However, if 102.5 mm SI is applied to rice at its most critical stage, i.e., RS, (Treatment T1), the yield and WUE increase by 34.2 and 34.9%, respectively.

The study also reveals that it is not necessary to maintain SAT all the time in field since it will enhance SP loss and thereby decrease the value of WUE. Many researchers earlier have reported that intermittent irrigation augments WUE and saves considerable amount of SI. This intermittent irrigation may be provided at one to several days after ponding water vanishes from rice field [12, 20]. In another approach, intermittent irrigation may be provided to rice at some level of critical moisture content, which varies from SAT to some degree of depletion of moisture from SAT level. Jensen et al. [9] and Allen et al. [2] have reported the depletion level as 20% from SAT when irrigation is required for rice. Since, the present study aims at deriving conclusion for the best WSI technique in upland rainfed regions where there is no other source of irrigation than management of rain water *in-situ* it is infeasible to provide SI to rice during CE and at times CD stage. However, the SR generated from the rice catchment during the above mentioned two stages may be harvested in a tank constructed at one end of the field and may be used as SI during RS when rainfall fails to meet the crop water demand. Kar et al. [11] have also reported that for effective utilization of soil moisture and to save irrigation, rice may be irrigated at a SMC within 0 to 20 mill bar water potential in sandy loam soil during RS. Extensive rice lands in China are now being irrigated to maintain SMC during RS at 60 to 100% SAT [19].

10.4.2.4   Surface Runoff

Surface runoff generated from the rice catchments for all treatments decrease as PE level increases. At 50% PE, values of SR are found to range from 117.6 to 341.4 mm for different treatments (Table 10.1). Amount of SR generated from the catchment increases for higher WSI treatments when the fields are frequently irrigated. Sometimes after irrigation is applied there may be high intensity of rainfall that causes SR to be more. The study reveals that at an average of 50% PE, there will be 117.6 mm SR that will be generated from the field for treatment T1 which if harvested properly can meet the average SI demand of 102.5 mm during RS.

10.4.2.5   Effective Rainfall

At 50% PE, values of effective rainfall (ER) for different treatments are found to range from 816.3 to 1157.7 mm. Total average seasonal rainfall during the rice growing season was observed to be 1157.7 mm that is 100% utilized by the treatment T0 as there was no provision of SR from the field. Of the various WSI techniques, treatment T1 is found to have highest ER of 1040.1 mm (89.9% of total seasonal rainfall) and T5 is found to have lowest ER of 816.3 mm (70.5% of total seasonal rainfall) (Table 10.1). Values of ER depend on SR generated from the field, that in turn depend on the SI applied to it. Since, treatment T1 requires less SI, it causes low SR to be generated from the field and so in-situ rainfall is more effectively used by the crop. Thus, the study reveals that of the different WSI techniques, treatment T1 is the best for effective utilization of rainfall, minimum demand of SI and produces more IUE as well as WUE. Hence, this treatment may be recommended for irrigation scheduling in upland rainfed rice regions of eastern India as well as in other similar agroclimatic regions of the world.

10.5 CONCLUSIONS

Using a water balance simulation model, various parameters like SP, AET, SI, and SR are computed on daily basis for upland rainfed rice field in eastern India. Treatments include both rainfed (T0- without any SI and SR) and different WSI techniques. The study reveals that yield of rice is seriously affected in the absence of SI. However, if SI is applied to rice during RS when SMC in the effective root zone is depleted by 20% from SAT (treatment T1) then yield, WUE as well as IUE are increased considerably. Treatment T1 is found to give maximum IUE and WUE of 29.85 and 2.90 kg/ha/mm, respectively among all the WSI treatments. This treatment uses only 102.5 mm of SI and where as about 682.6 mm of SI is required to maintain the rice field at SAT at all the time. Moreover, as compared to treatment T5, saving of irrigation is highest for treatment T1 than any other WSI treatments, which is found from simulation study as 85%. Thus, the study reveals that upland rainfed rice in eastern India must be irrigated when there is root zone depletion of SMC by 20% from SAT during the critical period of RS. This WSI technique may also be followed at similar agroclimatic regions of the world.

10.6 SUMMARY

A water balance simulation model is developed to estimate the different water balance parameters for various treatments of rainfed rice. The model uses 37 years (1977–2013) of daily meteorological, soil and crop data. The various treatments include rainfed (T_0) without any supplemental irrigation (SI) and water saving irrigation (WSI) techniques. The WSI techniques are (i) T1-SI applied at 20% management allowable deficit (MAD) of saturation (SAT) during the reproductive stage (RS) (ii) T2- SI applied at 20% MAD of SAT during critical growth (CD) and RS (iii) T3- SI applied at 20% MAD of SAT during CD, RS and maturity stage (MS) (iv) T4- SI applied at 20% MAD of SAT during all four stages, and (v) T5- SI applied at SAT in all the stages. At each irrigation, 50 mm of water is applied. Rice yields are simulated for various treatments in all years. Water requirement, irrigation use efficiency, water use efficiency and various water balance model parameters of rice for different treatments in all years are simulated.

The numeric parameters of yield and water balance model parameters are fitted to the normal distributions and values at various probabilities of exceedances are predicted. At 50% probability of exceedance, predicted yield of rice was 2280, 3060, 3125, 3250, 3400 and 4001 kg/ha for treatments T_0, T1, T_2, T3, T4 and T5, respectively. Values of SI of rice at the same probability level are predicted as 102.5, 166.4, 222.7, 330.9 and 682.6 mm for the treatment T_1, T2, T3, T4 and T5, respectively. The simulation results reveal that treatment T_1 is the best of all treatments with minimum water requirement of 1056.4 mm, highest water-use efficiency of 2.90 kg.(ha.mm)–1 and saves the highest irrigation water of 85.0% as compared to treatment T5.

ACKNOWLEDGEMENTS

The author is thankful to Dr. S. N. Panda, Professor, IIT, Kharagpur for allowing the experimental results of his project to be used in this chapter for validation of water balance model parameters.

KEYWORDS

- aqua-pro
- effective root zone
- evapotranspiration
- experiment
- irrigation use efficiency
- lysimeter
- model
- percolation
- piezometer
- ponding
- rainfall
- rainfed
- rice

- **rice simulation model**
- **saturation**
- **seepage**
- **simulation**
- **soil moisture content**
- **supplemental irrigation**
- **tensiometer**
- **treatments**
- **water balance**
- **water balance model**
- **water saving irrigation**
- **water use efficiency**
- **yield**

REFERENCES

1. Agrawal, A., 2000. *Optimal design of on-farm reservoir (OFR) for paddy-mustard cropping pattern using water balance approach.* MTech thesis, Indian Institute of Technology, Kharagpur, India.
2. Allen, R. G., Smith, M., Pereira, L. S., and Perrier, A., 1994. An update for the calculation of reference evapotranspiration. *ICID Bull.*, 43:35–92.
3. Bhuiyan, S. I., 1992. Water management in relation to crop production: case study on rice. *Outlook Agric.*, 21(4):293–299.
4. Dawe, D., Seckler, D., and Barker, R., 1998. Water supply and research for food security *Proceedings of the Workshop on Increasing Water Productivity and Efficiency in Rice-Based Systems.* July, IRRI, Los Banos, Philippines.
5. Doorenbos, J., and Kassam, A. H., 1979. *Yield Response to Water.* FAO Irrigation and Drainage Paper No. 33, Rome. Italy.
6. Doorenbos, J., and Pruitt, W. O., 1977. *Guidelines for Predicting Crop Water Requirements.* FAO Irrigation and Drainage Paper No. 24, Rome, Italy.
7. Guerra, L. C., Bhuiyan, S. I., Tuong, T. P., and Barker, R., 1998. Producing more rice with less water from irrigated systems. *SWIM Paper No. 5, International Water Management Institute,* Colombo, Sri Lanka, p. 1–24.
8. IRRI (International Rice Research institute), 1983. *Annual report for 1982.* Manila,
9. Jensen, J. R., Mannan, S. M. A., and Uddin, S. M. N., 1993. Irrigation requirement of transplanted monsoon rice in Bangladesh. *Agricultural Water Management,* 23:199–212.

10. Jones, P. G., and Thorton, P. K., 1993. A rainfall generator for agricultural applications in the tropics. *Agricultural and Forest Meteorology,* 63:1–19.

11. Kar, S., Varade, S. B., and Ghildyal, B. P., 1979. Soil physical conditions affecting root growth of upland rice. *Journal of Agricultural Sciences,* 93:719–726.

12. Mishra, H. S., Rathore, T. R., and Pant, R. C., 1990. Effect of intermittent irrigation on groundwater table contribution, irrigation requirement and yield of rice in Mollisols of the Tarai Region. *Agricultural Water management,* 18:231–241.

13. Mitra, B. N., 1979. Water management in rice fields: process studies and problems in India. *Seminar on Research Technology on Rural Poor. IDS*, Sussex, England.

14. Mizutani, M., Kalita, P. K., and Shinde, D., 1989. Effect of different rice varieties and mid- term drainage practice on water requirement in dry season paddy- observational studies on water requirement in lowland rice in Thailand. *Journal of Irrigation Engineering and Rural Planning,* 17:6–20.

15. Odhiambo, L. O., and Murty, V. V. N., 1996. Modeling water balance components in relation to field layouts in lowland paddy fields, I: model development. *Agricultural Water Management,* 30:185–199.

16. Pande, H. K., and Mitra, B. N., 1992. Water management for paddy. *Seminar on Irrigation Water management, Water Management Forum,* New Delhi, India: p. 425–445.

17. Panigrahi, B., Agrawal, A., Panda, S. N., and Mull, R., 2000. A simulation model for designing on-farm reservoirs for paddy cultivation. *Proceedings of the National Workshop on Rainwater and Groundwater Management for Sustainable Rice Ecosystem.* Indian Institute of Technology, Kharagpur, India, II: 51–62.

18. Panigrahi, B., Panda, S. N., and Mull, R., 1999. Dry and wet spell analysis for planning supplemental irrigation. *Proceedings of International Seminar on Civil and Environmental Engineering-New Frontiers and Challenges.* Asian Institute of Technology, Bangkok, Thailand, IV: 53–62.

19. SWIM Mission Report. 1997. Water Saving Techniques in Rice Irrigation. *SWIM Mission to Guilin Prefecture*, Guangxi Region, China.

20. Tripathi, R. P., Kushwaha, H. S., and Mishra, R. K., 1986. Irrigation requirement of rice under shallow water table conditions. *Agricultural Water Management,* 12:127–136.

21. Wickham, T. H., and Singh, V. P., 1978. Water management through wet soil. *In: Soil and Rice,* International Rice Research Institute, Los Banos, Philippines, p. 337–358.

22. Wolff, P., and Stein, Th. M., 1999. Water saving potentials of irrigated agriculture. *Journal of Applied Irrigation Science,* 33(2):153–173.

CHAPTER 11

PERFORMANCE OF PLANTATION CROPS UNDER CONSERVATION TRENCHES

C. R. SUBUDHI

Department of Soil and Water Conservation Engineering, College of Agricultural Engineering and Technology, Orissa University of Agriculture and Technology, Bhubaneswar-751003, Odisha, India, Phone: +91 9437645234, E-mail: rsubudhi5906@gmail.com

CONTENTS

11.1 Introduction 301
11.2 Literature Review 302
11.3 Materials and Methods 303
11.4 Results and Discussion 304
11.5 Conclusions 307
11.6 Summary 308
Acknowledgements 308
Keywords 309
References 309

11.1 INTRODUCTION

Kandhamal region of India receives rainfall around 1396 mm per year. Due to its uneven distribution, heavy downpour of rain at times results in sudden high runoff, which ultimately causes substantial soil loss. The uneven distribution of rainwater and movement of soil within the watershed results heavy loss to farmers. Therefore, conservation trenches for plantation crops help to conserve the soil and moisture and ultimately improves grain yield for the farmers. Kandhamal district suffers heavy soil loss due to hilly terrains and this study will help to check soil erosion to some extend in the sloppy areas.

The objectives of this study is to conserve moisture for establishment of plantation crop, to reduce soil erosion from upstream area, and to increase production of timber, fruit species, fuel wood and fodder.

11.2   LITERATURE REVIEW

In India for last 20 years, several researchers have studied the effects of cropping systems on runoff soil loss and crop productivity; and the influence of four land management systems on annual runoff and soil loss from the Vertisol watersheds in ICRISAT, Patencheru. Pathak et al. [6] indicated that the average annual run-off and soil loss was one-eleventh when compared to traditional flat land forms [6]. In a subsequent study in-situ soil, water conservation measures in North Western tract of India were found suitable for conserving soil and water [4]. The influence of four land configurations of rainwater was investigated during 1988–1991 on a Vertisol. Improved landform treatment (raised sunken bed) was able to reduce runoff by 6% and soil loss by 42% compared to the traditional landforms (flat beds) treatment [10]. Renovation of terrace and plantation of fruit and timber plants improved biomass production, net returns, growth of crop and productivity and reduces runoff in the range of 1.5–10.8 times, peak flow rate by 20 times and soil loss in the range of 1.2 to 5.2 times [8]. Effect of vegetative barrier like Vetiver has increased the rice yield, decreased the soil loss and the runoff compared to conventional practices by farmers [10]. There is a growing need for rain water management, since 96 M ha out of 142 M ha of net cultivated land of the country is rainfed. Scientific use of these resources will definitely increase the productivity & conservation of resources like soil and water [2]. Impact of different soil and water conservation techniques (contour bunding, terracing, land leveling, smoothening and gully plugging, sowing across the slope, vegetative barrier) have increased the production of *kharif* crops by 25–30%. Establishment of vegetative barrier with mechanical measures was more effective in controlling soil erosion (3.8 tons-ha–1) over conventional method (9.64 t.ha–1) and runoff thereby allowing more moisture available for crop growth [5]. The continuous contour V-ditch increased the crop yield significantly compared to no treatment [1].

11.3   MATERIALS AND METHODS

The experimental site lies in the Pila-Salki Watershed of Mahanadi Catchment (Figure 11.1). It falls under Sudreju revenue village of Khajuripada block in Phulbani district of Orissa. Based on Soil Conservation Department Govt. of Orissa, it is a part of watershed ORM 3-9-6-5. According to *watershed map classification reported by the Orissa Remote Sensing Application Center (Department of Science & Technology, Govt. of Odisha)*, the selected micro-watershed falls under sub-watershed No 17-07-31-01-01. This sub-watershed forms part of Survey of India Topographical Sheet Nos. 73D/2, 73D/6, 73D/3 and 73D/7. However,

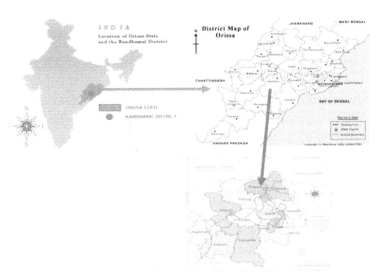

FIGURE 11.1   Location of experimental site.

the Micro-Watershed under study falls only under Topo Sheet No. 73D/6. These Micro-Watersheds are located at a distance of about 10 Km from Phulbani district headquarters on Phulbani–Sudrukumpa State Highway. During 2001–04, the on-farm trial was conducted at Sudreju under Dryland Agril Research Project, Orissa University of Agriculture & Technology, Phulbani, financed through National Agriculture Technology Project, Rainfed Rice Production System-7. Five following treatments were tested with 4 replications in a randomized block design:

T_1 – No treatment (control)
T_2 – Continuous V-ditches (CCVD) at 10 m horizontal intervals
T_3 – Continuous V-ditches at 20 m horizontal intervals
T_4 – V-ditches staggered at 5 m horizontal intervals
T_5 – V-ditches staggered at 10 m horizontal intervals.

The name of farmer is Kisore Pradhan. Mango variety Amrapalli at 5 m spacing was tried during Kharif season. The Niger, black gram and mustard were evaluated during Rabi season with 30 cm spacing. Weather was favorable for all crops. Field was contour surveyed and the slope was 4.15%. Soil loss was measured after the rainy season in the V-ditches. The soil was completely filled in 10 m continuous contour V Ditch. Hence, soil conserved was calculated using the size of the V-ditch before and after the rainy season.

11.4 RESULTS AND DISCUSSION

Monthly rainfall is presented in Table 11.1. It was observed that the year 2002 is a drought year with only 74% of mean rainfall, implying a deficit of 36% from mean annual rainfall. However, 2001 and 2003 are good years, receiving 39.6% and 4% more than the mean annual rainfall, respectively. The mean annual rainfall is 1396.14 mm. The fluctuation shows the rainfall is very erratic in all three years. Table 11.2 shows rate of growth of plantation crops. The rate of growth of mango was highest (3.02 cm/month) in T_2 and lowest (1.22 cm/month) in control during 2001–2003. Figure 11.2 shows the view of continuous contour V-ditch at 10 m interval in mango orchard.

TABLE 11.1 Monthly Rainfall (mm) During 2001, 2002 and 2003

Month	Monthly rainfall (normal) mm	2001		2002		2003	
		Actual mm	Deviation from normal %	Actual mm	Deviation from normal, %	Actual mm	Deviation from normal, %
January	9.18	0	-100	13.0	+41.6	0.0	-100
February	14.07	0	-100	0	-100	23.5	+67.0
March	21.70	56.0	+158.1	20.0	-7.8	12.5	-57.6
April	30.40	0	-100	32.0	+5.2	89.0	+192.7
May	57.48	48.0	-16.5	70.0	+21.8	7.0	-87.8
June	191.62	504.9	+163.5	149.0	-22.2	117.0	-38.9
July	353.62	797.6	+125.6	129.0	-63.5	237.0	-33.0
August	378.65	300.1	-20.7	329.0	-13.1	358.1	-5.4
September	218.57	124.7	-42.9	134.9	-38.3	350.1	+60.2
October	88.93	111.5	+25.4	11.0	-87.6	216.0	+142.9
November	27.48	6.9	-74.9	0	-100	0.0	-100
December	4.45	0	-100	0	-100	42.0	-843.8
Annual	1396.15	1949.7	+39.6	887.9	-36.4	1452.2	+4.0

TABLE 11.2 Yield, Growth Rate, Moisture Content and Soil Conserved in Different Treatments

	Treatments	Niger yield (q.ha^{-1}) (2001–02)	black gram yield (q.ha^{-1}) (2002–03)	Mustard yield (q.ha^{-1}) (2003–04)	Mean moisture content (%)	Mean growth rate of mango (cm/month)	Mean soil conserved (ton.ha^{-1})
T1	No treatment	2.33	6.12	4.17	3.67	1.22	0
T2	Continuous V-ditches at 10 m horizontal interval.	3.11	8.00	5.28	10.25	3.02	6.2
T3	Continuous V-ditches at 20 m horizontal interval.	2.44	7.12	4.85	5.59	2.47	3.2
T4	V-ditches staggered at 5 m horizontal interval.	2.51	7.37	5.15	8.47	2.42	5.5
T5	V-ditches staggered at 10 m horizontal intervals.	2.49	7.25	5.00	7.02	2.50	3.1
	SE (m)+	**0.13**	**0.57**	**0.05**	–	–	–
	CD (0.05)	**0.39**	**NS**	**0.17**	–	–	–

Note: One quintal, q = 100 kg.

FIGURE 11.2 View of continuous contour V-ditch with mango plants.

The grain-yield of Niger, black gram and mustard were 33.4%, 23.5% and 26.6% higher than the control, respectively (Table 11.2). This may be due to more soil and water conserved in the root zone of the crop as the moisture content in T_2 is more compared to all other treatments and lowest in control as there was no V-ditch (Table 11.2). The soil conserved in T_2 is 6.2 tons/ha followed by T_5 where soil conserved was 5.5 t/ha. Patil et al. [7] has obtained similar results and they reported lowest soil loss (1.51 t/ha) and highest survival percentage of cashew nut plantation in continuous contour trench compared to staggered trench (3.95 t/ha) and control (16.55 t/ha). Therefore, it can be concluded that 10 m CCVD can be recommended for uplands of degraded watershed at Kandhamal district of Odisha.

11.5   CONCLUSIONS

The present study reveals that grain yield of Niger, black gram and mustard were 33.4%, 23.5% and 26.6% higher than the control, respectively.

Though the cost of construction is little high, it is recommended to practice contour V-ditch at 10 m intervals, to conserve soil and moisture and to get more grain yield in degraded watershed of Kandhamal district. It is observed that in CCVD at 10 m interval, rate of mango growth was 3.02 cm/month in case of Amrapalli, which is 46% higher compared to control. Also we can conserve 6.2 t/ha of soil with 10 m CCVD, which is highest among all the treatments. It can be concluded that 10 meter CCVD can be recommended for upland of degraded watershed of Kandhamal district of Orissa.

11.6 SUMMARY

Kandhamal district situated in central part of Odisha receives an annual rainfall of 1396 mm and this region is highly prone to soil and runoff loss due to heavy rainfall during *kharif.* A trial was conducted during 2001–2004 to study the effects of conservation trenches on plantation crops. This trial was conducted on farmers field of Sudreju village of Kandhamal district, Odisha, India under National Agricultural Technology Project, Rainfed Rice Production System-7 with the objectives to: (i) conserve moisture for establishment of plantation crop; (ii) reduce erosion from upstream area; and (iii) increase production of timber, fruit species, fuel wood and fodder. The treatments were (i) No treatment/control, (ii) Continuous V-ditches at 10 m horizontal interval, (iii) Continuous V-ditches at 20 m horizontal interval, (iv) V-ditches staggered at 5 m horizontal interval, and (v) V-ditches staggered at 10 m horizontal interval. Mango varieties *Pusa Amrapalli* was tried during *kharif.* During *rabi,* black gram (*PU-30*) was tried in between mango rows. It is observed that in CCVD at 10 m interval, rate of mango growth was 3.02 cm/month in case of Amrapalli, which is 46% higher compared to control. The grain-yield of Niger, black gram and mustard were 33.4%, 23.5% and 26.6% higher than control, respectively. Though the cost of construction is little high, it is recommended to practice contour V-ditch at 10 m intervals to conserve soil and moisture and to get more grain yield in degraded watershed of Kandhamal district.

ACKNOWLEDGEMENTS

The authors acknowledge the help of Vice Chancellor, Orissa University of Agriculture and Technology; Director, Central Research Institute for

Dry land Agriculture, Hyderabad; and Dean of Research, University of Agriculture and Technology, Bhubaneswar for time-to-time guidance and financial help to carry out this project. The authors also acknowledge the help of Dry land Agriculture Research Project staff of Phulbani and staff of College of Agricultural Engineering and Technology, Orissa University of Agriculture and Technology, Bhubaneswar.

KEYWORDS

- black gram
- contour
- India
- mango
- Niger
- soil conservation
- trench
- V-ditch

REFERENCES

1. Anonymous, 2003. *Final progress report of NATP.* RRPS-7, DLAP, OUAT, Phulbani.
2. Arora, D. and Gupta, A. K., 2002. Effect of water conservation measures in a pasture on the productivity of Buffel grass. *Proceedings of Indian Association of Soil & Water Conservationists, Dehradun.* pp. 65–66.
3. Eswaran, V. B., 2002. Wasteland development. *Proceedings of Indian Association of Soil & Water Conservationists, Dehradun.* pp. 17–19.
4. Hadda, M. S., Khera, K. L., and Kukal, S. S., 2000. Soil and water conservation practices and soil productivity in North-Western sub-mountainous tract of India: A review. *Indian J. Soil Cons.,* 28:187–192.
5. Kumar, M., 2002. Impact of soil and water conservation on erosion loss and yield of Kharif crops under ravenous watershed. *Proceedings of Indian Association of Soil & Water Conservationists, Dehradun.* pp. 301–303.
6. Pathak, P., Mirande, S. M., and El-Swaify, S. A., 1985. Improved rainfed farming for semi-arid tropics: implications for soil and water conservation. In: El-Swaity, et.al. (Eds.), *Soil Erosion and Conservation.* Soil Conservation Society of America, USA, p. 338–354.

7. Patil, P. P., Gutal, G. B., Ganvir, B. N., and Bodake, P. S., 2004. Soil and moisture conservation practices for the hill slopes in Western Ghat of Maharashtra. *Extended abstracts of National Conferences on Resource Conserving Technologies for Social Upliftment.* p. 122–124.

8. Samra, J. S., 2002. Watershed management a tool for sustainable production. *Proceedings of Indian Association of Soil & Water Conservationists, Dehradun.* p. 1–10.

9. Sharma, V., Arora, S., Jalali, V. K., and Kher, D., 2004. Studies on constraints in adoption of improved soil and rainwater conservation practices in Kandi region of Jammu. *Journal of Research SKUAST*, 3(2):214–220.

10. Subudhi, C. R., Pradhan, P. C., and Senapati, P. C., 1999. Effect of grass bund on erosion loss and yield of rainfed rice, Orissa, India. *T. Vetiver Network*, 19:32–33.

CHAPTER 12

WATER REQUIREMENT OF CROPS

A. DALEI, C. R. SUBUDHI, and B. PANIGRAHI

*Department of Soil and Water Conservation Engineering,
College of Agricultural Engineering and Technology, Orissa
University of Agriculture and Technology, Bhubaneswar–751003,
Odisha, India, Phone: +91 9437645234,
E-mail: rsubudhi5906@gmail.com; kajal_bp@yahoo.co.in*

CONTENTS

12.1   Introduction..313
12.2   Review of Literature 315
12.3   Theoretical Considerations 318
    12.3.1   Penman-Monteith Equation 318
        12.3.1.1   Reference Surface 319
        12.3.1.2   Location..320
        12.3.1.3   Temperature..320
        12.3.1.4   Humidity 320
        12.3.1.5   Radiation 321
        12.3.1.6   Wind Speed 321
    12.3.2   Meteorological Factors Determining ETc 321
    12.3.3   Measurement..322
        12.3.3.1   Vapor Pressure Deficit (es – ea)..................322
        12.3.3.2   Extraterrestrial Radiation (Ra)....................322
        12.3.3.3   Solar or Shortwave Radiation (Rs) 323

12.3.3.4   Relative Shortwave Radiation (Rs/Rso)...... 323

12.3.3.5   Relative Sunshine Duration (n/N)............... 323

12.3.3.6   Albedo and Net Solar Radiation (Rns)........ 324

12.3.3.7   Net Longwave Radiation (Rnl) 324

12.3.3.8   Net Radiation (Rn) 324

12.3.3.9   Soil Heat Flux (G)...................................... 324

12.3.4   Estimating Missing Climatic Data 325

12.3.4.1   Estimating Missing Humidity Data............. 325

12.3.5   Introduction to Crop Evapotranspiration (ETc) 326

12.3.5.1   Direct Calculation 326

12.3.5.2   Crop Coefficient Approach 326

12.3.6   Irrigation Requirements of Crops............................. 328

12.4   Materials and Methods... 329

12.4.1   Study Area... 329

12.4.2   Crop Selection 329

12.4.3   Estimation of Daily
Evapotranspiration Using DSS_ET............................. 331

12.4.4   Ranking of Different ET0 Estimation Methods 331

12.4.4.1   Standard Error Estimate 339

12.4.5   Estimation of Correction Factor................................ 340

12.4.6   Estimation of Crop Water Requirement 340

12.4.7   Irrigation Requirement of Crops 340

12.5   Results and Discussions.. 341

12.5.1   Ranking of Different ET0 Estimation Methods 341

12.5.2   Estimation of Correction Factor................................ 342

12.5.3   Variations of Daily Rainfall and
Crop Evapotranspiration 342

12.5.4   Variations of Total Seasonal Rainfall and
Crop Evapotranspiration 346

12.5.5   Monthly Irrigation Requirement of Different Crops.... 350

12.6   Conclusions... 354

12.7   Summary 355

Keywords 356
References... 356

12.1   INTRODUCTION

Water requirement of crops in the form of Evapotranspiration (ETc) is the largest and one of the most basic components of the hydrologic cycle. It plays a very important role in the water and energy balance on earth's surface and also has a major role in agricultural and irrigation practices. It is the combination of soil evaporation and crop transpiration process. As reported by Almhab and Busu [3], about 70% of the water loss from the earth's surface occurs as evaporation. Thus, accurate estimation of ETc is of vital importance for many studies, such as hydrologic water balance, irrigation system design and management, water resources planning and management, etc. The evapotranspiration rate from a standard reference surface is the reference evapotranspiration (ET0). For effective planning and implementation of policies on irrigation projects, it is necessary to determine reference evapotranspiration which is further used in computing crop water use. It is essential for the development of modern irrigation management methodologies, optimum allocation of water and energy resources, and improved irrigation planning and management practices.

For estimating ETc from a well-watered agricultural crop first reference crop ET (ET0) from a standard surface is estimated and then an appropriate empirical crop coefficient for other crops is multiplied to determine the crop Evapotranspiration (ETc) for a particular crop. Various ET0 estimation methods have been developed earlier and these methods are mainly grouped into radiation, temperature, pan evaporation based and combination methods. Combination based ET estimation methods includes Penman vapor pressure deficit (VPD#1), Businger-van Bavel, Penman vapor pressure deficit (VPD#3), Penman-Monteith, 1972 Kimberly-Penman, FAO-24 Penman, FAO-24 Corrected Penman, FAO-PPP-17 Penman, 1982-Kimberly-Penman, CIMIS Penman and FAO-56 Penman-Monteith method. Radiation based methods includes Turc, Jensen-Haise, Priestly-Taylor and FAO-24 estimation methods. As reported by Dalei [6], Thornthwaite, SCS Blaney-Criddle, FAO-24 Blaney-Criddle, and Hargreaves come under temperature based methods and Christiansen Pan Evaporation, FAO-24 Pan and Pan Evaporation methods are included in evaporation based methods.

Due to wide application of ETc data, a number of indirect methods for estimation of ET0 have been developed based on the easily available location characteristics (elevation and latitude) and meteorological parameters (temperature, humidity, wind speed, solar radiation). However, not only data requirement varies from method to method, but also the performance and accuracy of different methods varies with different climatic condition and availability of data [10].

Therefore, it becomes impractical for many users to select the best ET0 estimation method for the available data and climatic condition. To overcome this problem, Reddy [13] developed a decision support system consisting of nine widely used ET0 estimation methods. This decision support system was further modified to include more ET0 estimation methods [14] and named as DSS_ET model.

This model was further improved by Bandyopadhyay et al. [4]. The DSS_ET model can be used to identify the best ET0 method for different climatic conditions. It is developed in Microsoft Visual Basic 6.0. It consists of a model base for estimating ET0 by different methods and ranking them and a user-friendly graphical interface. Generally, 22 and widely used ET methods are available in DSS_ET. These methods can be used for estimating monthly ET0 values for the time interval considered in this study. However, most of these methods can also be used for estimating daily or even hourly (Penman based methods) ET values. A user-friendly decision support system (DSS_ET) is used in the present study for the estimation of the reference evapotranspiration.

The aim of present study is to estimate the reference evapotranspiration by using the available methods and ranked them to find out which method gives the best suited value. using DSS_ET. These ET0 values can later be used for different purposes such as to derive irrigation water requirement of crops. The objectives of this study is as follows:

1. Compute the values of ETo by various estimation method, using DSS_ET decision support system.
2. To compare ETo estimated by various estimation methods and rank them to find out the alternative best methods and

3. To estimate the evapotranspiration (water requirement) of various crops grown in the area and determine their irrigation requirements.

12.2 REVIEW OF LITERATURE

It is essential now to develop strategies to optimize the use of water for crop production and to introduce effective water management practices. The new procedures and guidelines have been recently published in the FAO Irrigation and Drainage series and include the adoption of the Penman–Monteith approach as the new standard for determining reference crop evapotranspiration (ET0) calculations. Procedures have been developed to use the method also in conditions when no or limited data on humidity, radiation and wind speed are available. Procedures for estimating crop evapotranspiration are revised with an update of the crop coefficients that allow more accurate estimates for a wide range of crops and for various crop, soil and water management practices. Daily ET0 calculations are included by separating soil evaporation and crop transpiration estimates through the dual crop coefficient.

A user-friendly decision support system (DSS_ET) for estimation of reference evapotranspiration was developed by George et al. [9]. To evaluate the model with different data availability conditions various tests were conducted for five climatic stations such as Davis, Bellary, Kharagpur, Jagdalpur and Bombay. The results were compared with Penman Monteith method taking it as a standard method and the results indicated that DSS_ET model performed well under different data availability and climatic conditions. The Hargreaves and FAO-24 Blaney-Criddle methods ranked first for the Davis and Jagdalpur stations, respectively based on the weighted average standard error and the 1982 Kimberly-Penman method ranked first for Kharagpur station.

A new empirical equation for estimating hourly reference evapotranspiration which requires solar radiation, air temperature and relative humidity data were developed by Alexandris and Kerkides [2]. A regression analysis is performed in order to estimate the factors used in the empirical model. The performance of the model is evaluated against FAO-56 P-M and CIMIS Penman method. They concluded that the model performed as good as FAO-56 P-M and CIMIS Penman methods. Eight evapotranspiration estimation methods, i.e., Penman. Penman-Monteith, Pan Evaporation, Kimberly-Penman, Priestley-Taylor, Hargreaves, Samani-Hargreaves and Blaney-Criddle with 30 years of daily data, in the west coast of Peninsular Malaysia were developed by Jiabing et al. [11]. The Penman-Monteith, Blaney-Criddle and Pan methods estimate lower values of evapotranspiration with no significant difference among them. Penman method estimates reference evapotranspiration close to Penman-Monteith.

Pan coefficient by taking ratio of Penman-Monteith evapotranspiration to pan evaporation at 150 meteorological stations were evaluated by Xu et al. [16]. The results are compared to analyze the spatial distributions and temporal variations in the reference evapotranspiration as well as in the meteorological variables. Results indicate that the spatial distributions of reference evapotranspiration and pan evaporation are same. A significant decreasing trend in both the reference evapotranspiration and the pan evaporation was found for the whole catchment which is mainly caused due to a significant decrease in the net total radiation and to a lesser extent by a significant decrease in the wind speed over the catchment.

Performance of the ET0 methods on the basis of the three statistical indicators, i.e., modeling efficiency (ME), coefficient of residual mass (CRM), and root mean squared error (RMSE) expressed as a percentage of the arithmetic mean of observed values were studied by Bandyopadhyay et al. [4]. The performances of different methods were compared and it was found that the performance of different methods varies with different stations. The Hargreaves method performed better for all the stations. They concluded that under Indian conditions where no solar radiation measurements are available, the solar radiation can be estimated from air temperature extremes.

The performance of six commonly used reference evapotranspiration estimation methods such as Penman-Monteith, Priestley-Taylor, FAO-24 Radiation, Hargreaves, Blaney-Criddle and Class A pan by using weighing lysimeter data from a semiarid highland environment were studied by Benli et al. [5]. The performance of these methods was evaluated on the basis of Root Mean Square Errors (RMSE) and index of agreement for the daily data. The monthly averages as well as the mean absolute error (MAE) for the seasonal totals were computed to compare these methods. Ranking of methods was done where Penman-Monteith method ranked first. Daut et al. [8] estimated the solar radiation using a new method, which is a combination of Hargreaves method and linear regression. Estimated solar radiation data are compared and analyzed using coefficient of residual mass (CRM), root mean squared error (RMSE), Nash-Sutcliffe equation (NSE) and percentage error (e). The statistical analysis of the average monthly measured solar radiation data is compared with the estimated solar radiation data. The value of CRM was found to be closer to zero 'which indicate that proposed method is perfectly estimated, RMSE shows a low value, which indicates that, the method performance is well. Similarly the value of NSE is closer to 1, which indicates that the estimated solar radiations match perfectly with the measured data taken for the study. Finally the value of e (percentage error) is closer to zero indicates that the proposed method is acceptable and applicable.

Calibration of the Hargreaves (HG) and Priestley-Taylor (P-T) equations were carried on by Tabari and Talaee [15] on the basis of Penman-Monteith method in arid and cold climates of Iran during 1994–2005. The Hargreaves (HG) and Priestley-Taylor (P-T) equations are the simple equations and require only few weather data inputs for ET estimation. Regional calibration is required for acceptable performance of this method. Results indicate that calibration of these methods resulted in improvement of the original Hargreaves and Priestley- Taylor equations by reducing error of the ET estimates.

Reference crop evapotranspiration based on the meteorological data of Mengzhi weather station were calculated by Jiabing et al. [11] from 1961 to 2010 using FAO Penman-Moteith formula. Using the statistics test and regression analysis methods, the main meteorological factors that impact ET0 were discussed. The results showed that the annual ET0 in Mengzhi plain is decreased slowly over the recent 50 years, and reduced by 13.43 mm per decade. The influence capacity of meteorological factors on ET0 is as follows: annual average wind speed > relative humidity > sunshine hours > average maximum temperature. The maximum monthly ET0 appeared in April or May each year, and the minimum monthly ET0 occurred in December. ET0 declined significantly in April and May in recent 50 years; in other months, there was no significant increase or decline trend. The most important factor affecting ET0 from January to May and December is the average wind speed, and from June to September, ET0 was affected the most by the sunshine hours and in October and November by humidity. They studied actual evapotranspiration (ETc) and the crop coefficients (Kc) for the five tropical and temperate species of three-year-old bamboo plants, i.e., Bambusaoldhamii, Bambusa multiplex, Bambusa vulgaris, Phyllostachysaurea and Pseudosasa japonica using lysimeters for a period of more than one year under a tropical climate. The average ET rates for the bamboo species studied ranged from 4 to 7 mm day -1 with maximum values of between 10.7 and 17.1 mm day -1 during the wet season, and an average Kc of 1.1 to 1.9. The ET was correlated to weather parameters, especially minimum temperatures.

12.3   THEORETICAL CONSIDERATIONS

As a result of an expert consultation held in May 1990, the FAO Penman-Monteith method is now recommended as the sole standard method for the definition and computation of the reference evapotranspiration. The relatively accurate and consistent performance of the Penman-Monteith approach in both arid and humid climates has been indicated in both the ASCE and European studies. The analysis of the performance of the various calculation methods reveals the need for formulating a standard method for the computation of ETo. The FAO Penman-Monteith method is recommended as the sole standard method. It is a method with strong likelihood of correctly predicting ETo in a wide range of locations and climates and has provision for application in data-short situations. The FAO Penman-Monteith method requires radiation, air temperature, air humidity and wind speed data. Calculation procedures to derive climatic parameters from meteorological data and to estimate missing meteorological variables required for calculating ET0 are presented in this Part. The calculation procedures in this Publication allow for estimation of ET0 with the FAO Penman-Monteith method under all circumstances, even in the case of missing climatic data.

12.3.1   PENMAN-MONTEITH EQUATION

In 1948, Penman combined the energy balance with the mass transfer method and derived an equation to compute the evaporation from an open water surface from standard climatological records of sunshine, temperature, humidity, and wind speed. This so-called combination method was further developed by many researchers and extended to cropped surfaces by introducing resistance factors.

The resistance nomenclature distinguishes between aerodynamic resistance and surface resistance factors. The surface resistance parameters are often combined into one parameter, the 'bulk' surface resistance parameter, which operates in series with the aerodynamic resistance. The surface resistance, rs, describes the resistance of vapor flow through stomata openings, total leaf area and soil surface. The aerodynamic resistance, ra, describes the resistance from the vegetation upward and involves friction from air flowing over vegetative surfaces. Although the exchange process in a vegetation layer is too complex to be fully described by the two resistance factors, good correlations can be obtained between measured and calculated evapotranspiration rates, especially for a uniform grass reference surface.

12.3.1.1   Reference Surface

To obviate the need to define unique evaporation parameters for each crop and stage of growth, the concept of a reference surface was introduced. Evapotranspiration rates of the various crops are related to the evapotranspiration rate from the reference surface (ETo) by means of crop coefficients. In the past, an open water surface has been proposed as a reference surface. However, the differences in aerodynamic, vegetation control and radiation characteristics present a strong challenge in relating ET to measurements of free water evaporation. Relating ETo to a specific crop has the advantage of incorporating the biological and physical processes involved in ET from cropped surfaces. Grass, together with alfalfa, is a well-studied crop regarding its aerodynamic and surface characteristics and is accepted worldwide as a reference surface.

The FAO expert consultation on revision of FAO methodologies for crop water requirements accepted a hypothetical reference crop with an assumed crop height of 0.12 m, a fixed surface resistance of 70 s m^{-1} and an albedo of 0.23 as the reference surface. The reference surface closely resembles an extensive surface of green grass of uniform height, actively growing, completely shading the ground and with adequate water.

The Penman-Monteith form of the combination equation is:

$$ET_0 = \cfrac{\left(0.408\Delta\left(R_n - G\right) + \gamma\cfrac{900}{(T+273)}u_2\left(e_s - e_a\right) \right)}{\Delta + \gamma(1 + 0.34u_2)} \qquad (1)$$

where, Rn = net radiation, G = soil heat flux, (es-ea) = vapor pressure deficit of the air.

12.3.1.2 Location

In using the equation, altitude above sea level (m) and latitude of the location should be specified. These data are needed to adjust some weather parameters for the local average value of atmospheric pressure (a function of the site elevation above mean sea level) and to compute extraterrestrial radiation (Ra) and, in some cases, daylight hours (N). In the calculation procedures for Ra and N, the latitude is expressed in radian (i.e., decimal degrees times p /180). A positive value is used for the northern hemisphere and a negative value for the southern hemisphere.

12.3.1.3 Temperature

The (average) daily maximum and minimum air temperatures in degrees Celsius (°C) are required. Where only (average) mean daily temperatures are available, the calculations can still be executed but some underestimation of ETo will probably occur due to the non-linearity of the saturation vapor pressure – temperature relationship. Using mean air temperature instead of maximum and minimum air temperatures yields a lower saturation vapor pressure es, and hence a lower vapor pressure difference (es – ea), and a lower reference evapotranspiration estimate.

12.3.1.4 Humidity

The (average) daily actual vapor pressure, ea, in kilopascals (kPa) is required. The actual vapor pressure, where not available, can be derived from maximum and minimum relative humidity (%), psychrometric data (dry and wet bulb temperatures in °C) or dew point temperature (°C) .

12.3.1.5   Radiation

The (average) daily net radiation expressed in mega-joules per square meter per day (MJ m^{-2} day^{-1}) is required. These data are not commonly available but can be derived from the (average) shortwave radiation measured with a pyranometer or from the (average) daily actual duration of bright sunshine (hours per day) measured with a (Campbell-Stokes) sunshine recorder.

12.3.1.6   Wind Speed

The (average) daily wind speed in meters per second (ms^{-1}) measured at 2 m above the ground level is required. It is important to verify the height at which wind speed is measured, as wind speeds measured at different heights above the soil surface differ.

12.3.2   METEOROLOGICAL FACTORS DETERMINING ETc

Various meteorological factors that determine ETc are:

 a.   Solar radiation;
 b.   Air temperature;
 c.   Air humidity;
 d.   Wind speed;
 e.   Atmospheric parameters;
 f.   Atmospheric pressure (P);
 g.   Latent heat of vaporization;
 h.   Air temperature.

The methods for calculating evapotranspiration from meteorological data require various climatological and physical parameters. Some of the data are measured directly in weather stations. Other parameters are related to commonly measured data and can be derived with the help of a direct or empirical relationship.

This chapter discusses the source, measurement and computation of all data required for the calculation of the reference evapotranspiration by means of the FAO Penman-Monteith method.

12.3.3   MEASUREMENT

It is not possible to directly measure the actual vapor pressure. The vapor pressure is commonly derived from relative humidity or dew point temperature. It is better to utilize a dew point temperature that is predicted from daily minimum air temperature, rather than to use unreliable relative humidity measurements.

12.3.3.1   Vapor Pressure Deficit (es – ea)

The vapor pressure deficit is the difference between the saturation (es) and actual vapor pressure (ea) for a given time period. For time periods such as a week, ten days or a month es is computed using the Tmax and Tmin averaged over the time period and similarly the ea is computed, using average measurements over the period. When desired, es and ea for long time periods can also be calculated as averages of values computed for each day of the period.

12.3.3.2   Extraterrestrial Radiation (Ra)

The radiation striking a surface perpendicular to the sun's rays at the top of the earth's atmosphere, called the solar constant, is about 0.082 MJ m^{-2} min^{-1}. The solar radiation received at the top of the earth's atmosphere on a horizontal surface is called the extraterrestrial (solar) radiation, Ra. If the sun is directly overhead, the angle of incidence is zero and the extraterrestrial radiation is 0.0820 MJ m^{-2} min^{-1}. As seasons change, the position of the sun, the length of the day and, hence, Ra change as well. Extraterrestrial radiation is thus a function of latitude, date and time of day.

12.3.3.3   Solar or Shortwave Radiation (Rs)

The amount of radiation reaching a horizontal plane is known as the solar radiation, Rs. Because the sun emits energy by means of electromagnetic waves characterized by short wavelengths, solar radiation is also referred to as shortwave radiation. For a cloudless day, Rs is roughly 75% of extraterrestrial radiation. On a cloudy day, the radiation is scattered in the atmosphere, but even with extremely dense cloud cover, about 25% of the extraterrestrial radiation may still reach the earth's surface mainly as diffuse sky radiation.

12.3.3.4   Relative Shortwave Radiation (Rs/Rso)

The relative shortwave radiation is the ratio of the solar radiation (Rs) to the clear-sky solar radiation (Rso). Rs is the solar radiation that actually reaches the earth's surface in a given period, while Rso is the solar radiation that would reach the same surface during the same period but under cloudless conditions. The relative shortwave radiation is a way to express the cloudiness of the atmosphere; the cloudier the sky the smaller the ratio. The ratio varies between about 0.33 (dense cloud cover) and 1 (clear sky). In the absence of a direct measurement of Rn, the relative shortwave radiation is used in the computation of the net longwave radiation.

12.3.3.5   Relative Sunshine Duration (n/N)

The relative sunshine duration is another ratio that expresses the cloudiness of the atmosphere. It is the ratio of the actual duration of sunshine, n, to the maximum possible duration of sunshine or daylight hours N. In the absence of any clouds, the actual duration of sunshine is equal to the daylight hours ($n = N$) and the ratio is one, while on cloudy days n and consequently the ratio may be zero. In the absence of a direct measurement of Rs, the relative sunshine duration, n/N, is often used to derive solar radiation from extraterrestrial radiation. As with extraterrestrial radiation, the day length N depends on the position of the sun and is hence a function of latitude and date.

12.3.3.6 Albedo and Net Solar Radiation (Rns)

A considerable amount of solar radiation reaching the earth's surface is reflected. The fraction, a, of the solar radiation reflected by the surface is known as the albedo. The albedo is highly variable for different surfaces and for the angle of incidence or slope of the ground surface. It may be as large as 0.95 for freshly fallen snow and as small as 0.05 for a wet bare soil. A green vegetation cover has an albedo of about 0.20–0.25. For the green grass reference crop, is assumed to have a value of 0.23.

12.3.3.7 Net Longwave Radiation (Rnl)

The solar radiation absorbed by the earth is converted to heat energy. The terrestrial radiation is referred to as longwave radiation. The emitted longwave radiation (Rl, up) is absorbed by the atmosphere or is lost into space. The longwave radiation received by the atmosphere (Rl, down) increases its temperature and, as a consequence, the atmosphere radiates energy of its own. Part of the radiation finds it way back to the earth's surface. Consequently, the earth's surface both emits and receives longwave radiation. The difference between outgoing and incoming longwave radiation is called the net longwave radiation, Rnl.

12.3.3.8 Net Radiation (Rn)

The net radiation, Rn, is the difference between incoming and outgoing radiation of both short and long wavelengths. Rn is normally positive during the daytime and negative during the nighttime. The total daily value for Rn is almost always positive over a period of 24 hours, except in extreme conditions at high latitudes.

12.3.3.9 Soil Heat Flux (G)

The soil heat flux, G, is the energy that is utilized in heating the soil. G is positive when the soil is warming and negative when the soil is cooling.

12.3.4   ESTIMATING MISSING CLIMATIC DATA

The assessment of the reference evapotranspiration ETo with the Penman-Monteith method requires mean daily, ten-day or monthly maximum and minimum air temperature (Tmax and Tmin), actual vapor pressure (ea), net radiation (Rn) and wind speed measured at 2 m (u^2). If some of the required weather data are missing or cannot be calculated, it is strongly recommended that the user estimate the missing climatic data with one of the following procedures and use the FAO Penman-Monteith method for the calculation of ETo. Procedures to estimate missing humidity, radiation and wind speed data are given in this section.

12.3.4.1   Estimating Missing Humidity Data

Where humidity data are lacking, an estimate of actual vapor pressure, ea, can be obtained by assuming that dew-point temperature (Tdew) is near the daily minimum temperature (Tmin). The relationship between Tdew and Tmin holds for locations where the cover crop of the station is well watered. However, particularly for arid regions, the air might not be saturated when its temperature is at its minimum. Hence, Tmin might be greater than Tdew and a further calibration may be required to estimate dew-point temperatures. In these situations, "Tmin" in the above equation may be better approximated by subtracting 2–3°C from Tmin. In humid and sub-humid climates, Tmin and Tdew measured in early morning may be less than Tdew measured during the daytime because of condensation of dew during the night. After sunrise, evaporation of the dew will once again humidify the air and will increase the value measured for Tdew during the daytime. However, it is standard practice in 24-hour calculations of ETo to use Tdew measured or calculated during early morning.

This section has shown how solar radiation, vapor pressure and wind data can be estimated when missing. Many of the suggested procedures rely upon maximum and minimum air temperature measurements. Unfortunately, there is no dependable way to estimate air temperature when it is missing. Therefore it is suggested that maximum and minimum daily air temperature data are the minimum data requirements necessary to apply the FAO Penman-Monteith method.

12.3.5   INTRODUCTION TO CROP EVAPOTRANSPIRATION (ETc)

The crop coefficient approach is for calculating the crop evapotranspiration under standard conditions (ETc). The standard conditions refer to crops grown in large fields under excellent agronomic and soil water conditions. The crop evapotranspiration differs distinctly from the reference evapotranspiration (ETo). Crop evapotranspiration is calculated by multiplying ETo by Kc. Differences in evaporation and transpiration between field crops and the reference grass surface can be integrated in a single crop coefficient (Kc) or separated into two coefficients: a basal crop (Kcb) and a soil evaporation coefficient (Ke), i.e., Kc = Kcb + Ke. The approach to follow should be selected as a function of the purpose of the calculation, the accuracy required and the data available.

12.3.5.1   Direct Calculation

The evapotranspiration rate from a cropped surface can be directly measured by the mass transfer or the energy balance method. It can also be derived from studies of the soil water balance determined from cropped fields or from lysimeters. Crop evapotranspiration can also be derived from meteorological and crop data by means of the Penman-Monteith equation.

12.3.5.2   Crop Coefficient Approach

In the crop coefficient approach the crop evapotranspiration, ETc, is calculated by multiplying the reference crop evapotranspiration, ETo, by a crop coefficient, Kc: ETc = Kc ETo. The reference ETo is calculated using the FAO Penman-Monteith equation. Factors that affect crop coefficient are: Crop type, climate, soil evaporation, and crop growth stages.

There are two approaches of crop- coefficient used to estimate ETc. They are single crop coefficient approach and dual crop coefficient approach. In the single crop coefficient approach, the effect of crop transpiration and soil evaporation are combined into a single Kc coefficient. The coefficient integrates differences in the soil evaporation and crop transpiration rate between the crop and the grass reference surface. As soil evaporation may fluctuate daily as a result of rainfall or irrigation, the single crop coefficient expresses only the time-averaged (multi-day) effects of crop evapotranspiration.

In the dual crop coefficient approach, the effects of crop transpiration and soil evaporation are determined separately. Two coefficients are used: the basal crop coefficient (Kcb) to describe plant transpiration, and the soil water evaporation coefficient (Ke) to describe evaporation from the soil surface. The single Kc coefficient is replaced by:

$$Kc = Kcb + Ke \qquad (2)$$

where, Kcb =basal crop coefficient, and Ke =soil water evaporation coefficient.

The basal crop coefficient, Kcb, is defined as the ratio of ETc to ETo when the soil surface layer is dry but where the average soil water content of the root zone is adequate to sustain full plant transpiration. The soil evaporation coefficient, Ke, describes the evaporation component from the soil surface. If the soil is wet following rain or irrigation, Ke may be large. However, the sum of Kcb and Ke can never exceed a maximum value, Kc max, determined by the energy available for evapotranspiration at the soil surface. As the soil surface becomes drier, Ke becomes smaller and falls to zero when no water is left for evaporation. The estimation of Ke requires a daily water balance computation for the calculation of the soil water content remaining in the upper topsoil.

After the selection of the calculation approach, the determination of the lengths for the crop growth stages and the corresponding crop coefficients, a crop coefficient curve can be constructed. The curve represents the changes in the crop coefficient over the length of the growing season. The shape of the curve represents the changes in the vegetation and ground cover during plant development and maturation that affect the ratio of ETc to ETo. From the curve, the Kc factor and hence ETc can be derived for any period within the growing season. Shortly after the planting of annuals or shortly after the initiation of new leaves for perennials, the value for Kc is small, often less than 0.4. The Kc begins to increase from the initial Kc value, Kc_{ini}, at the beginning of rapid plant development and reaches a maximum value, Kc mid, at the time of maximum or near maximum plant development. During the late season period, as leaves begin to age and senesce due to natural or cultural practices, the Kc begins to decrease until it reaches a lower value at the end of the growing period equal to Kc end.

In dual crop coefficient, the single 'time-averaged' Kc curve incorporates averaged wetting effects into the Kc factor. The value for Kc mid is relatively constant for most growing and cultural conditions. However, the values for Kc_{ini} and Kc end can vary considerably on a daily basis, depending on the frequency of wetting by irrigation and rainfall. The dual crop coefficient approach calculates the actual increases in Kc for each day as a function of plant development and the wetness of the soil surface. As the single Kc coefficient includes averaged effects of evaporation from the soil, the basal crop coefficient, Kcb describing only plant transpiration, lies below the Kc value .The largest difference between Kc and Kcb is found in the initial growth stage where evapotranspiration is predominantly in the form of soil evaporation and crop transpiration is still small. Because crop canopies are near or at full ground cover during the mid-season stage, soil evaporation beneath the canopy has less effect on crop evapotranspiration and the value for Kcb in the mid-season stage will be nearly the same as Kc. Depending on the ground cover, the basal crop coefficient during the mid-season may be only 0.05–0.10 lower than the Kc value. Depending on the frequency with which the crop is irrigated during the late season stage, Kcb will be similar to (if infrequently irrigated) or less than the Kc value.

12.3.6 IRRIGATION REQUIREMENTS OF CROPS

The term water requirement (WR) of crops implies the total amount of water required at the field head regardless of its source to a mature crop.

It includes the evapotranspiration (ETc) needs, application losses and any other special needs. It however, does not include conveyance losses.

$$WR= ETc+ \text{Application losses}+ \text{Special needs} \qquad (3)$$

However in the main plot where crop is already in established condition, i.e., it has been sown already in the field, special needs for land preparation may be neglected. Assuming no application loss, we can assume water requirement (WR) is same as ETc. Irrigation requirement of crops (IR) can be expressed as:

$$IR= ETc -ER \qquad (4)$$

where, IR = irrigation requirement of crops, and ER = effective rainfall.

12.4   MATERIALS AND METHODS

12.4.1   STUDY AREA

Khurda district of Odisha, India is taken as the study area. The latitude of the study area is from 18°46' to 200 95' N and longitude of the area is 830°48' to 870°46' E and elevation is 42 m above mean sea level. The study was carried out by considering 30 years daily data like minimum and maximum air temperature, mean relative humidity, average wind speed, solar radiation, and rainfall obtained from the meteorological observatory located near to the study area and from the website http://global weather.tamu.edu/home/view/13292.

12.4.2   CROP SELECTION

Paddy is mostly cultivated during the kharif season in Odisha. In different parts of the state, all three verities paddy, i.e., short (Paddy I), medium duration (Paddy II) and long paddy (Paddy III) are almost equally cultivated during the kharif season. During rabi, six major crops are sown, which information is given in Table 12.1.

TABLE 12.1 Major Crop Information of the Study Area

CROPS	Total duration	Stages				KC value for different stages			
		Initial Stage(I)	Crop Dev. (II)	Mid Season (III)	Late Season (IV)	I	II	III	IV
Black gram	70	10	25	25	10	0.35	0.70	1.10	0.90
Green gram	60	10	20	20	10	0.35	0.70	1.10	0.90
Groundnut	137	25	30	40	25	0.45	0.75	1.05	0.70
Paddy-I	90	15	25	30	20	1.00	1.05	1.20	0.90
Paddy-II	120	15	50	25	30	1.00	1.05	1.20	0.90
Paddy-II	120	15	25	50	30	1.00	1.05	1.20	0.90
Paddy-III	150	15	30	60	45	1.00	1.05	1.20	0.90
Pea	90	15	25	35	15	0.45	0.80	1.15	1.05
Wheat	120	15	25	50	30	0.35	0.75	1.15	0.45

Total crop duration, duration of each crop stage, sowing and harvesting time of each crop are taken from Agriculture Hand Book of Odisha and Kc values of the selected crops are taken from FAO 24 publication of Doorenbos and Pruitt [7]. Various methods used to estimate the ET0 values are: (i) Penman-Monteith, (ii) Penman 1963, VPD #1, (iii) 1972 Kimberly-Penman, (iv) CIMIS Penman (v) 1982 Kimberly-Penman, (vi) Penman 1963, VPD #3 ((vii) FAO-PPP-17 Penman, (viii) Hargreaves et al. (1985), (ix) Businger-van Bavel, (x) Turc, (xi) FAO 24 Penman (c=1) and (xii) Priestly-Taylor. Details of equations and their applications to estimate ETo are available in FAO 56 publication of Allen et al. [1].

12.4.3 ESTIMATION OF DAILY EVAPOTRANSPIRATION USING DSS_ET

DSS_ET is a Decision Support System for estimation of crop evapotranspiration. The DSS_ET model [12] developed in Microsoft Visual Basic 6.0 is used in the study to estimate reference evapotranspiration. The DSS _ET developed for ET0 estimation includes a model base with decision-making capabilities, a graphical user interface and a database management system. The model base consists of 22 most commonly used and internationally accepted ET0 estimation methods based on combination theory, radiation, temperature and pan evaporation along with an algorithm based decision-making model. This model is used to identify the best ET0 estimation method for a given climatic condition. It identifies the data requirement of a method and if the available method satisfies the data requirement of the first-rank method (Penman-Monteith) as given in ASCE ranking. The system estimates the ET with that method: otherwise it searches for the next suitable method. Same procedure is repeated until a suitable method is identified for given location and data conditions. By using the available daily climatological data, the daily reference evapotranspiration (ET0) values were estimated for 30 years duration, using 12 available methods. Various DSS_ET input and output are shown in Figures 12.1a–h and Figures 12.2a–k, respectively.

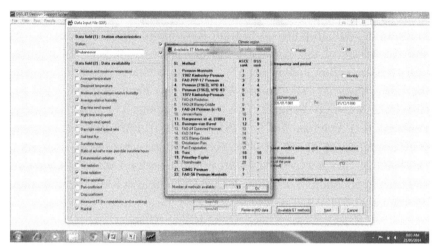

FIGURE 12.1A   DSS_ET input window 1.

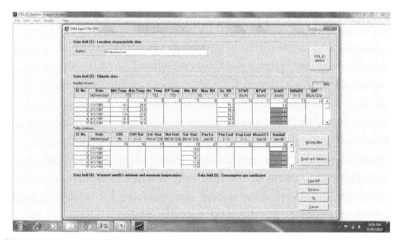

FIGURE 12.1B   DSS_ET input window 2.

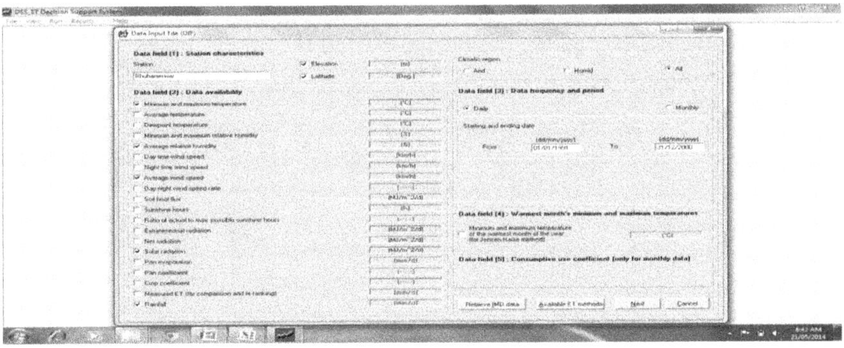

FIGURE 12.1C   DSS_ET input window 3.

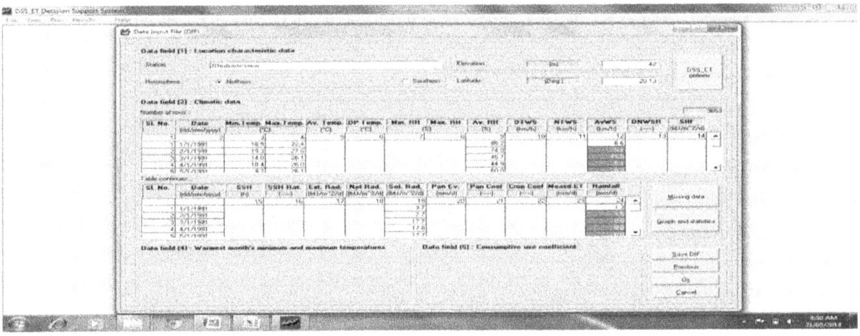

FIGURE 12.1D   DSS_ET input window 4.

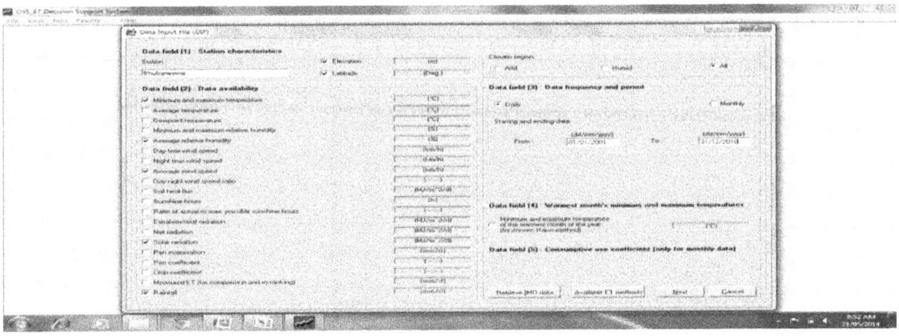

FIGURE 12.1E   DSS_ET input window 5.

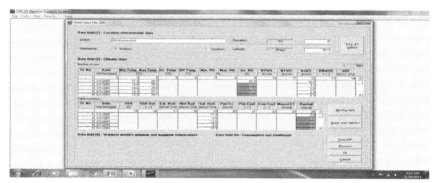

FIGURE 12.1F   DSS_ET input window 6.

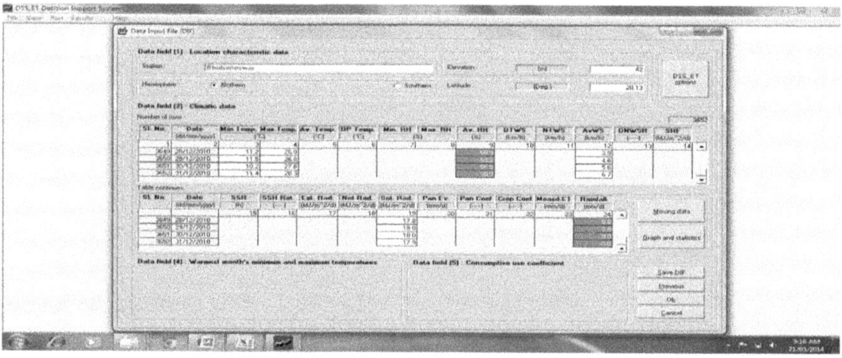

FIGURE 12.1G   DSS_ET input window 7.

FIGURE 12.1H   DSS_ET input window 8.

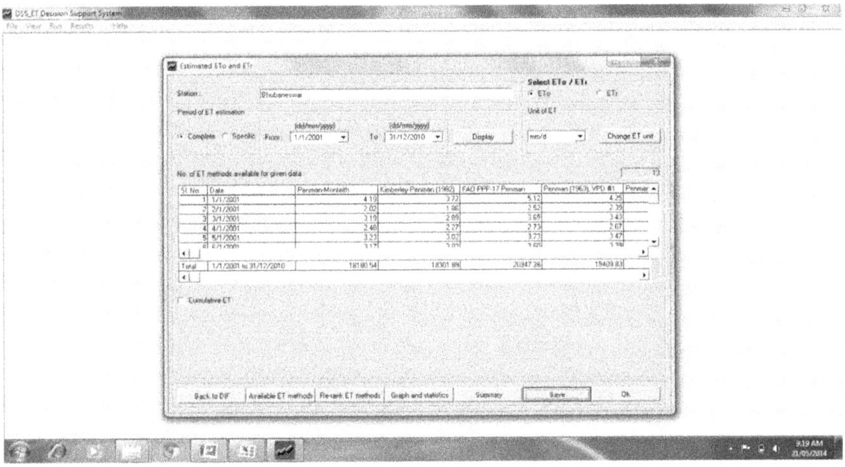

FIGURE 12.2A DSS_ET output window 9.

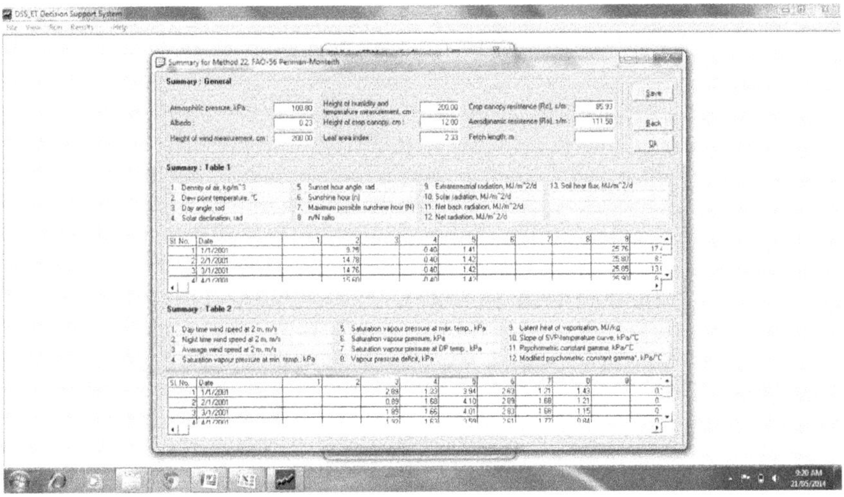

FIGURE 12.2B DSS_ET output window 10.

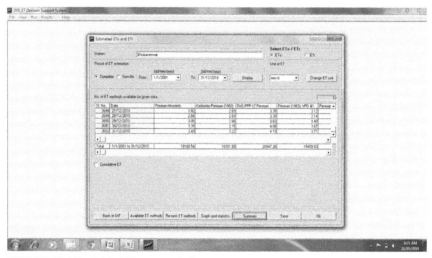

FIGURE 12.2C DSS_ET output window 11.

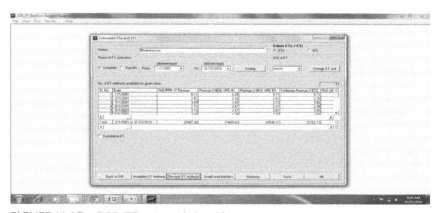

FIGURE 12.2D DSS_ET output window 12.

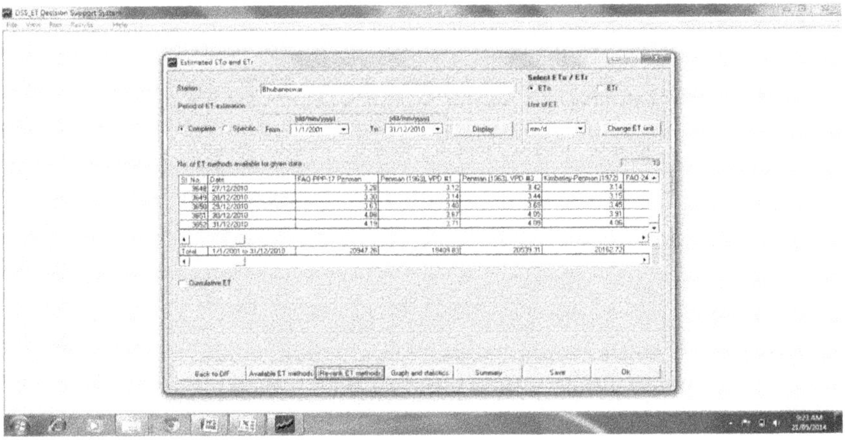

FIGURE 12.2E DSS_ET output window 13.

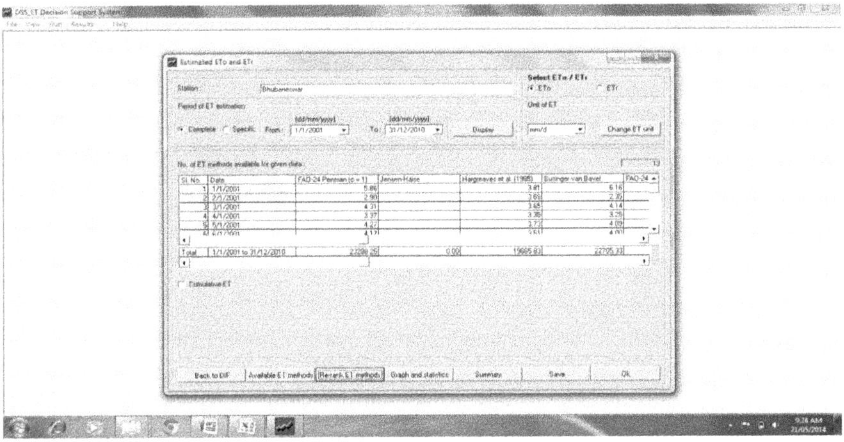

FIGURE 12.2F DSS_ET output window 14.

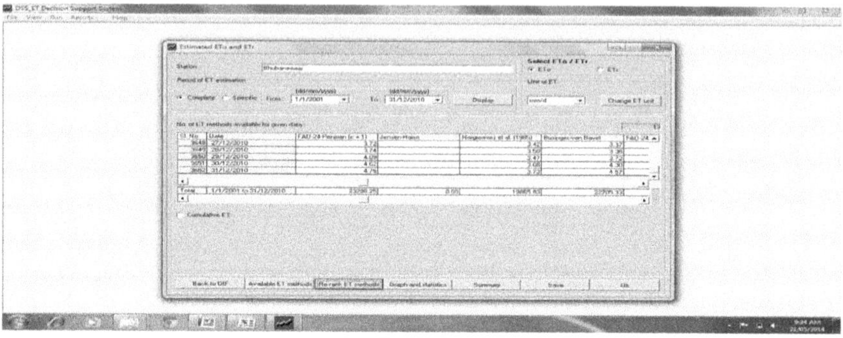

FIGURE 12.2G   DSS_ET output window 15.

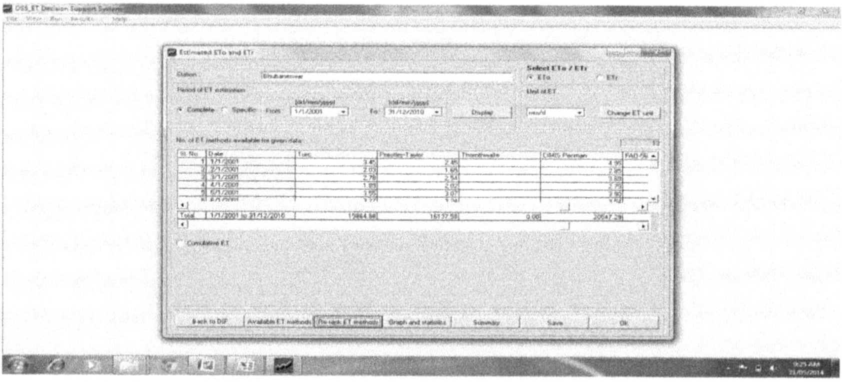

FIGURE 12.2H   DSS_ET output window 16.

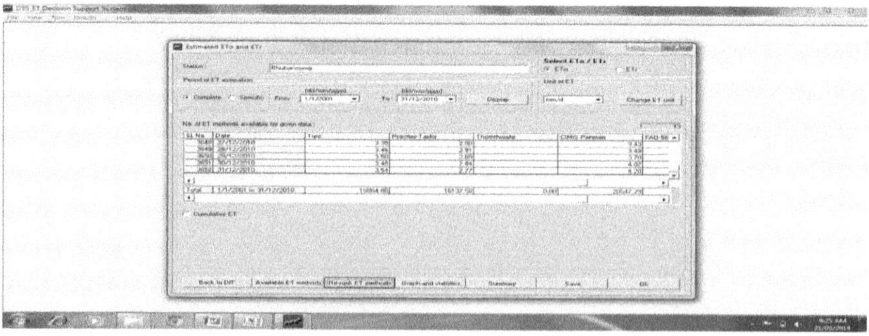

FIGURE 12.2I   DSS_ET output window 17.

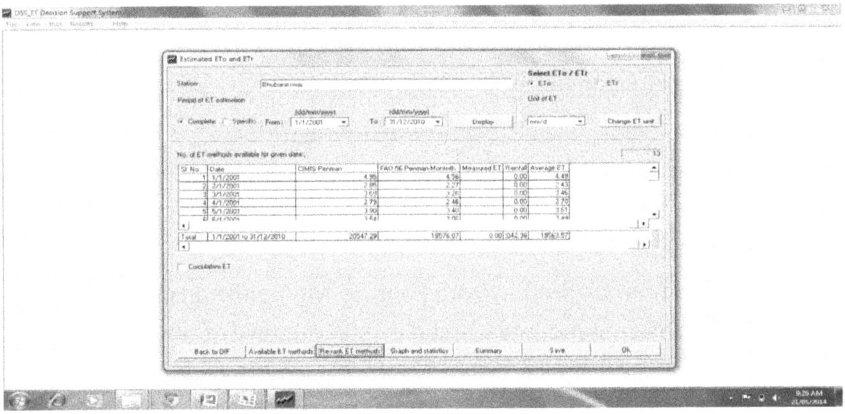

FIGURE 12.2J   DSS_ET output window 18.

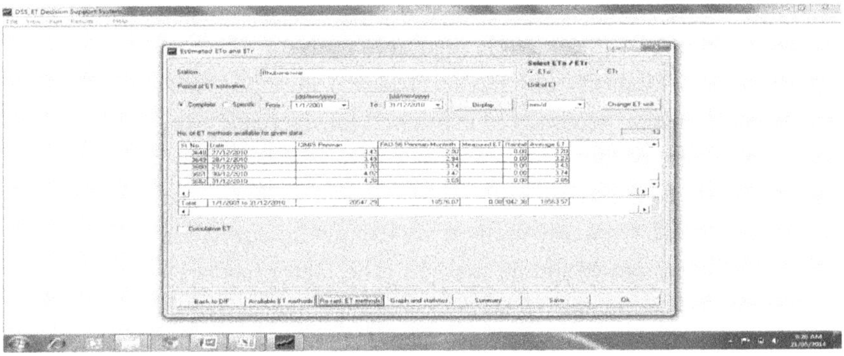

FIGURE 12.2K   DSS_ET output window 19.

12.4.4   RANKING OF DIFFERENT ET0 ESTIMATION METHODS

After estimation of reference crop evapotranspiration using 12 available methods, ranks were given to each method by comparing with reference ET0 data of standard FAO 56, Penman-Monteith method. An error estimation criterion, standard error estimate (SEE), is used to determine the error of each ET0 estimation methods.

12.4.4.1   Standard Error Estimate

Standard error of estimate, SEE is expressed as:

$$SEE = \left[\frac{\sum_{i=1}^{n} \left(Y_O - Y_E\right)^2}{n-1} \right]^{0.5} \tag{5}$$

where, Y_0 = evapotranspiration from Penman Monteith equation. YE = ET0 estimated by different methods, and n = total number of observations.

12.4.5   *ESTIMATION OF CORRECTION FACTOR*

Different evapotranspiration methods need various data for ET estimation. Among all these methods FAO 56, P-M method is considered as the best methods but data requirement is relatively high for this method. In case of limited data availability one has to choose a method other than FAO 56, PM method, hence, accuracy of estimated ET decreases. In study an effort has been made to develop correction factors for different methods other than FAO 56, P-M method. When the correction factor is multiplied with the estimated ET from any method we can get an equivalent result as that of FAO 56, Penman-Monteith method. The factor has been developed by considering 30 years daily ET0 data, and then average of the daily factors is taken as the correction factor for any particular method.

$$Correction\ factor = \frac{1}{n} \times \left(\frac{ET_{0p}}{ET_{oa}} \right) \tag{6}$$

where, ETop = reference evapotranspiration from FAO56, Penman-Monteith method and EToa = reference evapotranspiration from any other method.

12.4.6 ESTIMATION OF CROP WATER REQUIREMENT

Six rabi crops and three types of kharif paddy are considered in this study as the major crops of the study area. Thirty years reference crop evapotranspiration for all crops has been calculated by using FAO 56 P-M method. Then 30 years daily average is taken to determine ET0 for any day during the crop period. Daily crop water requirement was then calculated by multiplying crop coefficient (Kc) (Table 12.1) with the value of the estimated ET0 from FAO 56 P-M method.

12.4.7 IRRIGATION REQUIREMENT OF CROPS

Monthly irrigation requirement of selected crops has been calculated by deducting ETc values from the effective precipitation. Effective rainfall determination is always a top task. In this study, rainfall is converted to effective rainfall by following the rainfall effective rainfall relationship given in the website: file:///E:/CHAPTER%203%20%20EFFECTIVE%20RAINFALL.htm.

Special water requirements like, transplanting and sowing are finally included to find the total irrigation requirement of different crops.

12.5 RESULTS AND DISCUSSIONS

In this study, a trial has been made to determine the water requirement of six major crops including paddy. Twelve methods are used to estimate the daily reference evapotranspiration for 30-year period. All methods are then ranked with respect to the FAO56, Penman-Monteith (P–M) method. Then correction factors are derived for 12 methods in order to get equivalent result as that of FAO56 P-M method. Reference evapotranspiration was determined by taking 30 years daily average of ET0 for the crop period. Crop water requirement (ETc) and seasonal irrigation requirement were estimated for each crop using the result of reference ET0 for the crop period. In computation of ETc, daily average ETo computed by FAO 56 P-M method is considered.

TABLE 12.2 Ranking of ET0 Methods

Methods	Average factor	SEE	Rank
Penman-Monteith (ET0)	1.02	0.35	1
Penman 1963, VPD #1 (ET0)	0.94	0.69	2
1972 Kimberly-Panman (ET0)	0.92	0.90	3
CIMIS Penman (ET0)	0.89	0.98	4
1982 Kimberly-Panman (ET0)	1.02	1.00	5
Penman 1963, VPD #3 (ET0)	0.89	1.00	6
FAO-PPP-17 Penman (ET0)	0.89	1.26	7
Hargreaves et al. (1985) (ET0)	0.95	1.64	8
Businger-van Bavel (ET0)	0.82	2.22	9
Turc (ET0)	1.19	2.23	10
FAO-24 Penman (c=1) (ET0)	0.79	2.37	11
Priestly-Taylor (ET0)	1.16	2.47	12

12.5.1 RANKING OF DIFFERENT ET0 ESTIMATION METHODS

As FAO 56 P-M method is so far considered as the best ET0 estimation method, other methods are ranked with respect to FAO 56 P-M method. Ranking is based on value of standard error estimate. A method having lowest SEE value comes 1st in the ranking. SEE values and ranks of respective reference crop evapotranspiration methods are shown in Table 12.2.

From Table 12.2, it is evident that Penman-Monteith method has lowest SEE value and Priestly-Taylor method has highest ET0 value, hence, they are assigned 1st and last rank respectively. The main aim of the ranking is that a person can choose a better method among the 12 methods depending on the data availability when sufficient data is not available for FAO 56, P-M method.

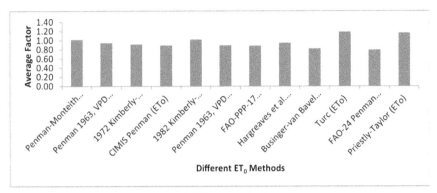

FIGURE 12.3   Correction factors of different methods.

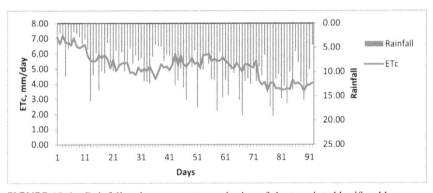

FIGURE 12.4   Rainfall and crop evapotranspiration of short variety kharif paddy.

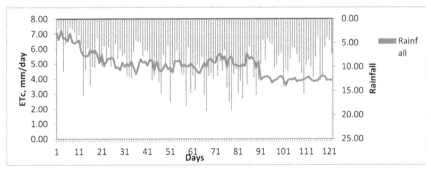

FIGURE 12.5   Rainfall and crop evapotranspiration of medium variety kharif paddy.

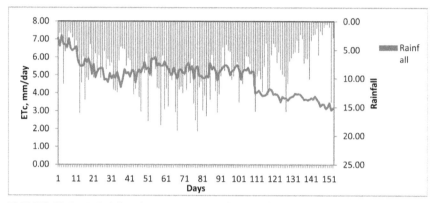

FIGURE 12.6 Rainfall and crop evapotranspiration of long variety kharif paddy.

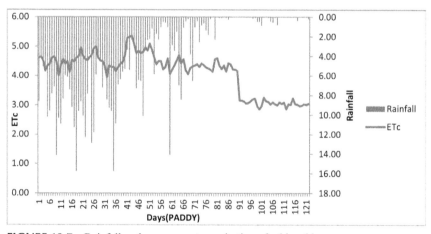

FIGURE 12.7 Rainfall and crop evapotranspiration of rabi paddy.

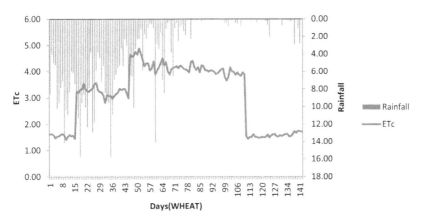

FIGURE 12.8   Rainfall and crop evapotranspiration of wheat in rabi.

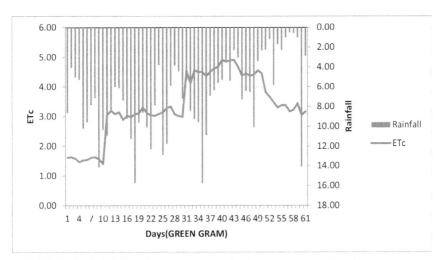

FIGURE 12.9   Rainfall and crop evapotranspiration of green gram in rabi.

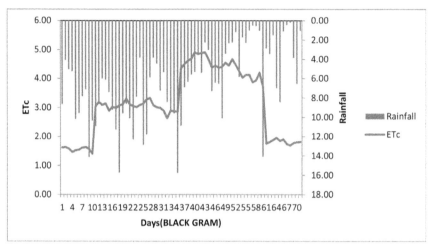

FIGURE 12.10   Rainfall and crop evapotranspiration of black gram in rabi.

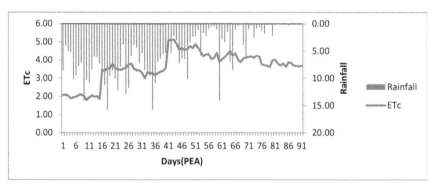

FIGURE 12.11   Rainfall and crop evapotranspiration of pea in rabi.

12.5.2   ESTIMATION OF CORRECTION FACTOR

Correction facts are calculated for different methods and are shown in Figure 12.3. These factors are multiplied with the estimated ET0 by any method to get an equivalent result as that of FAO 56, P-M method in a data scarce condition.

From Figure 12.3, it is observed that only four methods have average factor value above one and others have value less than one. For methods

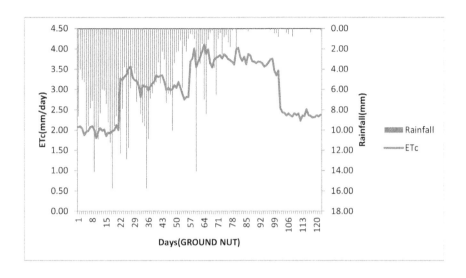

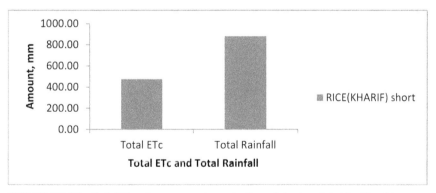

FIGURE 12.13   Rainfall and crop evapotranspiration of short variety kharif paddy.

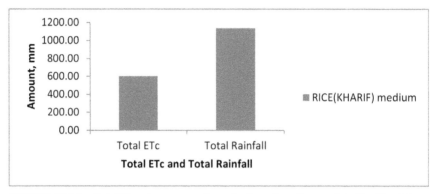

FIGURE 12.14   Rainfall and crop evapotranspiration of medium variety kharif paddy.

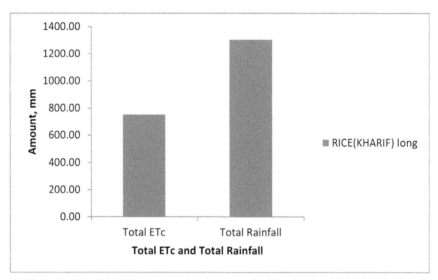

FIGURE 12.15   Rainfall and crop evapotranspiration of long variety kharif paddy.

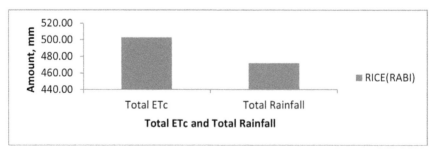

FIGURE 12.16   Rainfall and crop evapotranspiration of rabi paddy.

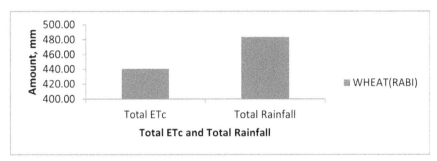

FIGURE 12.17 Rainfall and crop evapotranspiration of rabi wheat.

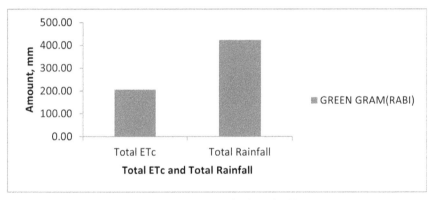

FIGURE 12.18 Rainfall and crop evapotranspiration of rabi green gram.

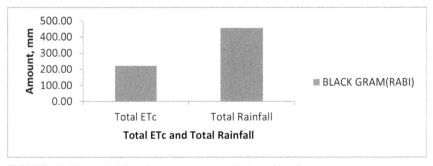

FIGURE 12.19 Rainfall and crop evapotranspiration of black gram.

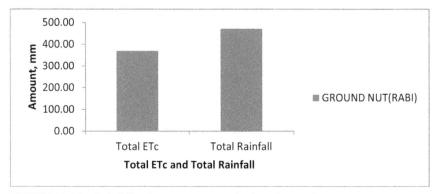

FIGURE 12.20   Rainfall and crop evapotranspiration of groundnut.

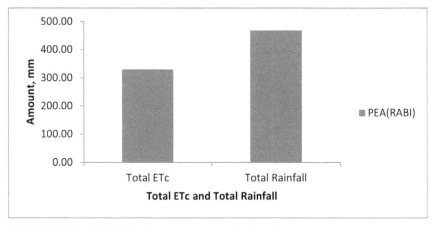

FIGURE 12.21   Rainfall and crop evapotranspiration of pea.

having factor value above one can be considered as the ET0 under estimating methods and methods having average factor value bellow one are
considered as ET0 over estimating methods.

FIGURE 12.22   Monthly irrigation requirement of short variety kharif paddy.

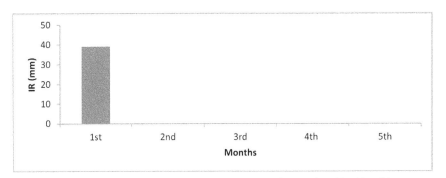

FIGURE 12.23   Monthly irrigation requirement of medium variety kharif paddy.

FIGURE 12.24   Monthly irrigation requirement of long variety kharif paddy.

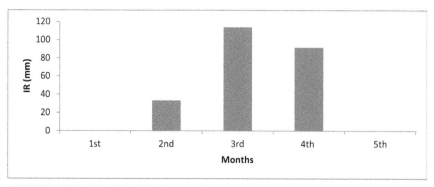

FIGURE 12.25 Monthly irrigation requirement of long variety rabi paddy.

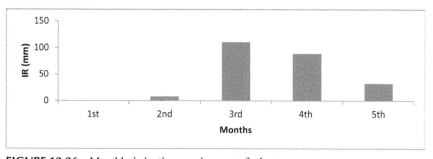

FIGURE 12.26 Monthly irrigation requirement of wheat.

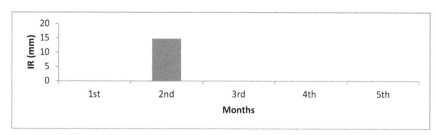

FIGURE 12.27 Monthly irrigation requirement of green gram.

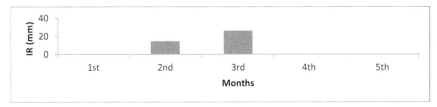

FIGURE 12.28   Monthly irrigation requirement of black gram.

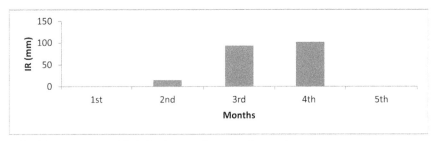

FIGURE 12.29   Monthly irrigation requirement of ground nut.

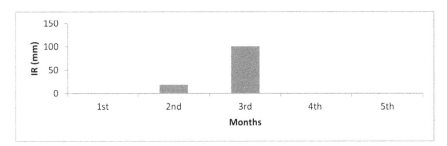

FIGURE 12.30   Monthly irrigation requirement of pea.

12.5.3 VARIATIONS OF DAILY RAINFALL AND CROP EVAPOTRANSPIRATION

The graphs representing crop water requirement/crop evapotranspiration and daily rainfall in the crop period for various crops are shown in Figures 12.4–12.12.

FIGURE 12.12   Rainfall and crop evapotranspiration of groundnut in rabi.

12.5.4 VARIATIONS OF TOTAL SEASONAL RAINFALL AND CROP EVAPOTRANSPIRATION

Total seasonal crop water requirement/crop evapotranspiration and total seasonal rainfall is shown in Figures 12.13–12.21 for various crops. The figures indicate that there is sufficient surplus water available during kharif but rabi crop needs irrigation because rainfall is not regular. For kharif long duration rice crop, water requirement is highest and for green gram, water requirement is the lowest.

12.5.5 MONTHLY IRRIGATION REQUIREMENT OF DIFFERENT CROPS

Monthly irrigation requirement of 3 kharif paddies and 6 rabi crops are shown in Figures 12.22–12.30.

For kharif paddies of three verities, irrigation is only required in 1st month of sowing. In other moths there is sufficient rainfall to meet the crop water demand. In rabi paddy, three months, i.e., 2nd, 3rd and 4th need irrigation. The irrigation amount varies from 30 to 108 mm. The 2nd month needs less irrigation whereas the 3rd month requires more irrigation. The 3rd month receives less rainfall, which falls in March and so the irrigation requirement is more in this month. For wheat these is four months of water requirement, in the 3rd moth of sowing highest irrigation water is required. For green gram, irrigation is required only in the second month, in other months no irrigation is required. For black gram irrigation is required in the second and third month of sowing. In other months sufficient rainfall is there to meet the crop water demand. For groundnut irrigation is required in 2nd 3rd and 4th month of sowing. Highest irrigation is required in 4th month after sowing. For pea, irrigation is required in the 2nd and 3rd month after sowing.

12.6   CONCLUSIONS

Evapotranspiration (ETc) is the largest and one of the most basic components of the hydrologic cycle. Accurate estimation of ETc is of vital importance for many studies, Such as hydrologic water balance, irrigation system design and management, water resources planning and management, etc. ETc is a complex and non-linear phenomenon as it depends on several interacting factors such as temperature, humidity, wind speed, radiation and type and growth stage of crop. Direct measurement of ET is a time consuming method and needs precisely and carefully planned experiments. Due to wide application of ETc data, a number of indirect methods for estimation of ETo have been developed based on the easily available location characteristics (elevation and latitude) and meteorological parameters (temperature, humidity, wind speed, solar radiation). The goal of present study was to estimate reference evapotranspiration by all applicable methods, ranking of methods and to obtain mean, variability of ETo and a factor for all methods. To achieve this goal, a user-friendly decision support system (DSS_ET) was used in the present study for the estimation of the reference evapotranspiration for the climatological station spread all over India. Out of the stations Bhubaneswar is considered, in this study. The long term daily climatic data of minimum and maximum air temperature, mean relative humidity, wind speed, solar radiation and sunshine hours obtained from website were used to estimate reference evapotranspiration. Thirteen methods were found applicable for the availability of data and climatic condition of the present study. Ranking of methods was done on the basis of standard error of estimate (SEE) to decide the best ETo estimation method. Mean and variability of ETo estimation method were also determined. In addition, a factor to convert ETo estimates of different methods to equivalent of FAO-56 P-M method was also determined for each methods for the chosen states. The following specific conclusions could be drawn from the present study:

Daily reference crop evapotranspiration was determined for the region using 12 methods. The performance of different methods was varied. Among the combination-based methods, Penman-Monteith method gave ET0 estimates close to standard FAO-56 P-M method than the other methods. In general combination based methods performed better than temperature and radiation based methods. Correction factors were determined for different methods with respect to FAO 56 Penman–Monteith method. The factors determined in this study can be used to convert ET0 estimates of other methods to equivalent of FAO-56 P-M estimates. Daily crop evapotranspiration were determined for major crops during kharif and rabi season. Monthly irrigation requirement were estimated for the selected crops during kharif and rabi season. The study indicates that there are irrigation requirements for all crops in both the seasons. However, rabi crops require more number and amount of irrigation than the kharif crops since in rabi the rainfall is less.

12.7   SUMMARY

In the present study, water requirement of different crops including cereals and pulses is estimated by using DSS_ET version 4.1 software. 30 years long term daily climatic data of minimum and maximum air temperature, relative humidity, average wind speed, solar radiation and rainfall were used in the study. The software DSS_ET developed for ET_0 estimation includes a model base with decision-making capabilities, a graphical user interface and a database management system. The model base consists of twelve most commonly used and internationally accepted ET_0 estimation methods based on combination theory, radiation, temperature and pan evaporation along with an algorithm based decision-making model. In this study, we have determined water requirement of six major crops of the state of Odisha, India. The well-established methods were used to estimate the daily reference evapotranspiration for 30-year period. For each crop, reference evapotranspiration was determined by taking 30 years daily average of ET_0 for the crop period. Daily crop water requirement was then calculated by multiplying crop coefficient (Kc) with the estimated ET_0 from FAO 56-P-M method. Water requirement of rice grown in *kharif* was found to be the highest and for green gram it was the lowest.

KEYWORDS

- **crop coefficient**
- **crop evapotranspiration**
- **effective rainfall**
- **estimation methods**
- **irrigation requirement**
- **paddy**
- **pan evaporation**
- **reference evapotranspiration**
- **solar radiation**
- **water requirement**

REFERENCES

1. Allen, R. G., Pereira, L. S., Raes, D., and Smith, M., 1998. *Crop evapotranspiration guidelines for computing crop water requirements.* Irrigation and Drainage Paper-56. Food and Agriculture Organization of the United Nations (FAO), Rome, Italy.
2. Alexandris, S. and Kerkides, P., 2003. New empirical formula for hourly estimations of reference evapotranspiration. *Agricultural Water Management,* 60:157–180.
3. Almhab, A. and Busu, I., 2008. *Decision support system for estimation of regional in arid areas: Application to the Republic of Yemen.* Mountain GIS e-Conference, 14–25 January, Department of remote Sensing, University Technology, Johor, Malaysia.
4. Bandyopadhyay, A., Bhadra, A. Raghuwanshi, N. S., and Singh, 2008. Estimation of monthly solar radiation from measured air temperature extremes. *Agricultural and forest Meteorology,* 148:1707–1718.
5. Benli, B., Bruggeman, A., Oweis, T., and Ustun, H., 2010. Performance of Penman-Monteith FAO-56 in a semiarid highland environment. *Journal of Irrigation and Drainage Engineering, ASCE,* 136(11):757–765.

6. Dalei, A., 2014. *Water Requirements of Major Crops of Khurda District of Odisha*. M. Tech. Thesis, Orissa University of Agriculture and Technology, Bhubaneswar, Odisha, India.
7. Doorenbos, J. and Pruitt, W. O., 1977. *Guidelines for predicting crop water requirements, Irrigation and Drainage paper 24*. Food and Agriculture Organization of the United Nations (FAO), Rome, Italy.
8. Daut, I., Irwanto, M., Irwan, Y. M., Gomesh, N., and Ahmad, N. S., 2011. Combination of Hargreaves method and linear regression as a new method to estimate solar radiation in Perlis, Northern Malaysia. *Solar Energy*, 85:2871–2880.
9. George, B. A., Reddy, B. R. S., Raghuwanshi, N. S., and Wallender, W. W., 2002. Decision support system for estimating reference crop evapotranspiration. *Journal of Irrigation and Drainage Engineering, ASCE*, 128(1):1–10.
10. Jensen, M. E., Burman, R. D., and Allen, R. G., 1990. *Evapotranspiration and irrigation water requirements*. ASCE Manuals No. 70, American Society of Civil Engineers (ASCE), NY, USA.
11. Jiabing, C., Yu, L., Tingwu. L., and Luis, S. P., 2007. Estimating reference evapotranspiration with the FAO Penman-Monteith equation using daily weather forecast messages. *Agricultural and Forest Meteorology*, 145: 22–35.
12. Lee, T. S., Najim, M. M. M. and Aminul, M. H., 2004. Estimating evapotranspiration of irrigated rice at the West coast of the Peninsular of Malaysia. *Journal of Applied Irrigation Science*, 39(1):103–117.
13. Reddy, B. R. S., 1999. Development of a decision support system for estimating evapotranspiration. MTech Thesis, Department of Agricultural and Food Engineering, Indian Institute of Technology Kharagpur, India.
14. Swarnakar, R. K. and Raghuwanshi, N. S., 2000). *DSS_ET user Manual*. Department of Agricultural and Food Engineering, Indian Institute of Technology, Kharagpur, India.
15. Tabari, H. and Talaee, P. H., 2011. Local calibration of the Hargreaves and Priestley-Taylor equations for estimating reference evapotranspiration in arid and cold climates of Iran based on the Penman-Monteith Model. *Journal of Irrigation and Drainage Engineering, ASCE*, 138(2):111–119.
16. Xu, C., Gong, L., Jiang, T., Chen, D., and Singh V. P., 2004. Analysis of spatial distribution and temporal trend of reference evapotranspiration and pan evaporation in Changjiang (Yangtze River) Catchment. *Journal of Hydrology*, 327:81–93.

APPENDICES

(Modified and reprinted with permission from: Goyal, Megh R., 2012. Appendices. Pages 317–332. In: *Management of Drip/Trickle or Micro Irrigation* edited by Megh R. Goyal. New Jersey, USA: Apple Academic Press Inc.)

APPENDIX A   CONVERSION SI AND NON-SI UNITS

To convert the Column 1 in the Column 2, Multiply by	Column 1 Unit SI	Column 2 Unit Non-SI	To convert the Column 2 in the Column 1 Multiply by

LINEAR

0.621 _____	kilometer, km (10^3m)	miles, mi _____	1.609
1.094 _____	meter, m	yard, yd _____	0.914
3.28 _____	meter, m	feet, ft _____	0.304
3.94×10^{-2} ___	millimeter, mm (10^{-3})	inch, in _____	25.4

SQUARES

2.47 _____	hectare, he	acre _____	0.405
2.47 _____	square kilometer, km^2	acre _____	4.05×10^{-3}
0.386 _____	square kilometer, km^2	square mile, mi^2 _____	2.590
2.47×10^{-4} ___	square meter, m^2	acre _____	4.05×10^{-3}
10.76 _____	square meter, m^2	square feet, ft^2 _____	9.29×10^{-2}
1.55×10^{-3} ___	mm^2	square inch, in^2 _____	645

CUBICS

9.73×10^{-3} ___	cubic meter, m^3	inch-acre _____	102.8
35.3 _____	cubic meter, m^3	cubic-feet, ft^3 _____	2.83×10^{-2}

6.10×10^{4}	cubic meter, m^3	cubic inch, in^3	1.64×10^{-5}
2.84×10^{-2}	liter, L (10^{-3} m^3)	bushel, bu	35.24
1.057	liter, L	liquid quarts, qt	0.946
3.53×10^{-2}	liter, L	cubic feet, ft^3	28.3
0.265	liter, L	gallon	3.78
33.78	liter, L	fluid ounce, oz	2.96×10^{-2}
2.11	liter, L	fluid dot, dt	0.473

WEIGHT

2.20×10^{-3}	gram, g (10^{-3} kg)	pound,	454
3.52×10^{-2}	gram, g (10^{-3} kg)	ounce, oz	28.4
2.205	kilogram, kg	pound, lb	0.454
10^{-2}	kilogram, kg	quintal (metric), q	100
1.10×10^{-3}	kilogram, kg	ton (2000 lbs), ton	907
1.10^{2}	mega gram, mg	ton (US), ton	0.907
1.10^{2}	metric ton, t	ton (US), ton	0.907

YIELD AND RATE

0.893	kilogram per hectare hectare	pound per acre	1.12
7.77×10^{-2}	kilogram per cubic meter	pound per fanega	12.87
1.49×10^{-2}	kilogram per hectare	pound per acre, 60 lb	67.19
1.59×10^{-2}	kilogram per hectare	pound per acre, 56 lb	62.71
1.86×10^{-2}	kilogram per hectare	pound per acre, 48 lb	53.75
0.107	liter per hectare	galloon per acre	9.35
893	ton per hectare	pound per acre	1.12×10^{-3}
893	mega gram per hectare	pound per acre	1.12×10^{-3}

0.446 _____ ton per hectare   ton (2000 lb) per _____ 2.24
                                         acre

2.24 _____ meter per second   mile per hour _____ 0.447

SPECIFIC SURFACE

10 _____ square meter per   square centimeter ____ 0.1
              kilogram   per gram

10^3 _____ square meter per   square millimeter ____ 10^{-3}
              kilogram   per gram

PRESSURE

9.90 _____ megapascal, MPa   atmosphere _____ 0.101

10 _____ megapascal   bar _____ 0.1

1.0 _____ megagram per   gram per cubic _____ 1.00
              cubic meter   cubic centimeter

2.09×10^{-2} ___ pascal, Pa   pound per square ____ 47.9
                                         feet

1.45×10^{-4} ___ pascal, Pa   pound per square ____ 6.90×10^3
                                         inch

To convert the	Column 1	Column 2	To convert the column
column 1 in the	Unit	Unit	2 in the column 1
Column 2,	SI	Non-SI	Multiply by
Multiply by			

TEMPERATURE

1.00 _____ Kelvin, K   centigrade, °C _____ 1.00
(K-273)   (C+273)

(1.8 C _____ centigrade, °C   Fahrenheit,°F _____ (F-32)/1.8
+ 32)

ENERGY

9.52×10^{-4} ___	Joule J	BTU _____	1.05×103
0.239 _____	Joule, J	calories, cal _____	4.19
0.735 _____	Joule, J	feet-pound _____	1.36
2.387×10^5 ___	Joule per square meter	calories per square ___ centimeter	4.19×10^4
10^5 _____	Newton, N	dynes _____	10^{-5}

WATER REQUIREMENTS

9.73×10^{-3} ___	cubic meter	inch acre _____	102.8
9.81×10^{-3} ___	cubic meter per hour	cubic feet per _____ second	101.9
4.40 _____	cubic meter per hour	galloon (US) per _____ minute	0.227
8.11 _____	hectare-meter	acre-feet _____	0.123
97.28 _____	hectare-meter	acre-inch _____	1.03×10^{-2}
8.1×10^{-2} ____	hectare centimeter	acre-feet _____	12.33

CONCENTRATION

1 _____	centimol per kilogram	milliequivalents _____ per 100 grams	1
0.1 _____	gram per kilogram	percents _____	10
1 _____	milligram per kilogram	parts per million ____	1

NUTRIENTS FOR PLANTS

2.29 _____	P	P_2O_5 _____	0.437
1.20 _____	K	K_2O _____	0.830
1.39 _____	Ca	CaO _____	0.715
1.66 _____	Mg	MgO _____	0.602

NUTRIENT EQUIVALENTS

Column A	Column B	Conversion A to B	Equivalent B to A
N	NH_3	1.216	0.822
	NO_3	4.429	0.226
	KNO_3	7.221	0.1385
	$Ca(NO_3)_2$	5.861	0.171
	$(NH_4)_2SO_4$	4.721	0.212
	NH_4NO_3	5.718	0.175
	$(NH_4)_2HPO_4$	4.718	0.212
P	P_2O_5	2.292	0.436
	PO_4	3.066	0.326
	KH_2PO_4	4.394	0.228
	$(NH_4)_2HPO_4$	4.255	0.235
	H_3PO_4	3.164	0.316
K	K_2O	1.205	0.83
	KNO_3	2.586	0.387
	KH_2PO_4	3.481	0.287
	Kcl	1.907	0.524
	K_2SO_4	2.229	0.449
Ca	CaO	1.399	0.715
	$Ca(NO_3)_2$	4.094	0.244
	$CaCl_2 \times 6H_2O$	5.467	0.183
	$CaSO_4 \times 2H_2O$	4.296	0.233
Mg	MgO	1.658	0.603
	$MgSO_4 \times 7H_2O$	1.014	0.0986
S	H_2SO_4	3.059	0.327
	$(NH_4)_2SO_4$	4.124	0.2425
	K_2SO_4	5.437	0.184
	$MgSO_4 \times 7H_2O$	7.689	0.13
	$CaSO_4 \times 2H_2O$	5.371	0.186

APPENDIX B   PIPE AND CONDUIT FLOW

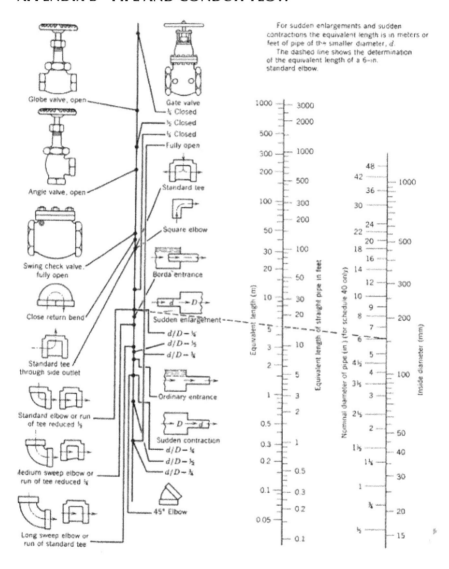

Friction loss (m per 100 m length of main line) of portable aluminum pipe with couplings: Based on Scobey's formula, for $K_S = 10$ m.

Flow		Pipe diameter, cm						
Liters/s	GPM	7.5	10	12.5	15	17.5	20	25
2.52	40	0.658	0.157	—	—	—	—	—
3.15	50	1.006	0.239	—	—	—	—	—
3.79	60	1.423	0.339	—	—	—	—	—
4.42	70	1.906	0.449	0.150	—	—	—	—
5.05	80	2.457	0.584	0.193	—	—	—	—
5.68	90	3.073	0.731	0.242	—	—	—	—
6.31	100	3.754	0.893	0.295	0.120	—	—	—
7.57	120	5.307	1.263	0.413	0.170	—	—	—
8.83	140	7.113	1.693	0.560	0.227	—	—	—
10.10	160	9.169	2.182	0.721	0.293	—	—	—
11.36	180	11.47	2.729	0.967	0.366	—	—	—
12.62	200	14.01	3.333	1.102	0.448	0.209	—	—
13.88	220	16.79	3.996	1.321	0.537	0.251	—	—
15.14	240	19.81	4.713	1.558	0.633	0.296	—	—
16.41	260	23.06	5.448	1.814	0.737	0.344	—	—
17.67	280	26.55	6.316	2.089	0.849	0.397	—	—
18.93	300	30.27	7.203	2.381	0.967	0.452	0.235	—
20.19	320	34.22	8.142	2.692	1.094	0.511	0.265	—
21.45	340	38.39	9.137	3.020	1.227	0.573	0.298	—
22.72	360	42.80	10.18	3.366	1.368	0.639	0.332	—
23.98	380	47.43	11.29	3.731	1.516	0.708	0.368	—
25.24	400	52.28	12.44	4.113	1.671	0.781	0.399	0.136
26.50	420	—	13.95	4.513	1.833	0.857	0.445	0.149
27.76	440	—	14.57	4.930	1.988	0.936	0.486	0.163
29.03	460	—	16.23	5.364	2.179	1.019	0.529	0.177
30.29	480	—	17.59	5.815	2.363	1.104	0.573	0.192
31.55	500	—	19.01	6.284	2.554	1.193	0.620	0.208
34.70	550	—	22.79	7.532	3.060	1.430	0.742	0.249
37.86	600	—	26.88	9.886	3.611	1.687	0.876	0.294
41.01	650	—	31.30	10.35	4.204	1.965	1.020	0.342

Flow		Pipe diameter, cm						
Liters/s	GPM	7.5	10	12.5	15	17.5	20	25
44.17	700	—	36.03	11.91	4.839	2.262	1.174	0.394
47.32	750	—	41.08	13.58	5.517	2.520	1.339	0.449
50.48	800	—	—	15.35	6.237	2.915	1.513	0.507
53.60	850	—	—	17.32	6.999	3.71	1.698	0.569
56.79	900	—	—	19.20	7.801	3.646	1.893	0.635
59.94	950	—	—	21.28	8.645	4.041	2.097	0.703
63.10	1000	—	—	23.45	9.530	4.454	2.312	0.775
69.49	1100	—	—	28.11	11.42	5.338	2.771	0.929
75.72	1200	—	—	31.75	13.58	6.298	3.269	1.096
82.03	1300	—	—	—	15.69	7.333	3.806	1.277
88.34	1400	—	—	—	18.06	8.441	4.382	1.470
94.65	1500	—	—	—	20.59	9.624	4.996	1.675
101.0	1600	—	—	—	23.28	10.88	5.648	1.894
107.3	1700	—	—	—	26.12	21.21	6.337	2.125
14.0	1800	—	—	—	—	13.61	7.064	2.369
120.0	1900	—	—	—	—	15.08	7.829	2.625
126.0	2000	—	—	—	—	16.62	8.630	2.894

Friction loss (m per 100 m length of lateral lines) of portable aluminum pipe with couplings: Based on Scobey's formula.

Flow, Liters/s	Pipe diameter, cm				
	5.0	7.5	10	12.5	15
	KS = 0.34	KS = 0.33		KS = 0.32	
1.26	—	—	—	—	—
1.89	0.32	—	—	—	—
2.52	2.53	—	—	—	—
3.15	4.40	0.565	0.130	—	—
3.79	6.85	0.858	0.198	—	—
4.42	9.67	1.21	0.280	—	—
5.05	12.9	1.63	0.376	0.122	—

Flow, Liters/s	Pipe diameter, cm				
	5.0	7.5	10	12.5	15
	KS = 0.34	KS = 0.33		KS = 0.32	
5.68	16.7	2.10	0.484	0.157	—
6.31	20.8	2.63	0.605	0.196	—
7.57	25.4	3.20	0.738	0.240	0.099
8.83	—	4.54	1.04	0.339	0.140
10.10	—	6.09	1.40	0.454	0.188
11.36	—	7.85	1.80	0.590	0.242
12.62	—	9.82	2.26	0.733	0.302
13.88	—	12.0	2.76	0.896	0.370
15.14	—	14.4	3.30	1.07	0.443
16.41	—	16.9	3.90	1.26	0.522
17.67	—	19.7	4.54	1.47	0.608
18.93	—	22.8	5.22	1.70	0.700
20.19	—	25.9	5.96	1.93	0.798
21.45	—	29.3	6.74	2.18	0.904
22.72	—	32.8	7.56	2.45	1.02
23.98	—	36.6	8.40	2.74	1.13
25.24	—	40.6	9.36	3.03	1.26
26.50	—	44.7	10.3	3.34	1.38
27.76	—	—	11.3	3.66	1.521
29.03	—	—	12.3	4.00	1.66
30.29	—	—	13.4	4.35	1.80
31.55	—	—	14.6	4.72	1.95
34.70	—	—	15.8	5.10	2.12
37.86	—	—	18.9	6.12	2.52
41.01	—	—	22.2	7.22	2.98
44.17	—	—	25.9	8.40	3.46
47.32	—	—	29.8	9.68	3.99
50.48	—	—	33.8	11.0	4.54
53.63	—	—		12.5	5.15
56.79	—	—		14.0	5.78
59.94	—	—		15.6	6.44
63.10	—	—		17.3	7.14

APPENDIX C   PERCENTAGE OF DAILY SUNSHINE HOURS: FOR NORTH AND SOUTH HEMISPHERES

Latitude	Jan	Feb	Mar	Apr	May	Jun	Jul	Aug	Sep	Oct	Nov	Dec
NORTH												
0	8.50	7.66	8.49	8.21	8.50	8.22	8.50	8.49	8.21	8.50	8.22	8.50
5	8.32	7.57	8.47	3.29	8.65	8.41	8.67	8.60	8.23	8.42	8.07	8.30
10	8.13	7.47	8.45	8.37	8.81	8.60	8.86	8.71	8.25	8.34	7.91	8.10
15	7.94	7.36	8.43	8.44	8.98	8.80	9.05	8.83	8.28	8.20	7.75	7.88
20	7.74	7.25	8.41	8.52	9.15	9.00	9.25	8.96	8.30	8.18	7.58	7.66
25	7.53	7.14	8.39	8.61	9.33	9.23	9.45	9.09	8.32	8.09	7.40	7.52
30	7.30	7.03	8.38	8.71	9.53	9.49	9.67	9.22	8.33	7.99	7.19	7.15
32	7.20	6.97	8.37	8.76	9.62	9.59	9.77	9.27	8.34	7.95	7.11	7.05
34	7.10	6.91	8.36	8.80	9.72	9.70	9.88	9.33	8.36	7.90	7.02	6.92
36	6.99	6.85	8.35	8.85	9.82	9.82	9.99	9.40	8.37	7.85	6.92	6.79
38	6.87	6.79	8.34	8.90	9.92	9.95	10.1	9.47	3.38	7.80	6.82	6.66
40	6.76	6.72	8.33	8.95	10.0	10.1	10.2	9.54	8.39	7.75	6.72	7.52
42	6.63	6.65	8.31	9.00	10.1	10.2	10.4	9.62	8.40	7.69	6.62	6.37
44	6.49	6.58	8.30	9.06	10.3	10.4	10.5	9.70	8.41	7.63	6.49	6.21
46	6.34	6.50	8.29	9.12	10.4	10.5	10.6	9.79	8.42	7.57	6.36	6.04
48	6.17	6.41	8.27	9.18	10.5	10.7	10.8	9.89	8.44	7.51	6.23	5.86
50	5.98	6.30	8.24	9.24	10.7	10.9	11.0	10.0	8.35	7.45	6.10	5.64
52	5.77	6.19	8.21	9.29	10.9	11.1	11.2	10.1	8.49	7.39	5.93	5.43
54	5.55	6.08	8.18	9.36	11.0	11.4	11.4	10.3	8.51	7.20	5.74	5.18

Latitude	Jan	Feb	Mar	Apr	May	Jun	Jul	Aug	Sep	Oct	Nov	Dec
NORTH												
56	5.30	5.95	8.15	9.45	11.2	11.7	11.6	10.4	8.53	7.21	5.54	4.89
58	5.01	5.81	8.12	9.55	11.5	12.0	12.0	10.6	8.55	7.10	4.31	4.56
60	4.67	5.65	8.08	9.65	11.7	12.4	12.3	10.7	8.57	6.98	5.04	4.22
SOUTH												
0	8.50	7.66	8.49	8.21	8.50	8.22	8.50	8.49	8.21	8.50	8.22	8.50
5	8.68	7.76	8.51	8.15	8.34	8.05	8.33	8.38	8.19	8.56	8.37	8.68
10	8.86	7.87	8.53	8.09	8.18	7.86	8.14	8.27	8.17	8.62	8.53	8.88
15	9.05	7.98	8.55	8.02	8.02	7.65	7.95	8.15	8.15	8.68	8.70	9.10
20	9.24	8.09	8.57	7.94	7.85	7.43	7.76	8.03	8.13	8.76	8.87	9.33
25	9.46	8.21	8.60	7.74	7.66	7.20	7.54	7.90	8.11	8.86	9.04	9.58
30	9.70	8.33	8.62	7.73	7.45	6.96	7.31	7.76	8.07	8.97	9.24	9.85
32	9.81	8.39	8.63	7.69	7.36	6.85	7.21	7.70	8.06	9.01	9.33	9.96
34	9.92	8.45	8.64	7.64	7.27	6.74	7.10	7.63	8.05	9.06	9.42	10.1
36	10.0	8.51	8.65	7.59	7.18	6.62	6.99	7.56	8.04	9.11	9.35	10.2
38	10.2	8.57	8.66	7.54	7.08	6.50	6.87	7.49	8.03	9.16	9.61	10.3
40	10.3	8.63	8.67	7.49	6.97	6.37	6.76	7.41	8.02	9.21	9.71	10.5
42	10.4	8.70	8.68	7.44	6.85	6.23	6.64	7.33	8.01	9.26	9.8	10.6
44	10.5	8.78	8.69	7.38	6.73	6.08	6.51	7.25	7.99	9.31	9.94	10.8
46	10.7	8.86	8.90	7.32	6.61	5.92	6.37	7.16	7.96	9.37	10.1	11.0

Mean daily maximum duration of bright sunshine hours (n) for different months and latitudes.

North South	Jan.– July	Feb.– Aug	Mar – Sept.	April– Oct.	May– Nov.	June– Dec.	July– Jan.	Aug.– Feb.	Sept.– Mar	Oct.– April	Nov.– May	Dec.– June
50	8.5	10.1	11.8	13.8	15.4	16.3	15.9	14.5	12.7	10.8	9.1	8.1
48	8.8	10.2	11.8	13.6	15.2	16.0	15.6	14.3	12.6	10.9	9.3	8.3
46	9.1	10.4	11.9	13.5	14.9	15.7	15.4	14.2	12.6	10.9	9.5	8.7
44	9.3	10.5	11.9	13.4	14.7	15.4	15.2	14.0	12.6	11.0	9.7	8.9
42	9.4	10.6	11.9	13.4	14.6	15.2	14.9	13.9	12.6	11.1	9.8	9.1
40	9.6	10.7	11.9	13.3	14.4	15.0	14.7	13.7	12.5	11.2	10.0	9.3
35	10.1	11.0	11.9	13.1	14.0	14.5	14.3	13.5	12.4	11.3	10.3	9.8
30	10.4	11.1	12.0	12.9	13.6	14.0	13.9	13.2	12.4	11.5	10.6	10.2
25	10.7	11.3	12.0	12.7	13.3	13.7	13.5	13.0	12.3	11.6	10.9	10.6
20	11.0	11.5	12.0	12.6	13.1	13.3	13.2	12.8	12.3	11.7	11.2	10.9
15	11.3	11.6	12.0	12.5	12.8	13.0	12.9	12.6	12.2	11.8	11.4	11.2
10	11.6	11.8	12.0	12.3	12.6	12.7	12.6	12.4	12.1	11.8	11.6	11.5
5	11.8	11.9	12.0	12.2	12.3	12.4	12.3	12.3	12.1	12.0	11.9	11.8
0	12.1	12.1	12.1	12.1	12.1	12.1	12.1	12.1	12.1	12.1	12.1	12.1

Mean daily percentage (P) of annual daytime hours for different latitudes.

Latitude	North South	Jan.– July	Feb.– Aug	March –Sept.	April– Oct.	May– Nov.	June– Dec.	July– Jan.	Aug.– Feb.	Sept.– March	Oct.– April	Nov.– May	Dec.– June
60°		0.15	0.20	0.26	0.32	0.38	0.41	0.40	0.34	0.28	0.22	0.17	0.13
58°		0.16	0.21	0.26	0.32	0.37	0.40	0.39	0.34	0.28	0.23	0.18	0.15
56°		0.17	0.21	0.26	0.32	0.36	0.39	0.38	0.33	0.28	0.23	0.18	0.16
54°		0.18	0.22	0.26	0.31	0.36	0.38	0.37	0.33	0.28	0.23	0.19	0.17
52°		0.19	0.22	0.27	0.31	0.35	0.37	0.36	0.33	0.28	0.24	0.20	0.17
50°		0.19	0.23	0.27	0.31	0.34	0.36	0.35	0.32	0.28	0.24	0.20	0.18
48°		0.20	0.23	0.27	0.31	0.34	0.36	0.35	0.32	0.28	0.24	0.21	0.19
46°		0.20	0.23	0.27	0.30	0.34	0.35	0.34	0.32	0.28	0.24	0.21	0.20
44°		0.21	0.24	0.27	0.30	0.33	0.35	0.34	0.31	0.28	0.25	0.22	0.20
42°		0.21	0.24	0.27	0.30	0.33	0.34	0.33	0.31	0.28	0.25	0.22	0.21
40°		0.22	0.24	0.27	0.30	0.32	0.34	0.33	0.31	0.28	0.25	0.22	0.21
35°		0.23	0.25	0.27	0.29	0.31	0.32	0.32	0.30	0.28	0.25	0.23	0.22
30°		0.24	0.25	0.27	0.29	0.31	0.32	0.31	0.30	0.28	0.26	0.24	0.23*
25°		0.24	0.26	0.27	0.29	0.30	0.31	0.31	0.29	0.28	0.26	0.25	0.24
20°		0.25	0.26	0.27	0.28	0.29	0.30	0.30	0.29	0.28	0.26	0.25	0.25
15°		0.26	0.26	0.27	0.28	0.29	0.29	0.29	0.28	0.28	0.27	0.26	0.25
10°		0.26	0.27	0.27	0.28	0.28	0.29	0.29	0.28	0.28	0.27	0.26	0.26
5°		0.27	0.27	0.27	0.28	0.28	0.28	0.28	0.28	0.28	0.27	0.27	0.27
0°		0.27	0.27	0.27	0.27	0.27	0.27	0.27	0.27	0.27	0.27	0.27	0.27

APPENDIX D   PSYCHOMETRIC CONSTANT (γ) FOR DIFFERENT ALTITUDES (Z)

$$\gamma = 10\text{--}3\ [(Cp.P) \div (\varepsilon.\lambda)] = (0.00163) \times [P \div \lambda]$$

γ, psychrometric constant [kPa C^{-1}]

c_p, specific heat of moist air = 1.013

[kJ kg^{-10}C^{-1}]

P, atmospheric pressure [kPa].

ε, ratio molecular weight of water

vapor/dry air = 0.622

λ, latent heat of vaporization [MJ kg^{-1}]

= 2.45 MJ kg^{-1} at 20°C.

Z (m)	γ kPa/°C	z (m)	γ kPa/°C	z (m)	γ kPa/°C	z (m)	γ kPa/°C
0	0.067	1000	0.060	2000	0.053	3000	0.047
100	0.067	1100	0.059	2100	0.052	3100	0.046
200	0.066	1200	0.058	2200	0.052	3200	0.046
300	0.065	1300	0.058	2300	0.051	3300	0.045
400	0.064	1400	0.057	2400	0.051	3400	0.045
500	0.064	1500	0.056	2500	0.050	3500	0.044
600	0.063	1600	0.056	2600	0.049	3600	0.043
700	0.062	1700	0.055	2700	0.049	3700	0.043
800	0.061	1800	0.054	2800	0.048	3800	0.042
900	0.061	1900	0.054	2900	0.047	3900	0.042
1000	0.060	2000	0.053	3000	0.047	4000	0.041

APPENDIX E   SATURATION VAPOR PRESSURE [ES] FOR DIFFERENT TEMPERATURES (T)

Vapor pressure function = e_s = [0.6108]*exp{[17.27*T]/[T + 237.3]}							
T °C	e_s kPa	T °C	e_s kPa	T °C	e_s kPa	T °C	e_s kPa
1.0	0.657	13.0	1.498	25.0	3.168	37.0	6.275
1.5	0.681	13.5	1.547	25.5	3.263	37.5	6.448
2.0	0.706	14.0	1.599	26.0	3.361	38.0	6.625
2.5	0.731	14.5	1.651	26.5	3.462	38.5	6.806
3.0	0.758	15.0	1.705	27.0	3.565	39.0	6.991
3.5	0.785	15.5	1.761	27.5	3.671	39.5	7.181
4.0	0.813	16.0	1.818	28.0	3.780	40.0	7.376

Vapor pressure function = $e_s = [0.6108]*exp\{[17.27*T]/[T + 237.3]\}$							
T °C	**e_s kPa**	**T °C**	**e_s kPa**	**T °C**	**e_s kPa**	**T °C**	**e_s kPa**
4.5	0.842	**16.5**	1.877	**28.5**	3.891	**40.5**	7.574
5.0	0.872	**17.0**	1.938	**29.0**	4.006	**41.0**	7.778
5.5	0.903	**17.5**	2.000	**29.5**	4.123	**41.5**	7.986
6.0	0.935	**18.0**	2.064	**30.0**	4.243	**42.0**	8.199
6.5	0.968	**18.5**	2.130	**30.5**	4.366	**42.5**	8.417
7.0	1.002	**19.0**	2.197	**31.0**	4.493	**43.0**	8.640
7.5	1.037	**19.5**	2.267	**31.5**	4.622	**43.5**	8.867
8.0	1.073	**20.0**	2.338	**32.0**	4.755	**44.0**	9.101
8.5	1.110	**20.5**	2.412	**32.5**	4.891	**44.5**	9.339
9.0	1.148	**21.0**	2.487	**33.0**	5.030	**45.0**	9.582
9.5	1.187	**21.5**	2.564	**33.5**	5.173	**45.5**	9.832
10.0	1.228	**22.0**	2.644	**34.0**	5.319	**46.0**	10.086
10.5	1.270	**22.5**	2.726	**34.5**	5.469	**46.5**	10.347
11.0	1.313	**23.0**	2.809	**35.0**	5.623	**47.0**	10.613
11.5	1.357	**23.5**	2.896	**35.5**	5.780	**47.5**	10.885
12.0	1.403	**24.0**	2.984	**36.0**	5.941	**48.0**	11.163
12.5	1.449	**24.5**	3.075	**36.5**	6.106	**48.5**	11.447

APPENDIX F   SLOPE OF VAPOR PRESSURE CURVE (Δ) FOR DIFFERENT TEMPERATURES (T)

$$\Delta = [4098.\ e^0(T)] \div [T + 237.3]^2$$
$$= 2504\{exp[(17.27T) \div (T + 237.2)]\} \div [T + 237.3]^2$$

T °C	Δ kPa/°C	T °C	Δ kPa/°C	T °C	Δ kPa/°C	T °C	Δ kPa/°C
1.0	0.047	13.0	0.098	25.0	0.189	37.0	0.342
1.5	0.049	13.5	0.101	25.5	0.194	37.5	0.350
2.0	0.050	14.0	0.104	26.0	0.199	38.0	0.358
2.5	0.052	14.5	0.107	26.5	0.204	38.5	0.367
3.0	0.054	15.0	0.110	27.0	0.209	39.0	0.375
3.5	0.055	15.5	0.113	27.5	0.215	39.5	0.384
4.0	0.057	16.0	0.116	28.0	0.220	40.0	0.393

T °C	Δ kPa/°C	T °C	Δ kPa/°C	T °C	Δ kPa/°C	T °C	Δ kPa/°C
4.5	0.059	16.5	0.119	28.5	0.226	40.5	0.402
5.0	0.061	17.0	0.123	29.0	0.231	41.0	0.412
5.5	0.063	17.5	0.126	29.5	0.237	41.5	0.421
6.0	0.065	18.0	0.130	30.0	0.243	42.0	0.431
6.5	0.067	18.5	0.133	30.5	0.249	42.5	0.441
7.0	0.069	19.0	0.137	31.0	0.256	43.0	0.451
7.5	0.071	19.5	0.141	31.5	0.262	43.5	0.461
8.0	0.073	20.0	0.145	32.0	0.269	44.0	0.471
8.5	0.075	20.5	0.149	32.5	0.275	44.5	0.482
9.0	0.078	21.0	0.153	33.0	0.282	45.0	0.493
9.5	0.080	21.5	0.157	33.5	0.289	45.5	0.504
10.0	0.082	22.0	0.161	34.0	0.296	46.0	0.515
10.5	0.085	22.5	0.165	34.5	0.303	46.5	0.526
11.0	0.087	23.0	0.170	35.0	0.311	47.0	0.538
11.5	0.090	23.5	0.174	35.5	0.318	47.5	0.550
12.0	0.092	24.0	0.179	36.0	0.326	48.0	0.562
12.5	0.095	24.5	0.184	36.5	0.334	48.5	0.574

APPENDIX G NUMBER OF THE DAY IN THE YEAR (JULIAN DAY)

Day	Jan	Feb	Mar	Apr	May	Jun	Jul	Aug	Sep	Oct	Nov	Dec
1	1	32	60	91	121	152	182	213	244	274	305	335
2	2	33	61	92	122	153	183	214	245	275	306	336
3	3	34	62	93	123	154	184	215	246	276	307	337
4	4	35	63	94	124	155	185	216	247	277	308	338
5	5	36	64	95	125	156	186	217	248	278	309	339
6	6	37	65	96	126	157	187	218	249	279	310	340
7	7	38	66	97	127	158	188	219	250	280	311	341
8	8	39	67	98	128	159	189	220	251	281	312	342
9	9	40	68	99	129	160	190	221	252	282	313	343
10	10	41	69	100	130	161	191	222	253	283	314	344
11	11	42	70	101	131	162	192	223	254	284	315	345
12	12	43	71	102	132	163	193	224	255	285	316	346
13	13	44	72	103	133	164	194	225	256	286	317	347
14	14	45	73	104	134	165	195	226	257	287	318	348

Day	Jan	Feb	Mar	Apr	May	Jun	Jul	Aug	Sep	Oct	Nov	Dec
15	15	46	74	105	135	166	196	227	258	288	319	349
16	16	47	75	106	136	167	197	228	259	289	320	350
17	17	48	76	107	137	168	198	229	260	290	321	351
18	18	49	77	108	138	169	199	230	261	291	322	352
19	19	50	78	109	139	170	200	231	262	292	323	353
20	20	51	79	110	140	171	201	232	263	293	324	354
21	21	52	80	111	141	172	202	233	264	294	325	355
22	22	53	81	112	142	173	203	234	265	295	326	356
23	23	54	82	113	143	174	204	235	266	296	327	357
24	24	55	83	114	144	175	205	236	267	297	328	358
25	25	56	84	115	145	176	206	237	268	298	329	359
26	26	57	85	116	146	177	207	238	269	299	330	360
27	27	58	86	117	147	178	208	239	270	300	331	361
28	28	59	87	118	148	179	209	240	271	301	332	362
29	29	(60)	88	119	149	180	210	241	272	302	333	363
30	30	—	89	120	150	181	211	242	273	303	334	364
31	31	—	90	—	151	—	212	243	—	304	—	365

APPENDIX H   STEFAN-BOLTZMANN LAW AT DIFFERENT TEMPERATURES (T):

$[\sigma^*(T_K)^4] = [4.903 \times 10^{-9}]$, MJ K^{-4} m^{-2} day^{-1}
where: $T_K = \{T[°C] + 273.16\}$

T	$\sigma^*(T_K)^4$	T	$\sigma^*(T_K)^4$	T	$\sigma^*(T_K)^4$
		Units			
°C	MJ m^{-2} d^{-1}	°C	MJ m^{-2} d^{-1}	°C	MJ m^{-2} d^{-1}
1.0	27.70	17.0	34.75	33.0	43.08
1.5	27.90	17.5	34.99	33.5	43.36
2.0	28.11	18.0	35.24	34.0	43.64
2.5	28.31	18.5	35.48	34.5	43.93
3.0	28.52	19.0	35.72	35.0	44.21
3.5	28.72	19.5	35.97	35.5	44.50
4.0	28.93	20.0	36.21	36.0	44.79
4.5	29.14	20.5	36.46	36.5	45.08

T	$\sigma^*(T_K)^4$	T	$\sigma^*(T_K)^4$	T	$\sigma^*(T_K)^4$
		Units			
°C	MJ m^{-2} d^{-1}	°C	MJ m^{-2} d^{-1}	°C	MJ m^{-2} d^{-1}
5.0	29.35	21.0	36.71	37.0	45.37
5.5	29.56	21.5	36.96	37.5	45.67
6.0	29.78	22.0	37.21	38.0	45.96
6.5	29.99	22.5	37.47	38.5	46.26
7.0	30.21	23.0	37.72	39.0	46.56
7.5	30.42	23.5	37.98	39.5	46.85
8.0	30.64	24.0	38.23	40.0	47.15
8.5	30.86	24.5	38.49	40.5	47.46
9.0	31.08	25.0	38.75	41.0	47.76
9.5	31.30	25.5	39.01	41.5	48.06
10.0	31.52	26.0	39.27	42.0	48.37
10.5	31.74	26.5	39.53	42.5	48.68
11.0	31.97	27.0	39.80	43.0	48.99
11.5	32.19	27.5	40.06	43.5	49.30
12.0	32.42	28.0	40.33	44.0	49.61
12.5	32.65	28.5	40.60	44.5	49.92
13.0	32.88	29.0	40.87	45.0	50.24
13.5	33.11	29.5	41.14	45.5	50.56
14.0	33.34	30.0	41.41	46.0	50.87
14.5	33.57	30.5	41.69	46.5	51.19
15.0	33.81	31.0	41.96	47.0	51.51
15.5	34.04	31.5	42.24	47.5	51.84
16.0	34.28	32.0	42.52	48.0	52.16
16.5	34,52	32.5	42.80	48.5	52.49

APPENDIX I   THERMODYNAMIC PROPERTIES OF AIR AND WATER

1. Latent Heat of Vaporization (λ)

$$\lambda = [2.501 - (2.361 \times 10^{-3})\, T]$$

where: λ = latent heat of vaporization [MJ kg^{-1}]; and T = air temperature [°C].

The value of the latent heat varies only slightly over normal temperature ranges. A single value may be taken (for ambient temperature = 20°C): λ = 2.45 MJ kg^{-1}.

2. Atmospheric Pressure (P)

$$P = P_o \left[\{T_{Ko} - \alpha(Z - Z_o)\} \div \{T_{Ko}\} \right]^{(g/(\alpha.R))}$$

Where: P, atmospheric pressure at elevation z [kPa]

P_o, atmospheric pressure at sea level = 101.3 [kPa]

z, elevation [m]

z_o, elevation at reference level [m]

g, gravitational acceleration = 9.807 [m s^{-2}]

R, specific gas constant = 287 [J kg^{-1} K^{-1}]

α, constant lapse rate for moist air = 0.0065 [K m^{-1}]

T_{Ko}, reference temperature [K] at elevation z_o = 273.16 + T

T, means air temperature for the time period of calculation [°C]

When assuming P_o = 101.3 [kPa] at z_o = 0, and T_{Ko} = 293 [K] for T = 20 [°C], above equation reduces to:

$$P = 101.3[(293^{-0}.0065Z)\ (293)]^{5.26}$$

3. Atmospheric Density (ρ)

$$\rho = [1000P] \div [T_{Kv}\ R] = [3.486P] \div [T_{Kv}],\ \text{and}\ T_{Kv} = T_K[1 - 0.378(ea)/P]^{-1}$$

where: ρ, atmospheric density [kg m^{-3}]

R, specific gas constant = 287 [J kg^{-1} K^{-1}]

$T_{Kv,}$ virtual temperature [K]

$T_{K,}$ absolute temperature [K]: T_K = 273.16 + T [°C]

e_a, actual vapor pressure [kPa]

T, mean daily temperature for 24-hour calculation time steps.

For average conditions (e_a in the range 1^{-5} kPa and P between 80^{-100} kPa), T_{Kv} can be substituted by: $T_{Kv} \approx 1.01\ (T + 273)$

4. Saturation Vapor Pressure function (e_s)

$$e_s = [0.6108]*\exp\{[17.27*T]/[T + 237.3]\}$$

where: e_s, saturation vapor pressure function [kPa]

T, air temperature [°C]

5. Slope Vapor Pressure Curve (Δ)

$$\Delta = [4098.\ e°(T)] \div [T + 237.3]^2$$

$$= 2504\{\exp[(17.27T) \div (T + 237.2)]\} \div [T + 237.3]^2$$

where: Δ, slope vapor pressure curve [kPa C^{-1}]

T, air temperature [°C]

e0(T), saturation vapor pressure at temperature T [kPa]

In 24-hour calculations, Δ is calculated using mean daily air temperature. In hourly calculations T refers to the hourly mean, T_{hr}.

6. Psychrometric Constant (γ)

$$\gamma = 10^{-3}\ [(Cp.P) \div (\varepsilon.\lambda)] = (0.00163) \times [P \div \lambda]$$

where: γ, psychrometric constant [kPa C^{-1}]

c_p, specific heat of moist air = 1.013 [kJ kg^{-10}C^{-1}]

P, atmospheric pressure [kPa]: equations 2 or 4

ε, ratio molecular weight of water vapor/dry air = 0.622

λ, latent heat of vaporization [MJ kg^{-1}]

7. Dew Point Temperature (T_{dew})

When data is not available, T_{dew} can be computed from e_a by:

$$T_{dew} = [\{116.91 + 237.3Log_e(ea)\} \div \{16.78 - Log_e(ea)\}]$$

where: T_{dew}, dew point temperature [°C]

e_a, actual vapor pressure [kPa]

For the case of measurements with the Assmann psychrometer, T_{dew} can be calculated from:

$$T_{dew} = (112 + 0.9T_{wet})[e_a \div (e^0\ T_{wet})]^{0.125} - [112 - 0.1T_{wet}]$$

8. Short Wave Radiation on a Clear-Sky Day (R_{so})

The calculation of R_{so} is required for computing net long wave radiation and for checking calibration of pyranometers and integrity of R_{so} data. A good approximation for R_{so} for daily and hourly periods is:

$$R_{so} = (0.75 + 2 \times 10^{-5}\, z)R_a$$

where: z, station elevation [m]

R_a, extraterrestrial radiation [MJ m^{-2} d^{-1}].

Equation is valid for station elevations less than 6000 m having low air turbidity. The equation was developed by linearizing Beer's radiation extinction law as a function of station elevation and assuming that the average angle of the sun above the horizon is about 50°.

For areas of high turbidity caused by pollution or airborne dust or for regions where the sun angle is significantly less than 50° so that the path length of radiation through the atmosphere is increased, an adoption of Beer's law can be employed where P is used to represent atmospheric mass:

$$R_{so} = (R_a)\, \exp[(-0.0018P) \div (K_t\, \sin(\Phi))]$$

where: K_t, turbidity coefficient, $0 < K_t < 1.0$ where $K_t = 1.0$ for clean air and

$K_t = 1.0$ for extremely turbid, dusty or polluted air.

P, atmospheric pressure [kPa]

Φ, angle of the sun above the horizon [rad]

R_a, extraterrestrial radiation [MJ m^{-2} d^{-1}]

For hourly or shorter periods, Φ is calculated as:

$$\sin \Phi = \sin \varphi \sin \delta + \cos \varphi \cos \delta \cos \omega$$

where: φ, latitude [rad]

δ, solar declination [rad] (Eq. (24) in Chapter 3)

ω, solar time angle at midpoint of hourly or shorter period [rad]

For 24-hour periods, the mean daily sun angle, weighted according to R_a, can be approximated as:

$$\sin(\Phi_{24}) = \sin[0.85 + 0.3\, \varphi \sin\{(2\pi J/365) - 1.39\} - 0.42\, \varphi^2]$$

where: Φ_{24}, average Φ during the daylight period, weighted according to R_a [rad]

φ, latitude [rad]

J, day in the year.

The Φ_{24} variable is used to represent the average sun angle during daylight hours and has been weighted to represent integrated 24-hour transmission effects on 24-hour R_{so} by the atmosphere. Φ_{24} should be limited to > 0. In some situations, the estimation for R_{so} can be improved by modifying to consider the effects of water vapor on short wave absorption, so that: $R_{so} = (K_B + K_D) R_a$ where:

$$K_B = 0.98 exp[\{(-0.00146P) \div (K_t \sin \Phi)\} - 0.091\{w/\sin \Phi\}^{0.25}]$$

where: K_B, the clearness index for direct beam radiation
K_D, the corresponding index for diffuse beam radiation
$K_D = 0.35 - 0.33 K_B$ for $K_B > 0.15$
$K_D = 0.18 + 0.82 K_B$ for $K_B < 0.15$
R_a, extraterrestrial radiation [MJ m^{-2} d^{-1}]
K_t, turbidity coefficient, $0 < K_t < 1.0$ where $K_t = 1.0$ for clean air and $K_t = 1.0$ for extremely turbid, dusty or polluted air.
P, atmospheric pressure [kPa]
Φ, angle of the sun above the horizon [rad]
W, perceptible water in the atmosphere [mm] = $0.14 e_a P + 2.1$
e_a, actual vapor pressure [kPa]
P, atmospheric pressure [kPa]

APPENDIX J   PSYCHROMETRIC CHART AT SEA LEVEL.\

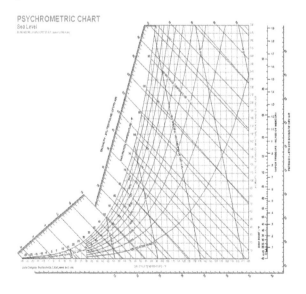

APPENDIX K

[<http://www.fao.org/docrep/T0551E/t0551e07.htm#5.5%20field%20 management%20practices%20in%20wastewater%20irrigation>]

1. **Relationship between applied water salinity and soil water salinity at different leaching fractions (FAO 1985)**

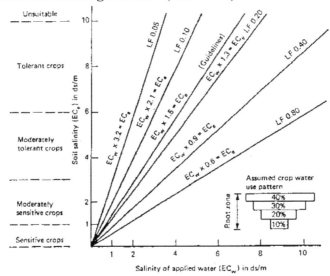

2. **Schematic representations of salt accumulation, planting positions, ridge shapes and watering patterns.**

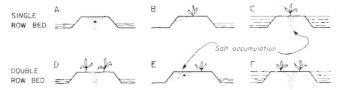

3. **Main components of general planning guidelines for wastewater reuse (Cobham and Johnson 1988)**

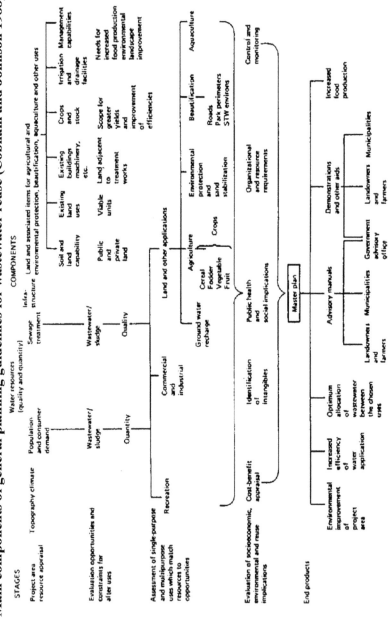

APPENDIX L VALUES OF *Kc* FOR FIELD AND VEGETABLE CROPS FOR DIFFERENT CROP GROWTH STAGES AND PREVAILING CLIMATIC CONDITIONS.

Crop		Relative humidity			
		RHmin > 70%		RHmin < 20%	
	Crop stage	Wind speed, m/sec			
	Initial 1	0–5	5–8	0–5	5–8
	Crop development 2	Values of K_c			
	Mid-season 3	0.95	0.95	1.0	1.05
	Late season/maturity 4	0.9	0.9	0.95	1.0
Barley	3	1.05	1.1	1.15	1.2
	4	0.25	0.25	0.2	0.2
Beans	3	0.95	0.95	1.0	1.05
(green)	4	0.85	0.85	0.9	0.9
Beans, dry /	3	1.05	1.1	1.15	1.2
pulses	4	0.3	0.3	0.25	0.25
Beets	3	1.0	1.0	1.05	1.1
	4	0.9	0.9	0.95	1.0
Carrots	3	1.0	1.05	1.1	1.15
	4	0.7	0.75	0.8	0.85
Sweet corn	3	1.05	1.1	1.15	1.2
(maize)	4	0.95	1.0	1.05	1.1
Cotton	3	1.05	1.15	1.2	1.25
	4	0.65	0.65	0.65	0.7
Crucifers	3	0.95	1.0	1.05	1.1
(cabbage,	4	0.80	0.85	0.9	0.95
cauliflower, broccoli)					
Cucumber	3	0.9	0.9	0.95	1.0
	4	0.7	0.7	0.75	0.8
Lentil	3	1.05	1.1	1.15	1.2
	4	0.3	0.3	0.25	0.25
Melons	3	0.95	0.95	1.0	1.05
	4	0.65	0.65	0.75	0.75
Millet	3	1.0	1.05	1.1	1.15
	4	0.3	0.3	0.25	0.25
Oats	3	1.05	1.1	1.15	1.2
	4	0.25	0.25	0.2	0.2

Crop	Relative humidity				
		RHmin > 70%		RHmin < 20%	
	Crop stage	Wind speed, m/sec			
	Initial 1	0–5	5–8	0–5	5–8
	Crop development 2	Values of K_c			
	Mid-season 3	0.95	0.95	1.0	1.05
	Late season/maturity 4	0.9	0.9	0.95	1.0
Onion (dry)	3	0.95	0.95	1.05	1.1
	4	0.75	0.75	0.8	0.85
Onion (green)	3	0.95	0.95	1.0	1.05
	4	0.95	0.95	1.0	1.05
Peanuts (Groundnut)	Mid-season 3	0.95	1.0	1.05	1.1
	Late season/maturity 4	0.55	0.55	0.6	0.6
Peas	3	1.05	1.1	1.15	1.2
	4	0.95	1.0	1.05	1.1
Potato	3	1.05	1.1	1.15	1.2
	4	0.7	0.7	0.75	0.75
Radish	3	0.8	0.8	0.85	0.9
	4	0.75	0.75	0.8	0.85
Safflower	3	1.05	1.1	1.15	1.2
	4	0.25	0.25	0.2	0.2
Sorghum	3	1.0	1.05	1.1	1.15
	4	0.5	0.5	0.55	0.55
Soybeans	3	1.0	1.05	1.1	1.15
	4	0.45	0.45	0.45	0.45
Spinach	3	0.95	0.95	1.0	1.05
	4	0.9	0.9	0.95	1.0
Sugarbeet	3	1.05	1.1	1.15	1.2
	4	0.9	0.95	1.0	1.0
Sunflower	3	1.05	1.1	1.15	1.2
	4	0.4	0.4	0.35	0.35
Tomato	3	1.05	1.1	1.2	1.25
	4	0.6	0.6	0.65	0.65
Wheat	3	1.05	1.1	1.15	1.2
	4	0.25	0.25	0.2	0.2

Note: Values of Kc in this table are for field and vegetable crops; values of Kc for other crops are reported by Doorenbos and Pruitt (1977).

APPENDIX M   CROP TOLERANCE AND YIELD POTENTIAL OF CROPS AFFECTED BY IRRIGATION WATER SALINITY (ECw) OR SOIL SALINITY (ECe)

Field crops	100%		90%		75%		50%		0% Maximum	
	EC_e	EC_w	EC_e	EC_w	EC_e	EC_w	EC_e	EC_w	EC_e	EC_w
Barley (Hordeum vulgare)	8.0	5.3	10	6.7	13	8.7	18	12	28	19
Cotton (Gossypium hirsutum)	7.7	5.1	9.6	6.4	13	8.4	17	12	27	18
Sugarbeet (Beta vulgaris)	7.0	4.7	8.7	5.8	11	7.5	15	10	24	16
Sorghum (Sorghum bicolor)	6.8	4.5	7.4	5.0	8.4	5.6	9.9	6.7	13	8.7
Wheat (Triticum aestivum)	6.0	4.0	7.4	4.9	9.5	6.3	13	8.7	20	13
Wheat, durum (Triticum turgidum)	5.7	3.8	7.6	5.0	10	6.9	15	10	24	16
Soybean (Glycine max)	5.0	3.3	5.5	3.7	6.3	4.2	7.5	5.0	10	6.7
Cowpea (Vigna unguiculata)	4.9	3.3	5.7	3.8	7.0	4.7	9.1	6.0	13	8.8
Groundnut (Peanut) (Arachis hypogaea)	3.2	2.1	3.5	2.4	4.1	2.7	4.9	3.3	6.6	4.4
Rice (paddy) (Oriza sativa)	3.0	2.0	3.8	2.6	5.1	3.4	7.2	4.8	11	7.6
Sugarcane (Saccharum officinarum)	1.7	1.1	3.4	2.3	5.9	4.0	10	6.8	19	12
Corn (maize) (Zea mays)	1.7	1.1	2.5	1.7	3.8	2.5	5.9	3.9	10	6.7
Flax (Linum usitatissimum)	1.7	1.1	2.5	1.7	3.8	2.5	5.9	3.9	10	6.7
Broadbean (Vicia faba)	1.5	1.1	2.6	1.8	4.2	2.0	6.8	4.5	12	8.0

Field crops	100%		90%		75%		50%		0% Maximum	
	EC_e	EC_w	EC_e	EC_w	EC_e	EC_w	EC_e	EC_w	EC_e	EC_w
Bean (*Phaseolus vulgaris*)	1.0	0.7	1.5	1.0	2.3	1.5	3.6	2.4	6.3	4.2
Vegetables										
Squash, zucchini (courgette) (Cucurbita pepo melopepo)	4.7	3.1	5.8	3.8	7.4	4.9	10	6.7	15	10
Beet, red (Beta vulgaris)	4.0	2.7	5.1	3.4	6.8	4.5	9.6	6.4	15	10
Squash, scallop (Cucurbita pepo melopepo)	3.2	2.1	3.8	2.6	4.8	3.2	6.3	4.2	9.4	6.3
Broccoli (Brassica oleracea botrytis)	2.8	1.9	3.9	2.6	5.5	3.7	8.2	5.5	14	9.1
Tomato (Lycopersicon esculentum)	2.5	1.7	3.5	2.3	5.0	3.4	7.6	5.0	13	8.4
Cucumber (Cucumis sativus)	2.5	1.7	3.3	2.2	4.4	2.9	6.3	4.2	10	6.8
Spinach (Spinacia oleracea)	2.0	1.3	3.3	2.2	5.3	3.5	8.6	5.7	15	10
Celery (Apium graveolens)	1.8	1.2	3.4	2.3	5.8	3.9	9.9	6.6	18	12
Cabbage (Brassica oleracea capitata)	1.8	1.2	2.8	1.9	4.4	2.9	7.0	4.6	12	8.1
Potato (Solanum tuberosum)	1.7	1.1	2.5	1.7	3.8	2.5	5.9	3.9	10	6.7
Corn, sweet (maize) (Zea mays)	1.7	1.1	2.5	1.7	3.8	2.5	5.9	3.9	10	6.7
Sweet potato (Ipomoea batatas)	1.5	1.0	2.4	1.6	3.8	2.5	6.0	4.0	11	7.1
Pepper (Capsicum annuum)	1.5	1.0	2.2	1.5	3.3	2.2	5.1	3.4	8.6	5.8

Field crops	100%		90%		75%		50%		0% Maximum	
	EC_e	EC_w	EC_e	EC_w	EC_e	EC_w	EC_e	EC_w	EC_e	EC_w
Lettuce (Lactuca sativa)	1.3	0.9	2.1	1.4	3.2	2.1	5.1	3.4	9.0	6.0
Radish (Raphanus sativus)	1.2	0.8	2.0	1.3	3.1	2.1	5.0	3.4	8.9	5.9
Onion (Allium cepa)	1.2	0.8	1.8	1.2	2.8	1.8	4.3	2.9	7.4	5.0
Carrot (Daucus carota)	1.0	0.7	1.7	1.1	2.8	1.9	4.6	3.0	8.1	5.4
Bean (Phaseolus vulgaris)	1.0	0.7	1.5	1.0	2.3	1.5	3.6	2.4	6.3	4.2
Turnip (Brassica rapa)	0.9	0.6	2.0	1.3	3.7	2.5	6.5	4.3	12	8.0
Wheatgrass, tall (Agropyron elongatum)	7.5	5.0	9.9	6.6	13	9.0	19	13	31	21
Wheatgrass, fairway crested (Agropyron cristatum)	7.5	5.0	9.0	6.0	11	7.4	15	9.8	22	15
Bermuda grass (Cynodon dactylon)	6.9	4.6	8.5	5.6	11	7.2	15	9.8	23	15
Barley (forage) (Hordeum vulgare)	6.0	4.0	7.4	4.9	9.5	6.4	13	8.7	20	13
Ryegrass, perennial (Lolium perenne)	5.6	3.7	6.9	4.6	8.9	5.9	12	8.1	19	13
Trefoil, narrow leaf birds foot (Lotus corniculatus tenuifolium)	5.0	3.3	6.0	4.0	7.5	5.0	10	6.7	15	10
Harding grass (Phalaris tuberosa)	4.6	3.1	5.9	3.9	7.9	5.3	11	7.4	18	12
Fescue, tall (Festuca elatior)	3.9	2.6	5.5	3.6	7.8	5.2	12	7.8	20	13

Field crops	100%		90%		75%		50%		0% Maximum	
	EC_e	EC_w	EC_e	EC_w	EC_e	EC_w	EC_e	EC_w	EC_e	EC_w
Wheatgrass, standard crested (Agropyron sibiricum)	3.5	2.3	6.0	4.0	9.8	6.5	16	11	28	19
Vetch, common (Vicia angustifolia)	3.0	2.0	3.9	2.6	5.3	3.5	7.6	5.0	12	8.1
Sudan grass (Sorghum sudanense)	2.8	1.9	5.1	3.4	8.6	5.7	14	9.6	26	17
Wildrye, beardless (Elymus triticoides)	2.7	1.8	4.4	2.9	6.9	4.6	11	7.4	19	13
Cowpea (forage) (Vigna unguiculata)	2.5	1.7	3.4	2.3	4.8	3.2	7.1	4.8	12	7.8
Trefoil, big (Lotus uliginosus)	2.3	1.5	2.8	1.9	3.6	2.4	4.9	3.3	7.6	5.0
Sesbania (Sesbania exaltata)	2.3	1.5	3.7	2.5	5.9	3.9	9.4	6.3	17	11
Sphaerophysa (Sphaerophysa salsula)	2.2	1.5	3.6	2.4	5.8	3.8	9.3	6.2	16	11
Alfalfa (Medicago sativa)	2.0	1.3	3.4	2.2	5.4	3.6	8.8	5.9	16	10
Lovegrass (Eragrostis sp.)	2.0	1.3	3.2	2.1	5.0	3.3	8.0	5.3	14	9.3
Corn (forage) (maize) (Zea mays)	1.8	1.2	3.2	2.1	5.2	3.5	8.6	5.7	15	10
Clover, berseem (Trifolium alexandrinum)	1.5	1.0	3.2	2.2	5.9	3.9	10	6.8	19	13
Orchard grass (Dactylis glomerata)	1.5	1.0	3.1	2.1	5.5	3.7	9.6	6.4	18	12

Field crops	100%		90%		75%		50%		0% Maximum	
	EC_e	EC_w	EC_e	EC_w	EC_e	EC_w	EC_e	EC_w	EC_e	EC_w
Foxtail, meadow (Alopecurus pratensis)	1.5	1.0	2.5	1.7	4.1	2.7	6.7	4.5	12	7.9
Clover, red (Trifolium pratense)	1.5	1.0	2.3	1.6	3.6	2.4	5.7	3.8	9.8	6.6
Clover, alsike (Trifolium hybridum)	1.5	1.0	2.3	1.6	3.6	2.4	5.7	3.8	9.8	6.6
Clover, ladino (Trifolium repens)	1.5	1.0	2.3	1.6	3.6	2.4	5.7	3.8	9.8	6.6
Clover, strawberry (Trifolium fragiferum)	1.5	1.0	2.3	1.6	3.6	2.4	5.7	3.8	9.8	6.6
Fruit crops										
Date palm (phoenix dactylifera)	4.0	2.7	6.8	4.5	11	7.3	18	12	32	21
Grapefruit (Citrus paradisi)	1.8	1.2	2.4	1.6	3.4	2.2	4.9	3.3	8.0	5.4
Orange (Citrus sinensis)	1.7	1.1	2.3	1.6	3.3	2.2	4.8	3.2	8.0	5.3
Peach (Prunus persica)	1.7	1.1	2.2	1.5	2.9	1.9	4.1	2.7	6.5	4.3
Apricot (Prunus armeniaca)	1.6	1.1	2.0	1.3	2.6	1.8	3.7	2.5	5.8	3.8
Grape (Vitus sp.)	1.5	1.0	2.5	1.7	4.1	2.7	6.7	4.5	12	7.9
Almond (Prunus dulcis)	1.5	1.0	2.0	1.4	2.8	1.9	4.1	2.8	6.8	4.5
Plum, prune (Prunus domestica)	1.5	1.0	2.1	1.4	2.9	1.9	4.3	2.9	7.1	4.7
Blackberry (Rubus sp.)	1.5	1.0	2.0	1.3	2.6	1.8	3.8	2.5	6.0	4.0
Boysenberry (Rubus ursinus)	1.5	1.0	2.0	1.3	2.6	1.8	3.8	2.5	6.0	4.0
Strawberry (Fragaria sp.)	1.0	0.7	1.3	0.9	1.8	1.2	2.5	1.7	4	2.7

Source: Maas and Hoffman (1977) and Maas (1984).

INDEX

A

Abbotsford, 6, 7, 10, 20, 24, 26
  recharge system, 24
Aerial photography, 223, 225
Aerodynamic resistance, 319
Afforestation/forest densification, 269
Aggregation method, 121
Agricultural
  crops, 228
  development, 126, 260, 272
  land, 119, 158, 162, 229, 257–259,
  268, 271
  management parameters, 144
  production, 24, 228, 232, 269
  watersheds, 133, 134, 137, 161, 162
Agricultural Non-Point Source Pollution
  (AGNPS) model, 131–137, 141–146,
  150–158, 161, 162
Agriculture, 8, 42, 107, 133, 158–161,
  188, 227, 228, 237, 238, 255, 257,
  271–274, 282
  production, 8, 228
  seasons, 107
  use (crop classification), 216, 237
Agroforestry, 268, 269, 274
Agro-horticulture, 229, 269
Agronomical/biological erosion, 160
Air
  humidity, 321
  temperature, 6–9, 52, 192–195,
  205–208, 315–322, 325, 326, 329,
  355, 356
Albedo and Net Solar Radiation (Rns),
  324
Algorithm, 35–37, 112, 136, 331, 356
All India Soil and Land Use Survey
  map, 136

Alluvial
  formations, 253
  plain, 255, 256, 260
  soils, 136
Animal feedlots, 137
Antecedent moisture conditions, 143
Aqua-pro, 48, 52, 297
  moisture sensor, 52
  soil water sensor, 52
Aquatic habitat, 188
Aquifer, 4–8, 25, 104–110, 126, 127
  storage capability, 105
  system, 104
Artificial neural network (ANN) model,
  31–33, 37–39
  technique, 38
Atmosphere radiates energy, 324
Atmospheric
  conditions, 14
  deposition, 194
  parameters, 321
  pressure, 8, 320, 321
Augmentation, 122
Automatic Weather Station, 52
Australian water balance model, 72,
  78–89, 99, 100

B

Bare soil, 48
Basin
  length, 244–247, 263, 274
  perimeter, 244, 263, 274
  shape factor, 246
  width, 244, 274
Bathymetric data, 174
Best management practices (BMPs),
  134, 158, 161
Bhairabbanki watersheds, 145, 150

Bifurcation ratio, 233, 244, 245, 263, 264, 272–274
Biological
  engineering measures, 160
  erosion, 160
  measures, 158, 161
  oxygen demand, 196
  properties, 65
Black gram, 304–309, 346, 353, 354
Boreholes, 109
Boundary conditions, 14, 48, 49, 75, 104, 174, 194
Brier score, 72, 76, 90, 91, 97–100
Buffer operations, 251
Building sluices, 170

C

Calibration, 3, 6, 15–17, 22–26, 32, 54, 76–80, 93–99, 144–146, 150, 151, 156–162, 179, 200, 317, 325
  period, 94, 96, 99
  trials, 145
California, 187, 192, 209–211
Campbell-Stokes, 321
Carbon dioxide, 220
Channel parameter calculation, 144, 146
Channel transport capacity, 137
Charnokite, 122, 123
Chemical
  laboratory, 243, 244, 265
  methods, 4
  nutrients, 163
  oxygen demand, 137
  transport, 44, 137, 163
    process, 44
Circulatory ratio, 272–274
Circulatory ratio, 246
City population, 173
Cityscapes, 171
Climate change, 188–193, 205–209, 211
Climateorological data, 287
Climatic conditions, 156, 161, 162, 191, 263, 314, 315
Climatic variability, 4
Coefficient of,

determination, 55, 59, 60, 66, 146, 199, 202
  residual mass (CRM), 316, 317
Compactness coefficient, 247
Comprehensive bootstrapping techniques, 32
Computational technology, 32
Conservation practice factor, 144, 161
Construction of recharge structures, 114, 116, 121, 125
Continuous Rank Probability Score (CRPS), 72, 76, 92, 99
Contour bunding, 159–163, 302
Conventional methods, 104, 105, 227
Conventional modeling, 32
Cost-effective assessment, 229
CREAMS model, 138
Critical growth stage, 45, 51, 58, 283
Crop
  coefficient, 46, 66, 285, 313, 315, 326–328, 341, 357
  evapotranspiration, 315, 326, 328, 354, 356
  growing season, 188
  management factor, 144
  production, 109, 287, 315
  selection, 329
  stress factor, 66
  water requirement, 341, 356
Cropping intensity, 126, 127, 271
Cropping pattern, 42, 227, 268
Cross-validation error, 37
Cultivation, 42, 43, 221, 287
Culturable waste lands, 260
Curve number (CN), 138
Cyclic experience, 55

D

Dactylis glomerata L, 8
Data driven models, 32
Data transmission, 222
Database management system, 223, 331, 356
Database query, 251

Davis Tahoe Environmental Research Center (TERC), 196
Deccan volcanic province, 105
Decision support system, 127, 331
Delineation, 127, 225, 226, 234
Dharwar landscape, 136
Digital elevation model (DEM), 116, 144
Dike heights, 287
Dissolved organic phosphorus (DOP), 196
Distributed Model Inter-comparison Project (DMIP), 74
Domestic and agricultural, 111
Dominant drainage pattern, 117
Drainage and Surface Water Body Map, 242
Drainage density, 104, 112, 114–122, 127, 230, 233, 244, 248, 249, 263, 264, 272–274
Drainage-efficiency, 170
Drexel University, 182
Drum lysimeter, 52
Dry land cropping, 270
Dupuit's approach, 47
Dynamic and non-linear systems, 32
Dynamics and physical process, 189

E

Ecological balance, 132
Economic conditions, 218
Economic development, 9
Ecosystem, 43, 50, 172, 188, 203, 208, 211, 255
Effective root zone, 43–54, 58, 63, 66, 67, 283–289, 293, 296, 297
Electrical resistivity, 5
Electromagnetic radiation, 222
Electro-magnetic spectrum, 223
Elongation ratio, 233, 244, 247, 263, 272–274
Empirical model, 315
Empirical shape parameter, 23
Ensemble construction, 87

Ensemble streamflow prediction (EPS), 74
Environmental
    management, 218
    processes, 133
    protection, 132
    reasons, 109
Epilimnion layer, 199
Equivalent height, 193
ERDAS/IMAGINE software, 136
Erodibility factor, 163
Evaporation, 48, 85–87, 189–193, 200–211, 253, 288, 313–319, 325–331, 356
Evapotranspiration, 13, 14, 44–52, 66, 67, 84, 86, 170, 284–289, 297, 313–331, 339–350, 354–357
Experimental farm, 50, 288
Experimental requirements, 25
Extraterrestrial (solar) radiation, 322

F

False color composites (FCCs), 263, 272, 274
Fault zone, 122
Faulty land, 132
Fertilization, 24
Fertilizers, 133
    availability factor, 144
Field water balance, 44–49, 54, 55, 65–67
Financial support, 66, 100
Finite element method (FEM), 48
Flood
    area, 108
    damage reduction, 84
    plain granite gneiss, 122, 123
    plain upland, 122
    plain, 108–113, 119, 122, 127, 227
    plain, 84, 260
    protection, 191, 211
Food engineering, 41, 50, 66, 71, 131, 288
Forest, 120, 124, 217, 218, 238, 257–260, 267
Forest plantation, 268, 270

G

General theory, 37
Generalization
  capability, 37
  performance, 37
Genuchten-Mualem model, 12, 27
Geo morphology layer, 122
Geographical area, 105, 126, 132, 237, 260, 269
Geographical information center, 274
Geographical information system, 133, 162, 215, 223, 229, 234, 274
  environment, 105, 106, 228, 231, 232
  platform, 104
  techniques, 134
Geoinformatics, 219, 224, 274
Geologic conditions, 111
Geological
  conditions, 113, 126
  formations, 106
  lithological developments, 245
  structure, 104
  survey, 119
Geology, 108, 225, 231–233, 241
Geometry, 48, 140
Geomorphic option, 144–146
Geomorphic units, 227, 230, 232, 273
Geo-morphological
  aspects, 105
  parameters, 116
  unit, 121
Geomorphology, 112, 119, 127, 220, 224, 226, 230–236, 268
Geophysical fluid dynamics laboratory (GFDL), 193, 209
Geophysical methods, 5
Geo-processing functions, 219
Geo-referencing satellite data products, 236
Germination, 47–50, 285, 288
  period, 48, 50, 288
Global circulation model, 209, 211
Global warming, 191, 207, 208, 220
Government and non-government organizations, 158

Granitic zones, 117
Graphical comparison, 55
Graphical representation, 156
Green roofs, 171, 177–184
Ground truth, 136, 216, 237–241, 242, 274
Groundwater
  analytical analysis, 4
  balance, 109
  chemical methods, 4
  derivation from the water budget, 4
  direct measurements, 4
  flow, 84, 104, 111
  geophysical methods, 4
  management, 25, 26, 105
  quality, 125
  recharge estimation, 4
  recharge structures, 104–109, 114, 125
  roundwater resources, 104, 105
  system, 104
Groups, 116, 143, 173, 177–183, 243, 253

H

Hard engineering methods, 171
Hargreaves (HG), 317
Harvesting structure, 59, 61, 65, 253, 254, 266
Heat budget, 189, 193, 194, 200, 209
Heat transfer, 26, 171, 189, 198, 211
HEC-HMS, 71, 78, 84, 87–89, 99, 100
Herbicides, 133
Heterogeneity, 23
Hidden layer, 34, 35, 38, 39
Hidden nodes in hidden layers, 35
Hierarchy process, 112
Hills and plateaus, 261
Homogeneous, 44, 49, 84
Horticulture, 229, 231, 255, 269
Human and natural systems, 218
Humidity measurements, 322
Hydraulic
  conductivity, 12, 15, 16, 23, 25, 26, 107, 176, 293

functions, 49
geometry, 140
gradients, 11
parameters, 15, 49, 54
properties, 12, 45, 49, 51, 108
structure, 134
Hydro-geological basin parameters, 104
Hydro-geomorphic parameters, 121
Hydro-meteorological inputs, 211
Hydrodynamic, 189, 190, 198, 211
Hydrogeological characteristics, 4
Hydrogeomorphological
  characteristics, 260
  map, 242, 251, 254, 260, 266, 273, 274
  studies, 230
  units, 242, 260
Hydrograph, 85, 127, 140, 144, 146, 161
Hydrograph modeling, 85
Hydrologic
  budgets, 189, 192, 193
  component, 189
  computation options, 144
  conditions, 4, 156
  cycle, 5, 189, 313, 355
  engineering center, 84
  modeling system (HEC-HMS), 71, 84, 161, 163
  processes, 32
  system, 72, 99
Hydrological
  analysis, 117
  computations, 145
  ensemble prediction, 74
  models, 16, 17, 73, 78, 83, 87, 92, 98, 99
  observations, 73
  properties, 232
  state, 37
  system, 73
  water quality process, 83
Hydrology, 74, 100, 131, 137, 138
  hydrograph generation, 138
  peak runoff rate, 138
  runoff, 138
Hydropower generation, 188, 204

HYDRUS-1D, 5, 6, 11, 13, 15, 25–27
HYDRUS-1D calibration, 15
HYDRUS-1D parameter estimation
  module, 15
HYDRUS-2D, 41, 44–48, 54, 65–67
HYDRUS-2D model, 45–48, 54, 65
Hyperbolic tangent function, 36
Hypotheses testing techniques, 173
Hypsographic information, 197

I

Indian Institute of Technology, 41, 50, 52, 66, 71, 100, 131, 288
Industrial and agricultural, 109
Infiltration capacity, 86, 230, 248
Infiltration function, 86
Input data to DLM-WQ, 194
  flows and pollutant loadings, 195
  lake data, 195
  meteorological data input, 194
Input parameters, 16, 34, 37, 38, 144–146, 156
Intensive agriculture, 229, 268, 271, 274
Inverse modeling, 27, 54
Irrigation, 43, 49, 50, 58, 65, 105, 109, 126, 127, 171, 183, 188, 191, 237, 253–257, 282–287, 290, 294–297, 313–315, 327–329, 341, 351–357
Irrigation projects, 313
Irrigation requirement of crops, 328, 329, 341
Isohyetal lines, 195, 198, 209, 211
Isohyetal map, 195
Isotopes, 4

K

Kharif, 107, 111, 114, 126, 127, 136, 228, 237, 254, 271, 274, 303, 308, 329, 341–348, 351–357
  season, 107, 329

L

Laboratory analysis data, 244
Laboratory measurements, 10, 27
Lake

dynamics, 189, 211
physical data, 194
reservoir heat budget, 211
reservoir water budget, 211
stability, 211
surface water level, 202
surface, 189, 191, 195, 202, 206–209
Tahoe Interagency Monitoring
Program (LTIMP), 195, 211
water temperature, 189, 192, 194,
197, 200
Land
capability, 127
productivity index, 228
resources development plan, 217, 254
slope, 138, 161, 233, 274
transformations, 240
Land Use Change on Hydrology by
Ensemble Modeling (LUCHEM), 74
Land use/land cover (LU/LC), 134
map, 217, 240, 257
puincha micro-watershed, 258
Latent heat of vaporization, 321
Lateral sediment inflow rate, 141
Lateritic soils, 136
Latxaga watershed, 133
Levenberg–Marquardt algorithm, 36–39
Linear programming (LP), 72, 76, 87,
99, 100, 107, 108
Linear transfer functions, 36
Lithological control, 117
Load simulation program in C++, LSPC,
211
Local authorities, 183
Long-term estimate, 27
Longwave radiation, 193–195, 200, 323,
324
Low cost, 6–9, 25, 27, 221
Low impact development (LID), 184
Low-flow augmentation, 188
Lysimeter, 4, 67, 289–297, 318, 316,
326

M

Magnitude flows, 170

Magnitude of storm discharge, 170
Management schemes, 188
Manning's roughness coefficient, 144,
146, 161
Marginal agricultural lands, 221
Mathematical models, 44, 133, 162, 189
lake heat budget, 194
lake hydrologic budget, 193
Mathematical structures, 33
Mean square error (MSE), 74
Mean stream length, 245
Meteorological
center, 52
data, 14, 26, 192–194, 198, 232, 317,
318, 321
factors, 317, 321
hydrological stations, 174
parameters, 49, 52, 195, 314, 355
station, 195
MIKE SHE, 71, 78, 81, 83, 87–89, 99,
100
Mining area, 227
Mitigation, 183, 252
MLP-ANN technique, 38
Model
calibration, 144, 145
evaluation, 55
performance, 16, 55, 146, 182
topology, 75
validation, 145, 150, 156, 179
Modeling efficiency (ME), 316
Modeling hidden layer, 39
Monsoon season, 42, 106, 108, 115, 119
Morphological characteristics, 117
Morphometric analysis, 226, 230, 233,
234, 263, 274
Morphometric parameters, 230, 244,
263, 273
Morphometry, 224, 268
Mualem hydraulic model, 49, 54
Multi criteria analysis, 106, 121
Multi layer, 39
Multi Layer Perceptron-Artificial Neural
Network (MLP-ANN) Technique, 33
Multi-category probability predictions,
91

Multi-criteria analysis, 127
Multilinear regression, 287
Multi-spectral satellite imageries, 119

N

Nala bund, 251, 254, 275
Nash Sutcliffe Efficiency (NSE), 76, 90, 94, 98–100, 317
National Geodetic Vertical Datum (NGVD), 197
Natural
 randomness, 73
 Resources Conservation Service, 136
 resources, 132, 133, 219, 220, 224, 226, 229, 232, 233, 237
 storage structures, 114
 water cycle processes,, 170
Nayagarh district, 107, 119, 126
Net Longwave Radiation (Rnl), 324
Net Radiation (Rn), 324
Nevada, 191, 192, 209, 211
Nevada Division of Environmental Protection (NDEP), 191
Niger, 304–309
Nitrogen (N), 137
Non-linear optimization, 35
Nonmetallic roads, 241, 242
Non-monsoon season, 42
Non-point source pollution, 133, 136, 163
Nonpoint Source Pollution (AGNPS) model, 133
NPS pollution, 134
Numerical analysis, 85
Numerical model, 6, 11, 23, 27, 182, 183
Nutrient flow rates, 137
Nutrient imbalance, 43

O

Observation nodes, 48, 54
Optimally used land, 269
Optimization, 32, 34, 36, 65, 106, 109, 113, 127, 144
 model, 65, 106, 113
 procedures, 109

techniques, 34, 106
Organize matter, 243
Overland flow, 14, 84–86, 137, 138, 145, 163, 233
Overland runoff, 137, 138
Overlaying analysis, 251, 255, 275
Oxidize organic and oxidizable inorganic compounds, 137

P

Paddy, 229, 329, 341–352, 354, 357
Pan evaporation, 316, 357
Parasite blooms, 189
Particulate organic phosphorus (POP), 196
PCSWMM model, 176, 179
Pediplain, 262, 270, 271, 275
Pedons, 243, 244, 265, 275
Pedotransfer function, 27
Penman-Monteith method, 13, 46, 285, 313–318, 322–326, 339–342, 356
Percolation, 6, 44, 45, 52, 66, 67, 86, 106, 232, 233, 251–254, 272, 275, 283, 284, 289, 297
Performance measure, 36
Pervious pavements, 172, 174, 182
Pesticides, 133
Phosphorous, 137
Photographic emulsions, 223
Physical
 based modeling, 25, 27
 laws, 32
 mechanism, 32
 processes, 73, 319
 properties, 51, 117
Physio-chemical analysis, 243
Physiographic analysis, 243
Piezometer, 67, 289, 291, 297
Piezometric observations, 112
Piezometric studies, 111
Plantation, 119, 221, 255–259, 268, 270, 271, 302–308
Ponding
 depth, 43–47, 52, 55, 58, 65, 67, 285
 phase, 46, 285

situations, 48
water, 43, 46, 47, 55, 285, 286, 294
Priestley-Taylor (P-T), 317
PRISM data, 197, 198, 211
Problem solving environment, 105
Puincha micro-watershed, 224
Pyranometer, 321

Q

Quantitative estimation, 44
Quasi-Newton method, 36

R

Rabi, 105–111, 114, 119, 126, 127, 136,
    228, 237, 271, 275, 308, 329, 341–
    349, 352–356
    season, 105–111, 114, 119, 126, 127,
    356
Radiation, 9, 311–316, 321, 322
Radiation based methods, 313
Rain garden, 174
Rain Harvesting, 174
Rain water harvesting, 178, 184
Rainfall-recharge relationship, 108
Rainfall-runoff, 32–39, 84–86, 144, 145,
    156
Rainfed, 43–46, 50, 54, 61, 65, 67, 237,
    282–290, 292, 294–297, 302
    agriculture, 65
    ecosystem, 67
Rainwater harvesting, 171, 179, 182,
    183, 231
Rank probability score (RPS), 72, 76,
    91, 99
Rank probability skill score, 76, 99
Recreational values, 171
Reference evapotranspiration, 67, 313,
    316, 326, 331, 340, 355–357
Relative shortwave radiation (Rs/Rso),
    323
Relative sunshine duration (n/N), 312,
    323
Reproductive phase, 43, 58
Reservoir
    flow, 39

inflow, 39
    management, 32, 211
    operation, 85, 211
Residual moisture, 52, 61, 107, 287
Resistance factors, 319
Rice (Oryza sativa L.), 42
Rice field, 43–66, 282, 284–289,
    291–296
Rice plants, 43
Richards' equation, 5, 11, 45
Robust prediction, 27
Root mean square error (RMSE), 55, 67,
    199, 316, 317
Root zone depth, 44, 49, 52, 58, 63,
    286–289
ROSETTA, 12, 15, 17, 25, 27
Ruggedness number, 249
Runoff volume, 137, 138, 144, 145

S

Saaty's scaling ratios, 112
SACRAMENTO, 72, 78, 86–89, 99, 100
Sacramento model, 86
Satellite
    data, 134, 223, 228, 232–236, 252
    image, 240–244, 275
    imagery, 144
Saturation soil water, 46
Scatter diagram, 100
Scatter plot, 96
SCS triangular hydrograph, 140
Sediment
    availability, 143
    chemical transport, 137
    discharge, 141, 143
    runoff, 137
    transport, 131, 137, 141–143
    yield, 133, 158–163
Seepage, 44–52, 63–67, 110, 283, 284,
    289, 298
Seepage and percolation, 281, 292
Seepage face conditions, 49
Selected models classes, 88
    physically based distributed models,
    88

MIKE SHE, 88
SWAT, 88
lumped conceptual models, 88
AWBM, 88
HEC-HMS, 88
SACRAMENTO, 88
SIMHYD, 88
TANK, 88
Selection of,
activation function, 36
stopping criteria, 36
training algorithm, 35
Sensitivity analysis, 16, 17, 20, 23, 27,
145, 156, 158, 161–163
Sewerage systems, 170
Sigmoidal type, 36, 38
transfer function, 38
Silt loam, 136
Silvi-pasture, 229, 275
Simplified Daily Conceptual Rainfall-
Runoff (SIMHYD), 86
Simulation, 41–51, 53, 57, 59, 63–67,
73–81, 83, 85, 87–100, 174, 181, 196,
198, 209, 281, 283, 289
hydrus-2D model, 48
Boundary and Initial Conditions,
48
Field Water Balance Model
Simulation, 48
Space and Time Discretization, 48
Slope map, 119, 224, 232, 233, 243,
251–255, 265, 273, 275
Slope plantation, 218, 267, 271
Societal and environmental concern, 132
Socioeconomics, 158
Socioeconomic groups, 178
Software, 6, 44, 45, 84, 85, 115, 119,
121, 127, 136, 162, 163, 198, 219,
236, 239, 245, 252, 254, 266, 268,
273, 274, 356
Soil and water assessment tool (SWAT),
71, 84
Soil and water conservation measure,
275

Soil characteristics, 59, 84, 136, 232,
272
Soil conservation, 159, 228, 231, 269,
270, 273, 309
measures, 159, 228, 269, 270
service, 136
Soil data measurements, 25
Soil erodibility factor, 141, 161
Soil erosion range, 159
Soil heat flux (G), 312, 324
Soil map, 136, 144, 226, 227, 232, 273
Soil moisture, 5–9, 14–27, 43–47, 52,
57, 86, 221, 252, 254, 283–286, 293,
295
accounting model (SMAR), 86
content, 6, 23, 57, 283, 298
Soil properties, 15, 65
Soil reservoir, 45
Soil resource inventory, 228
Soil resource map, 217, 243, 244, 255,
265, 275
Soil surface, 14, 23, 44, 49, 51, 319,
321, 327, 328
Soil textural, 136
Soil water, 5, 18, 19, 43, 45, 49–58,
65–67, 287, 326, 327
balance model, 45, 52
content, 5, 19, 43, 49, 50, 52, 54, 55,
57, 58, 65, 66, 327
data, 54
ponding depth, 55
measurements, 52
research, 44
Solar or shortwave radiation (Rs), 323
Solar radiation, 6, 8, 10, 13, 52,
314–317, 322–329, 355–357
Soluble pollutants, 137
Soluble reactive phosphorous (SRP),
196
Source pollution, 132, 133, 162
Spatial
change, 114
datasets, 235
distribution, 115
resolution, 197
variation, 119

Spillway design, 84
Split-sampling method, 37
Standard deviation, 10, 11
Standard error estimate (SEE), 339
Statistic chart, 24
Statistic parameters, 55, 146, 156
Storm water, 170, 171
Strahler's system, 117
Stream frequency, 233, 244, 250, 263, 272–275
Stream length, 117, 244–246, 263, 273, 275
Stream order, 117, 245, 275
Structure of ANN model (MLP network), 33
Subdendratic pattern, 117
Sub-humid climate, 135
Sub-humid sub-tropical region, 161
Sun's illumination, 228
Supplemental irrigation, 42, 45, 51, 52, 59, 67, 283, 284, 289, 290, 296, 298
Surface condition constants (SCC), 146
Surface drainage, 67
Surface runoff from rice field, 45
    check dams, 252
    nala bunds, 252
    renovation of water bodies, 252
    water harvesting structure, 252
Sustainable urban development, 184
Sustainable urban drainage systems (SUDS), 171, 184
    evaluation, 170, 174, 179
    green roofs, 171
    pervious pavements, 172
    rainwater harvesting, 171
    technologies, 171, 174, 176, 181
    urban green space, 172
Sustainable water management, 211
    Software used in GIS, 239
        digitizing software, 239
        editing and labeling, 239
        designing and overlaying software, 239
        text and table addition software, 239

system and peripheral used in GIS, 239
    Black and white CCAL COMP scanner, 239
    Pentium-IV dual core micro processor, 239
    HP-810 INKJET color printer, 239

T

Tabular and statistical analysis, 217, 251, 252, 275
Tahoe City precipitation, 200, 207, 209
Tahoe Environmental Research Center, 211
Tarafeni watershed, 145, 159–162
Telecommunication, 222
Temporal and spatial meteorology, 189
Temporal data, 223
Tensiometer, 48, 52, 67, 289, 298
Terrace systems, 137
Terrain data, 223
Texture ratio, 233, 247, 263, 273
Thematic layers, 121–233
Thematic map, 229–232, 236–239, 250, 268, 273
    integration, 217, 250
Thematic resource maps, 235
Tomography, 5
Topo sheets, 216, 236
Topographic and thematic map, 222
Topographical elevation, 114
Topography, 45, 50, 107, 134, 231, 234, 248, 269, 270
Topology buildings, 250
Topo-sheet, 136, 234, 240, 273, 275
Topsoil surface, 49
Traditional concept, 282
Traditional drainage systems, 170
Training algorithm, 35, 38, 75
Trans-boundary aquifer, 6
Transformation scheme, 177
Transmission of energy, 222
Trapezoidal bund, 49
Tree plantation, 218, 267, 271
Triangular elements, 48

Triangular hydrograph, 140

U

United State Geological Survey, 211
United States Natural Resources
    Conservation Services, 211
University of California, UC, 212
Upland Charnokite rank, 122
Upland granite gneiss, 122
Urban flooding, 171
    problem, 171
Urban green, 176–179, 181, 183
    space, 179, 182–184
USDA Agricultural Research Service
    (USDA-ARS), 136

V

Vadose zone, 5, 6, 11–14, 15, 23–27
    modeling, 5, 11, 25
        boundary conditions, 14
        calibration, 15
        governing equations,11
Validation period, 81–83, 93–99
Van Genuchten-Mualem (VGM) model,
    12, 45
    model, 12, 16, 45
    parameters, 15, 17, 25
Vapor pressure deficit, 311, 322
V-ditch, 303–309
Vegetation, 119
    period, 24, 26
Verification dataset, 91
Versatile technology, 222
Vertical
    boundaries, 49
    depth, 189
    direction, 5, 198
    percolation, 44–47, 52, 63–66
Village and Road Network Map, 240
Village boundary map, 236
Vital natural resources, 132
Volumetric water content, 12, 13

W

Waste land, 158, 221, 238, 260, 275
Water balance, 44–48, 52–57, 65–67, 86,
    87, 114, 191, 196, 202, 209, 283–285,
    288–292, 296–298, 313, 327, 355
    components, 44, 55, 65, 66
    model, 44–49, 52–57, 65, 66, 86,
        283, 284, 288, 289, 291, 296–298
    parameters, 44–48, 53–55, 65, 66,
        283, 284, 288–292, 296
    rice, 284
    simulations, 54
        actual evapotranspiration, 46
        field water balance model, 45
        germination period of rice, 48
        hydrus-2D model, 44
        ponding depth,46
        surface runoff,47
        vertical percolation and lateral
            seepage,47
        water balance model for rice field,
            45
Water body, 260
Water conflicts, 188
Water conservation plan, 233
Water content, 5, 12, 45, 49, 54, 55, 58,
    327
Water harvesting structure, 43, 54, 61,
    127, 251, 252, 272, 275
Water impoundments, 137
Water management, 43, 44, 50, 59, 65,
    173, 191, 208, 209, 224, 232, 257,
    260, 282, 286, 302, 315
Water pollution, 137
Water quality, 83, 133, 159, 189, 194,
    196, 210, 212, 219
Water requirement, 43, 108, 114, 282,
    283, 289, 297, 314, 315, 328, 329,
    341, 354, 356, 357
Water resource development plan, 217,
    252, 266, 275
Water saving irrigation, 43, 67, 282,
    283, 296, 298

techniques, 43, 283, 285, 287, 292, 295, 296
Water system, 32, 232
    optimization framework, 32
Water use efficiency, 43, 282, 283, 290, 296, 298
Water utilization, 184
Watershed
    analysis, 133, 162
    characterization and prioritization, 225
    condition, 144, 158
    degradation, 220
    grids, 143
    hydrology, 212

    management, 133, 219, 220, 224, 226
    morphometry analysis, 244
Weather station, 6–13, 20, 25–27, 317
Weathered zone, 122
Weight matrices, 34
Weir height, 43, 45, 47, 50, 52, 57–60, 67
Wind speed, 9, 311, 321
Wind velocity, 52
Winter crop, 61

Z

z test, 177
zooplankton sub-model, 190

9 781774 636022